MICROBIAL FOOD CONTAMINATION

Edited by
Charles L. Wilson, Ph.D.
Samir Droby, Ph.D.

CRC Press
Boca Raton London New York Washington, D.C.

Library of Congress Cataloging-in-Publication Data

Microbial food contamination / edited by Charles L. Wilson and Samir Droby.
p. cm.
Includes bibliographical references and index.
ISBN 0-8493-2229-4 (alk. paper)
1. Food—Microbiology. 2. Food contamination. I. Wilson, Charles L.
II. Droby, Samir

QR115 .M456 2000
664′.001′579—dc21

00-060871

This book contains information obtained from authentic and highly regarded sources. Reprinted material is quoted with permission, and sources are indicated. A wide variety of references are listed. Reasonable efforts have been made to publish reliable data and information, but the author and the publisher cannot assume responsibility for the validity of all materials or for the consequences of their use.

Direct all inquiries to CRC Press LLC, 2000 N.W. Corporate Blvd., Boca Raton, Florida 33431.

International Standard Book Number 0-8493-2229-4
Library of Congress Card Number 00-060871
Printed in the United States of America 1 2 3 4 5 6 7 8 9 0
Printed on acid-free paper

Preface

A U.S./Israel workshop on "Microbial Food Contamination" was held at the U.S. Fish and Wildlife Service, National Training Center, Shepherdstown, WV, November 8–12, 1998. The 62 participants consisted of scientists, administrators, regulatory officials, and industry representatives from the U.S., Israel, and Singapore. (Co-organizers of the Workshop were Charles Wilson and Art Miller of the U.S. and Samir Droby and Nachman Paster of Israel.) The workshop took a holistic view of the microbial food contamination problem by including bacterial, viral, and fungal contamination of food and the interaction of these various contaminants. International experts lectured on the incidences and nature of microbial food contaminants, the human health effects of microbial food contaminants, and the detection and monitoring of microbial food contaminants. On the last day of the Workshop, consideration was given to how microbial food contaminants affect international trade.

This book contains chapters written by the presenters at this Workshop. These authors are at the forefront of developments in food contamination in the U.S. and Israel. The scope of the book is unique in that consideration is given not only to bacterial contamination of food, but also to a wide range of other contaminants and their interactions. Also, the impact of food contamination on international trade is addressed.

The participants at this Workshop are appreciative of the support given by the United States–Israel Binational Agricultural Research and Development Fund (BARD), without which the Workshop would not have been possible.

Charles L. Wilson
Samir Droby

United States–Israel Binational Agricultural Research and Development Fund

BARD, the United States–Israel Binational Research and Development Fund, supports mutually beneficial, mission-oriented, strategic, and applied research of agricultural problems conducted cooperatively by American and Israeli scientists. Most BARD projects focus on increasing agricultural productivity, particularly in hot and dry climates, and emphasize plant and animal health, food quality and safety, and environmental issues.

Since 1979, BARD has funded over 850 research projects. Proposals undergo extensive peer and panel review evaluations, which are carried out simultaneously and independently in both countries. This assures that only top-quality proposals (approximately 35 annually) with the highest scientific merit, most feasible objectives, and greatest potential benefits to the agriculture of *both* countries and appropriate collaborative arrangements are supported. BARD has funded projects in most of the 50 states.

BARD has earned a fine reputation and built a bridge of cooperation between U.S. scientists and farmers and their Israeli counterparts. The benefits of this synergistic relationship have had significant impact on the agriculture of the U.S., Israel, the Middle East, and worldwide. In addition to funding research proposals, BARD also funds two to three workshops annually that are aimed at identifying research needs in areas of new development, defining directions and opportunities regarding contemporary agricultural needs, and providing information necessary to address policy issues.

BARD workshops are also intended to promote increased contact between BARD scientists and other scientists throughout the world who work in academic and industrial communities in areas related to the binational agricultural interests of BARD. Microbial food contamination has been identified as such an area and worthy of BARD support. The workshop proposal submitted to BARD underwent peer review and was ultimately recommended for funding. The Board recognized the uniqueness of this workshop in bringing together experts from different fields working in the area of microbial food contamination. We trust that this publication will bring the presentations and deliberations of the workshop to many interested parties and will contribute to solutions to this problem of increasing importance.

Edo Chalutz
Executive Director, BARD

About the Editors

Charles L. Wilson, Ph.D., is currently a Research Plant Pathologist with the U.S. Department of Agriculture, Agricultural Research Service, in Kearneysville, WV. His career spans over 40 years, during which he has been at the forefront of research and thinking on the biological control of plant diseases and weeds. In the early part of his career, Dr. Wilson was a professor at the University of Arkansas (1958–1968), where he received the Arkansas Alumni's highest award for distinguished teaching and research. He then served on the faculty at Ohio State University from 1970 to 1979. Dr. Wilson joined the USDA ARS Appalachian Fruit Research Station in 1980, where he initiated a research program to find biologically based alternatives to synthetic fungicides for the control of fruit diseases. This internationally recognized research program has yielded a variety of innovative technologies for the control of postharvest diseases of fruits and vegetables that include the use of antagonistic microorganisms, natural plant-derived fungicides, and the use of induced resistance.

Dr. Wilson is exploring the use of natural plant- and animal-derived antimicrobials for the control of plant diseases and foodborne human pathogens. He is also doing research on the incorporation of natural antimicrobials into food packaging to reduce microbial food contamination and decay. In conjunction with investigators in his laboratory and the Volcani Center in Israel, Dr. Wilson has authored a series of international patents involving the use of antagonistic yeasts to control postharvest diseases. In addition to contributing to our basic understanding of the mode of action of antagonistic yeasts for the control of postharvest diseases, Dr. Wilson has worked closely with industry to bring about the large-scale testing and commercialization of this technology.

In 1984, Dr. Wilson received the Washington Academy of Sciences distinguished service award in the biological sciences for "pioneering research in understanding and manipulating plant diseases." He was also elected a Fellow of the Academy. In 1988, he received the ARS-NAA Scientist of the Year Award for "innovative research on biological control of postharvest diseases of fruit." In 1994, Dr. Wilson was named a Fellow of the American Phytopathology Society, and in 1996 he received the Award of Excellence from the Federal Laboratory Consortium for Technology Transfer for his role in developing one of the first EPA-registered "biofungicides" for the control of postharvest diseases of fruits and vegetables.

Dr. Wilson has authored or co-authored over 185 scientific publications, 12 patents, and two books on gardening. He has previously co-edited books for Academic Press (*Exotic Plant Pests and North American Agriculture*) and CRC Press (*Biological Control of Postharvest Diseases — Theory and Practice*).

Samir Droby, Ph.D., is a Senior Research Scientist at the Agricultural Research Organization (ARO), Volcani Center, Bet Dagan, Israel, and a group leader at the Department of Postharvest Science of Fresh Produce. He received his B.Sc. (1980), M.Sc. (1982), and Ph.D. (1985) in Plant Pathology and Microbiology from the Hebrew University of Jerusalem and joined the Department of Postharvest Science at the Volcani Center in 1987 after spending two years as a postdoctoral fellow at the Department of Plant Pathology, University of California, Riverside. His current research interests are biological control of postharvest diseases of fruits and vegetables, mode of action of biocontrol agents, and host-pathogen interactions.

Throughout his professional career, Dr. Droby has been a research scientist in the field of postharvest pathology, in general, and in developing nonchemical control methods for postharvest diseases of fruits and vegetables, in particular. In conjunction with American colleagues in the USDA Laboratory at the Appalachian Fruit Research Station, Kearneysville, WV, he was instrumental in developing a commercial product used to biologically control postharvest diseases of citrus and pome fruit. Currently, he is engaged in developing another biological product, based on the use of a naturally occurring yeast antagonist, for the control of pre- and postharvest decay of table and wine grapes. His work in the field of biological control has gained international recognition, and he has been invited by several universities and research institutions around the world to lecture about various aspects of his work.

Dr. Droby has published more than 100 papers in international journals, chapters in books, and proceedings in symposia. He is a member of the board of directors of the Phytopathological Society of Israel and a member of the American Phytopathological Society and the International Society of Plant Pathology.

The Contributors

Charles W. Bacon
Toxicology and Mycotoxin Research Unit
Agricultural Research Service
U.S. Department of Agriculture
Richard B. Russell Agricultural Research Center
Athens, Georgia

Larry R. Beuchat
Center for Food Safety and Quality Enhancement
University of Georgia
Griffin, Georgia

D. Bhatnagar
Food and Feed Safety Research Unit
Southern Regional Research Center
U.S. Department of Agriculture
Agricultural Research Service
New Orleans, Louisiana

Edo Chalutz
Executive Director
BARD, United States–Israel Binational Agricultural Research and Development Fund
Bet Dagan, Israel

T.E. Cleveland
Food and Feed Safety Research Unit
Southern Regional Research Center
U.S. Department of Agriculture
Agricultural Research Service
New Orleans, Louisiana

Dean O. Cliver
WHO Collaborating Center for Food Virology
Department of Population Health and Reproduction
School of Veterinary Medicine
University of California, Davis
Davis, California

P.J. Cotty
Food and Feed Safety Research Unit
Southern Regional Research Center
U.S. Department of Agriculture
Agricultural Research Service
New Orleans, Louisiana

Samir Droby
Department of Postharvest Science
Agricultural Research Organization
The Volcani Center
Bet Dagan, Israel

Ahmed El Ghaouth
Faculte des Sciences et Techniques
Universite de Nouakchott
Nouakchott, Mauritania

I. Eshkenazi
Department of Molecular Microbiology and Biotechnology
Tel-Aviv University
Ramat-Aviv, Israel

Pina M. Fratamico
U.S. Department of Agriculture
Agricultural Research Service
Eastern Regional Research Center
Wyndmoor, Pennsylvania

S.J. Fritschel
Qualicon, Inc.
Wilmington, Delaware

Daniel Y.C. Fung
Department of Animal Sciences and Industry
Kansas State University
Manhattan, Kansas

Michael J. Goldblatt
Director, Defense Sciences Office
Defense Advanced Research Projects Agency (DARPA)
Arlington, Virginia

V.K. Juneja
Food Safety Research Unit
U.S. Department of Agriculture
Agricultural Research Service
Eastern Regional Research Center
Wyndmoor, Pennsylvania

Isaac Klinger
Kimron Veterinary Institute
Veterinary Services and Animal Health
Ministry of Agriculture
Bet Dagan, Israel

Michael D. McElvaine
U.S. Department of Agriculture
Office of Risk Assessment and Cost-Benefit Analysis
Washington, D.C.

Filmore I. Meredith
Toxicology and Mycotoxin Research Unit
Agricultural Research Service
U.S. Department of Agriculture
Richard B. Russell Agricultural Research Center
Athens, Georgia

T. Neufeld
Department of Molecular Microbiology and Biotechnology
Tel-Aviv University
Ramat-Aviv, Israel

William P. Norred
Toxicology and Mycotoxin Research Unit
Agricultural Research Service
U.S. Department of Agriculture
Richard B. Russell Agricultural Research Center
Athens, Georgia

Ronald D. Plattner
Bioactive Agents Research
Agricultural Research Service
U.S. Department of Agriculture
National Center for Agricultural Utilization Research
Peoria, Illinois

James K. Porter
Toxicology and Mycotoxin Research Unit
Agricultural Research Service
U.S. Department of Agriculture
Richard B. Russell Agricultural Research Center
Athens, Georgia

Carrie E. Regenstein
Director, University Computing Center
Assistant Dean for Educational Technology
University of Rochester
Rochester, New York

Joe M. Regenstein
Professor of Food Science
Cornell Kosher Food Initiative
Department of Food Science
Cornell University
Ithaca, New York

Ronald T. Riley
Toxicology and Mycotoxin Research Unit
Agricultural Research Service
U.S. Department of Agriculture
Richard B. Russell Agricultural Research Center
Athens, Georgia

J. Rishpon
Department of Molecular Microbiology and Biotechnology
Tel-Aviv University
Ramat-Aviv, Israel

V. Sacks
Department of Molecular Microbiology and Biotechnology
Tel-Aviv University
Ramat-Aviv, Israel

Avishay-Abraham Stark
Department of Biochemistry
Faculty of Life Sciences
Tel-Aviv University
Ramat-Aviv, Israel

D.W. Thayer
Food Safety Research Unit
U.S. Department of Agriculture
Agricultural Research Service
Eastern Regional Research Center
Wyndmoor, Pennsylvania

Kenneth A. Voss
Toxicology and Mycotoxin Research Unit
Agricultural Research Service
U.S. Department of Agriculture
Richard B. Russell Agricultural Research Center
Athens, Georgia

H. Michael Wehr
U.S. Food and Drug Administration
Center for Food Safety and Applied Nutrition
Office of Constituent Operations
Washington, D.C.

Charles L. Wilson
Appalachian Fruit Research Station
U.S. Department of Agriculture
Agricultural Research Service
Kearneysville, West Virginia

Michael Wisniewski
Appalachian Fruit Research Station
U.S. Department of Agriculture
Agricultural Research Service
Kearneysville, West Virginia

Table of Contents

Control of Microbial Food Contaminants

Microbial Food Contamination and International Trade

1 Foodborne Human Pathogens and Their Toxic Metabolites in Israel

Isaac Klinger

CONTENTS

I. INTRODUCTION

Historical evidence that the problem of food safety has always troubled mankind is available from since the beginning of human culture. Foodborne infections and intoxication (FBII) were common in ancient times, but without rational explanations for the appearance of some symptoms after the consumption of certain foods ancient peoples based their food laws on religious beliefs. Trial and error were the main tools for reaching conclusions about the safety of foods, and taboos were the primary tool for prevention of FBII. The most comprehensive code of ancient laws is *The Bible*, particularly the book of Leviticus. Certain foods were described as "unhealthy" and unfit for human consumption, particularly those of animal origin. Some animals were defined as "unclean" and were therefore unsuitable for consumption. The consumption of the blood of slaughtered animals was totally prohibited, as the blood was considered to be the soul of the animal. Believers were allowed to consume the flesh but not the soul. All other agricultural products were declared suitable. Similar laws were also prevalent in ancient Egypt. According to Plutarch, pigs were declared "unhealthy animals;" therefore, pork consumption was prohibited. In ancient Rome, similar laws were issued for the prohibition of goat meat consumption, whereas the population of ancient Rome ate pork meat enthusiastically. In Japan, beginning in the 7th century when Buddhism became the state religion, the meat of farm animals was totally prohibited for human consumption.

Much is known about food laws promulgated in Europe during the Middle Ages. Such laws were developed especially in urban communities, primarily to protect commercial or other interests. Some of the laws had a hygienic basis, such as a law passed in the town of Strasbourg (1322) which stated that meat from sick animals could only be sold to hospitals for use as food for sick people. Only when the study of food microbiology began to develop at the end of the 19th century could the pattern of FBII agents be understood.

A wide range of microorganisms and their toxic metabolites may be transmitted to humans in edible materials used as food. The variety of foods causing FBII depends largely on the dietary habits of various ethnic groups, reflecting the preferences of particular populations for certain products and cooking habits. Traditionally, attention has been given to those agents that cause disturbances of the gastrointestinal tract; however, in recent years, information has accumulated regarding an increasing number of agents that are involved in episodes of FBII but have an effect on other body systems.

In general, food of an animal origin is more involved in episodes of FBII than is food of a plant origin. The most likely reasons are that animals can act as active or passive vectors of FBII agents, and the products themselves offer nutrients for the growth of certain FBII agents. But, with the increase in population densities and the trend of recycling marginal water in agriculture, urban effluents may gain access to food of a plant origin normally considered to be safe but which transfers FBII from latent or clinically sick excreters. Another trend in recent years is that certain microorganisms not considered to be pathogens are gaining importance as FBII agents. These are the so-called "new" or "emerging" pathogens. This finding is mainly due to our improved understanding of infection chains and to improvements in diagnostic techniques.

The agents involved in FBII consist of a range of microorganisms that includes fungi, bacteria, viruses, parasitic protozoa, helminthic parasites, their metabolites, and others not yet fully understood. Virtually each material used for human food harbors a unique assortment of microorganisms originating from contaminants gathered on the product prior to or during harvesting or in the post-harvesting period; therefore, every agricultural product has a typical spectrum of contaminants reflected by the exposure pattern of the product. The information gathered in recent decades about possible FBII agents that may be present in a product indicates treatments that should be employed for ensuring a safe product. There is no demand that materials used as human food should be free of microorganisms. Most agricultural products have a certain load of microorganisms usually considered as saprophytes and as such are tolerated. As a result, microbiological standards are recommended for most products. Some of these contaminants, however, may gain importance due to new scientific evidence on their role as FBII agents. Reports on outbreaks of FBII or emerging foodborne pathogens usually catch the public's attention, and the natural reaction of the public is purchase withdrawal, thus causing very severe damage to entire agricultural sectors.

II. PRODUCTS OF ANIMAL ORIGIN

A. Poultry Meat

Poultry meat is considered to be one of the most common foods that cause FBII. Industrially slaughtered poultry may harbor some 27 different species of bacteria, some of which are considered as pathogens. Post-slaughter treatment is usually limited to rinsing with potable water, which does not eliminate the contamination. Poultry meat is blamed as the most common vector that introduces FBII agents such as *Salmonella*, *Campylobacter*, *Listeria*, and others into the food-preparing establishment.

B. Eggs

At one time, eggs were considered the perfect food in the perfect wrapping; however, in the last decade the transovarial transmission of *Salmonella enteritidis* PT4 inside eggs has been reported in several countries. This episode caused the egg industry severe economic damage. No treatment was described for the product itself, except to reduce the multiplication rate of bacteria present in the egg to a certain extent by reducing the storage temperature and the marketing period.

C. Red Meat

Beef, sheep, and goat meat was considered to be safe because the hides of the slaughtered animals are removed during the harvesting of the edible tissues. When working in the abattoir is in accordance with hazard analysis and critical control point (HACCP) concepts, the contamination of the processed carcasses with FBII agents can be avoided. However, in recent years, due to contamination of beef with a certain strain of a bacteria considered to be a normal inhabitant of the animal intestines, *Escherichia coli* (O157:H7), the bacterium was identified as being an agent causing very severe disease among humans, particularly high-risk consumer groups. This contamination, as well as contamination involving a new group of pathogens called the Prions (emerging pathogens not recognized before), is blamed as being the agent of TSE in humans and has caused the beef industry the most severe economical damage in the history of agriculture.

D. Milk

Since the beginning of modern microbiology, milk has been recognized as a vector for several diseases such as tuberculosis, brucellosis, and salmonellosis. Milk was the first food item of animal origin to be treated by exposure to heat treatment sufficient to kill pathogenic agents while maintaining the nutritional value of the product at the same time. Most countries in the developed world have adopted regulations prohibiting the marketing of nonpasteurized milk; however, in those countries where such regulations have not been not issued or are not enforced, episodes of FBII originating from various milk products have been reported.

E. Fish and Seafood

Fish may be contaminated by various means from the surrounding water. Some FBII agents are permanent residents of the marine environment and should be recognized as such. Another way contaminants gain access to products is through contamination of the marine environment by urban effluents.

2 Innovative Technology To Reduce Microbial Food Contamination

Michael J. Goldblatt

CONTENTS

I. INTRODUCTION

One of the most exciting new technologies for reducing microbial food contamination is the Microsphère® technology being developed by Bernàrd Technologies, Inc. (BTI). The Microsphère® system is active packaging with a twist — it is also intelligent. Rather than taking something out of the packaged environment, such as oxygen, the Microsphère® system generates, on demand, controlled amounts of chlorine dioxide gas, an extremely potent biocide, in the package environment. The Microsphère® system has been designed to be cost effective, compatible with existing packaging manufacturing equipment and practices, and transparent to both the food manufacturer and the end-use customer. Because of the technology's flexibility, packaging material can be tailored to the individual antimicrobial requirements as dictated by the material being packaged.

II. BIOCIDAL USES OF CHLORINE DIOXIDE

Chlorine dioxide is very effective against a broad range and a large variety of microbes, including HIV, *Escherichia coli*, and poliovirus. Some examples of microorganisms known to be controlled by chlorine dioxide include:

- Viruses: HIV, poliovirus, rotavirus, herpesvirus, ECHO virus
- Bacteria: *E. coli, Salmonella, Staphylococcus*
- Spore formers: *Bacillus* spp., *Clostridium* spp.
- Molds: *Chaetomium, Aspergillus*
- Fungi: *Botrytis, Alternaria, Colletotrichum*
- Protozoa: *Cryptosporidium, Giardia*

Chlorine dioxide gas was discovered in 1811. Chlorine dioxide was unexploited until new technologies became available that permitted on-site, on-demand generation. It was then approved and used to bleach wheat flour and to remove objectionable tastes and odors from municipal waters. A short time later, its potential to bleach wood pulp was realized, and its primary industrial use was born. In the 1950s and 1960s, research articles began to appear that discussed the extensive antimicrobial activity of chlorine dioxide.

Chlorine dioxide has received more attention in recent years, due to a large extent to Howard Alliger and Aaron Rosenblatt, and is employed with greater frequency for disinfecting drinking water; reducing microbial contamination on fresh food, produce, and meats; sanitizing medical and food equipment; and for viral kill and slime control in cooling tower waters.[1–3] It has a unique ability to break down phenolic compounds and remove phenolic tastes and odors from the water. Another favorable feature is said to be its lack of reaction with ammonia. That characteristic is of great importance to humans and the environment. Chlorine dioxide is used in the treatment of drinking water, as well as in wastewater, and for elimination of cyanides, sulfides, aldehydes, and mercaptans. Ingols et al.[4] showed the oxidation capacity of ClO_2 in terms of available chlorine to be greater than that of chlorine. Finally, chlorine dioxide does not form trihalomethanes or chlorophenols.[5]

In 1976, the U.S. Environmental Protection Agency (EPA) discovered that trihalomethanes (THMs) were being produced in drinking water as a by-product of chlorination. The EPA considers THMs to be carcinogenic.[21] Calabrese et al.[6] and Bull[7] have also reported on the health effects of chlorine dioxide in water. The books *Chlorine Dioxide: Chemistry and Environmental Impact of Oxychlorine Compounds*[9] and *Ozone/Chlorine Dioxide Oxidation Products of Organic Materials*[10] are good reference sources for additional information about chlorine dioxide. Hoehn et al.[11] have published a study that addresses odors reportedly associated with the use of chlorine dioxide as a drinking water disinfectant. The EPA's Division of Drinking Water began a long-term program to discover replacements for chlorine in drinking water. Of the three leading candidates, chlorine dioxide was judged to be the best overall compound on the basis of its high antimicrobial activity and its ability to remain in solution and most importantly the fact that it does not produce chlorinated organics such as THMs.

In further work, Rosenblatt et al.[12] and Ridenour et al.[13] demonstrated greater sporicidal activity for chlorine dioxide than for chlorine. The greater sporicidal activity of chlorine dioxide is explained by greater utilization of its oxidation capacity involving a full charge of five electrons. Harakeh et al.[14] evaluated the effectiveness of solutions of chlorine dioxide against *Yersinia enterocolitica* and *Klebsiella pneumoniae*. Berman et al.[15] showed good antiviral activity by chlorine dioxide at pH 10 in less than 15 seconds.

The studies of Ridenour and Armbruster,[13] Rosenblatt et al.,[12] and Benarde et al.[1] indicate that chlorine dioxide is an effective water disinfectant for achieving the destruction of bacteria. Chlorine dioxide is also a potent virucide[3,16–18] and is potentially effective against a recently recognized public health hazard, waterborne *Cryptosporidium* oocytes.[19] Their data show that its effectiveness is not lessened by increases in the pH of water.

Recently, chlorine dioxide gas has been recognized as a means for sterilizing medical devices. Studies have shown that even at low concentrations (20 mg/L) chlorine dioxide is an effective sterilant. Additionally, Rosenblatt and Knapp[12,20] have reported on the importance of relative humidity for microbial inactivation and concluded that 50% or higher is optimal for sterilization, and Jen and Woodworth[23] extended these studies.

In a major step toward safety and localized generation of chlorine dioxide from the solid state, Bernàrd Technologies, Inc., in the early 1990s, began developing its Microsphère® product line. This product line includes a number of proprietary, controlled-release, solid-state antimicrobial and deodorizing products that form localized Microatmosphere® environments.

III. DESCRIPTION OF MICROSPHÈRE® SOLID-STATE CHEMISTRY

Chlorine dioxide chemistry is centered on the conversion of sodium chlorite into chlorine dioxide without producing free chlorine. This conversion occurs when ions of chlorite, from sodium chlorite, are acidified with various acid groups. A description of the compounds likely to be encountered in chlorine dioxide chemistry is provided in Table 2.1 for clarity.

Generated chlorine dioxide normally uses sodium chlorite and an acid to drive the reaction as completely as possible toward the production of chlorine dioxide. "Activation" of the Microsphère® system is designed to complete the conversion of chlorite to chlorine dioxide. The mode of action depicted in Figure 2.1 is initiated when moisture from the surface environment reacts with a hydrophobic phase component in the Microsphère® material to form an acid which in turn reacts with a hydrophilic phase to convert ionomeric chlorite to chlorine dioxide. The proprietary chemistry, which enables the sustained controlled release of chlorine dioxide, is contained in micron-sized inorganic particles that are combined with polymeric components to form film material.

IV. CHEMICAL STRUCTURE AND MODE OF ACTIVITY OF CLO_2 CHEMISTRY

The chlorite ion and chlorine dioxide are chemically very similar and often referred to as the same entity. The mode of action of chlorine dioxide is not well understood. Chlorine dioxide exists as a free radical in nature.

TABLE 2.1
Chlorine Dioxide Chemistry

Compound	Symbol	Description
Sodium chlorite (chlorite ion)	$NaClO_2$ (ClO_2^-)	This compound is the primary precursor to chlorine dioxide in use today. It is the molecule that is "converted" to chlorine dioxide. The chlorite ion is also the primary by-product of the reaction of chlorine dioxide with other compounds and is then converted to simple chloride, a component of table salt.
Sodium chlorate (chlorate ion)	$NaClO_3$ (ClO_3^-)	This is an oxychlorine compound used in some liquid processes to generate gaseous chlorine dioxide. Its ion can also be a minor component of the by-product of the reaction of chlorine dioxide in solution. Chlorate is not generated by the chemistries in Microsphère®.
Chlorous acid	$HClO_2$	This is a weak acid that is intermediate in the reaction path between sodium chlorite and chlorine dioxide. It is thought to have high antimicrobial activity, especially when combined with chlorine dioxide itself. It is produced and maintained only under certain conditions of pH and concentration in aqueous systems. Since Microsphère® is a solid-based system, chlorous acid is not formed.
Chlorine dioxide	ClO_2	This is a powerful oxidizer that exists as a gas in nature but is 40 times more soluble in water than in air; therefore, air concentrations are extremely low.

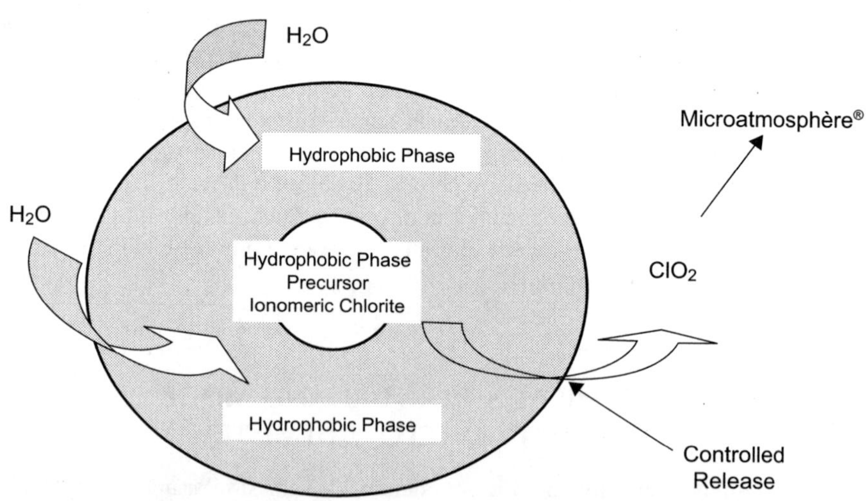

FIGURE 2.1 Microsphère® mode of action.

The chlorite molecule is converted to chlorine dioxide but probably goes through at least one intermediate compound, which is then converted to chlorine dioxide. Under various conditions of concentration and pH, the rate of conversion can vary. Once the chlorine dioxide locates and extracts an electron, it is reduced back toward the chlorite ion.

The activity of the Microsphère® system is believed to stem from the source of the electron extracted by the chlorine dioxide component. At least four specific amino acids readily react with chlorine dioxide: two aromatic amino acids, tryptophane and tyrosine, and two sulfur-bearing amino acids, cysteine and methionine. The ring structures of tryptophane and tyrosine are a rich source of electrons, which can be captured by strong oxidizers such as chlorine dioxide. The sulfur-bearing amino acids have an electronegative character and also readily give up electrons.

An oxidative attack on these amino acids causes tremendous disruption to cellular functions, as these molecules are the building blocks of proteins and essential constituents to life. The mechanism by which oxidation of amino acids causes structural disruption of the protein chain or, in the case of the sulfur-containing amino acids, a disruption of the disulfide bonds linking several protein chains is illustrated in Figure 2.2. In either case, the enzyme will become dysfunctional and the associated biochemical reaction will cease. The disulfide bonds, responsible for the structural integrity of the polypeptide molecule, are broken, allowing for the two chains to separate. Because proteins must have a precise tertiary structure, this separation "denatures" the enzyme and renders it inactive. This can lead directly to microbial death.

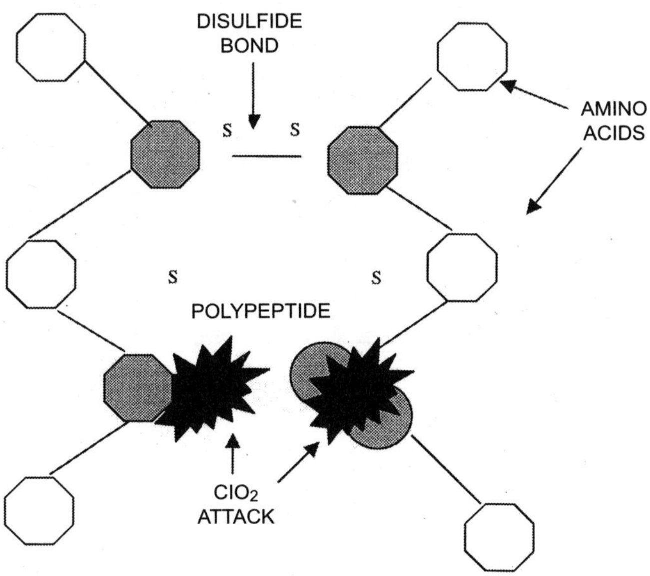

FIGURE 2.2 Oxidative attack by chlorine dioxide on polypeptide chains.

V. MICROSPHÈRE® MATERIAL CHARACTERISTICS

The Microsphère® system is a unique patented dry powder which is stable when stored dry and is activated when exposed to threshold levels of humidity in ambient air. The controlled release of chlorine dioxide can be custom tailored to occur over a time range varying from hours to months at concentrations from less than 1 ppm to hundreds of parts per million. The chemistry can also be tailored to respond to humidities as low as 20% or as high as 75%. As an example, Figure 2.3 illustrates the controlled release of chlorine dioxide as a function of time obtained with the Microsphère® system.

In its inactive state, Microsphère® has the following characteristics:

- Single-component white powder or granules, with no odor
- Solid-state matrix of precursor ingredients to produce a Microatmosphere® environment
- Nonflammable, non-explosive
- Functions as a bacteriostat, microbicide, and deodorant when activated
- Stable shelf-life which allows long-term inventory
- Environmental compatibility for disposal
- Toxic Substances Control Act (TSCA)-listed raw materials

Activated Microsphère® films have the following characteristics:

- They exhibit high activity against a broad spectrum of microorganisms, including vegetative or actively growing cells and spores.
- They are active over a broad range of pH.
- They do not produce chlorine and/or chlorinated by-products.
- Antimicrobial activity accelerates as the humidity increases.
- They provide continuous extended chlorine dioxide generation.
- They generate levels of chlorine dioxide (parts per million) required for antimicrobial activity.

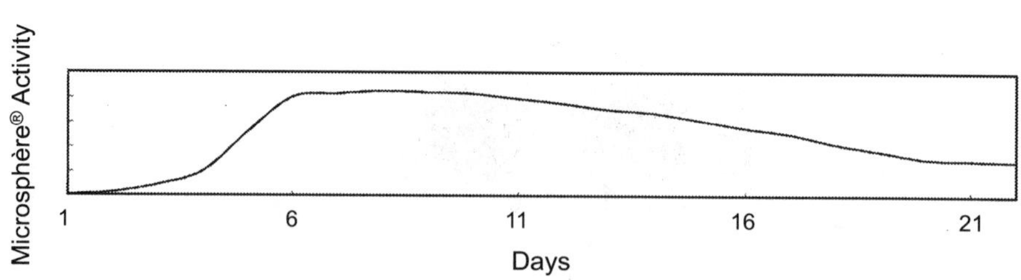

FIGURE 2.3 Controlled release of chlorine dioxide, as a function of time, obtained with the Microsphère® system.

VI. SAFETY OF CHLORINE DIOXIDE

Many evaluations have shown ClO_2 compounds to be non-toxic. Toxicology tests include ingestion of ClO_2 in drinking water, additions to tissue culture, injections into the blood, seed disinfection,[22] insect egg disinfection, injections under the skin of animals and into the brains of mice, burns administered to over 1500 rats, and injections into the stalks of plants. Standard tests include Ames mutation, Chinese hamster, rabbit's eye, skin abrasion, pharmacodynamics, and teratology.[24]

Seed corn was coated with a polymeric film containing the Microsphère® system and then planted in regional experimental farm plots. Mold contamination was reduced to the same level as for Captan™-treated seed, and germination proceeded normally with yield reduction.[25] With the prospect of changing from chlorine to chlorine dioxide in our water supply, the EPA and American Water Works have commissioned over 100 papers and studies on the toxicity of ClO_2. Many controlled animal studies on the effects of ingesting sodium chlorite and chlorine dioxide have been conducted using 1- to 1000-mg/L concentrations. Metabolically, both ClO_2 and ClO_2^- are rapidly reduced following ingestion. Radioactive chlorine tests show that most of the tagged chlorine is excreted from the urine in the form of Cl^- ion with a small amount of ClO_2^-. The no observed effect level (NOEL) from animal ingestion involving ClO_2 and ClO_2^- ranges up to 100 ppm.[26–28] The half-life for the elimination of ClO_2 and ClO_2^- from the plasma is less than half that of HOCl (hypochlorite).[8]

In one study, human volunteers drank up to 24 ppm ClO_2 or ClO_2^- in solution and showed no adverse effects.[29] Several studies examined the effects on reproductive toxicity or teratology. There is no evidence of fetal malformation or birth defects at ClO_2 concentrations of up to 100 ppm through drinking or skin route.[30–32] With prolonged feeding, toxicity is produced mainly in the red blood cell. Rats fed up to 1000 mg/L chronically for 6 months showed no significant hematological changes. After 9 months, however, red blood cell counts, hematacrit, and hemoglobin were decreased in all treatment groups. Lack of toxicity on a long-term but low-level basis is dramatically illustrated by a study where rats[33] were fed ClO_2 in high doses over a 2-year period.

VII. REGULATORY STATUS OF CHLORINE DIOXIDE IN U.S.

Several U.S. governmental agencies — the Food and Drug Administration (FDA), the EPA, and the U.S. Department of Agriculture (USDA) — regulate and have approved the use of chlorine dioxide and its precursor, sodium chlorite. Sodium chlorite is approved as an ingredient in adhesives used for packaging, transportation, and holding of foods. Chlorine dioxide may be safely used as a sanitizing solution on food-processing equipment and utensils. Sodium chlorite may be used as a slimicide in the manufacture of paper and paperboard that contact food. Chlorine dioxide may also be used to bleach cherries and flour and is approved as an antimicrobial agent in water used for poultry processing and for the washing of uncut,

TABLE 2.2
Comparison of Minimum Bactericidal Concentrations (2.5 Minutes)

	Microorganisms				
Reagents	***E. coli* (ppm)**	***S. aureus* (ppm)**	**MRSA (ppm)**	***B. subtilis* (spore) (ppm)**	***A. niger* (ppm)**
GA	100,000	100,000	100,000	100,000	100,000
PN	10,000	>10,000	>10,000	>10,000	>10,000
EtOH	500,000	500,000	500,000	>500,000	>500,000
CHG	100	10	1,000	1000	>10,000
BAC	100	10	100	1000	10,000
PVP–I	10	100	100	>1000	1000
SH	10	10	10	>1000	1000
CD	1	1	1	100	10

Note: GA = glutaraldehyde; PN = phenol; EtOH = absolute ethyl alcohol; CHG = chlorhexidine digluconate; BAC = benzalkonium chloride; PVP-I = povidone iodine; SH = sodium hypochlorite; and CD = chlorine dioxide.

unpeeled fruits and vegetables, cut potatoes (French fries), and red meats. The USDA has approved the use of chlorine dioxide as a sanitizer for all surfaces; for treating boilers, steam lines, and cooling systems; and for cleaning floors and walls. Chlorine dioxide is permitted as a chemical germicide for disinfecting hemodialysis equipment or supplies for safe reuse. The EPA has exempted sodium chlorite and chlorine dioxide from the requirement of a tolerance for residues as a seed soak treatment in the growing of the raw agricultural commodities crop groups, and as an alternate or supplemental disinfectant or oxidant in drinking water.

VIII. COMPARATIVE BACTERICIDAL EFFECTIVENESS OF DIFFERENT BIOCIDES, INCLUDING CHLORINE DIOXIDE

Takayama et al.[35] illustrate (Table 2.2) the bactericidal efficacy of chlorine dioxide relative to other commonly used disinfectants. The disinfectants used in the study were chlorine dioxide, glutaraldehyde, phenol, absolute ethyl alcohol, chlorhexidine digluconate, benzalkonium chloride, povidone iodine, and sodium hypochlorite. The table provides the minimum bactericidal concentrations in parts per million for a 2.5-minute exposure for five different organisms. The minimum bactericidal concentrations for chlorine dioxide are significantly lower than for any of the other disinfectants shown.

The Microsphère® technology differs from traditional biocides in that it employs a solid-state chemistry that can be incorporated into packaging material to create on demand a vapor gas-phase biocidal microenvironment within the package. Prior to packaging a food product, the solid-state precursor chemistry remains inactive. Upon exposure to moisture in the food, the solid-state precursor chemistries become activated and generate the biocidal microenvironment that creates an active Microatmosphere® environment of chlorine dioxide that controls a broad spectrum of microbes without requiring direct surface contact.

IX. CURRENT DEVELOPMENT ACTIVITIES

BTI is currently pursuing regulatory approval and commercialization of a variety of products incorporating its technology to control harmful microorganisms and help reduce the potential for disease transmission. Of relevance to the food industry, BTI is now finishing the development of incorporating its technology into packaging material commonly used in the food industry. This will provide the industry with environmentally compatible, sustained-release packaging that will release chlorine dioxide at a specific concentration and rate to reduce product spoilage and increase transport and distribution shelf-life by inhibiting unwanted development and colonization of microorganisms. Recent advances in the chemistry of the Microsphère® technology have led to the creation of Microlite™ technology, a light-activated system for the generation of chlorine dioxide from the solid state. As with the Microsphère® system, the Microlite™ system can readily be incorporated into the manufacture of common packaging material.

REFERENCES

1. Benarde, M.A., Israel, B.M., Olivieri, V.P., and Granstrom, M.L. 1965. Efficiency of chlorine dioxide as a bactericide. *Appl. Microbiol.,* 13:776–780.
2. Roller, S.D. 1978. Some Aspects of the Mode of Action of Chlorine Dioxide on Bacteria, Master's thesis, The Johns Hopkins University, Baltimore, MD.
3. Chen, Y.S., and Vaughn, J.M. 1990. Inactivation of human and simian rotavirus by chlorine dioxide. *Appl. Environ. Microbiol.,* 56:1363–1366.
4. Igols, R.S. and Ridenour, G.M. 1948. Chemical properties of chlorine dioxide in water treatment. *J. Am. Water Works Assoc.,* 40:1207–1227.
5. Symons, J.M. et al. 1976. *Interim Treatment Guides for the Control of Chloroform and other Trihalomethanes.* Water Supply Research Division, Municipal Environmental Research Laboratory, U.S. Environmental Protection Agency, Cincinnati, OH.
6. Calabrese, E.J., Moore, G.S., and Tuthill, R.W. 1978. Health effects of chlorine dioxide as a disinfectant in potable water: a literature survey. *J. Environ. Health,* 41:26–31.
7. Bull, R.J. 1980. Health effects of alternate disinfectants and their reaction products. *J. Am. Water Works Assoc.,* 72:299–303.
8. Bull, R.J., Health effects of drinking water disinfectants and disinfectant by-products. *Environ. Sci. Tech.,* Vol. 16, Nov. 10, 1982.
9. Masschelein, W.J. 1979. *Chlorine Dioxide: Chemistry and Environmental Impact of Oxychlorine Compounds,* R.G. Rice, Ed. Ann Arbor Science Publishers, Ann Arbor, MI.

10. Rice, R.G. and Cotruvo, J.A., Eds. 1978. *Ozone/Chlorine Dioxide Oxidation Products of Organic Materials*. International Ozone Institute, Cleveland, OH.
11. Hoehn, R.C. et al. 1990. Household odors associated with the use of chlorine dioxide. *J. Am. Water Works Assoc.,* 82:166–172.
12. Rosenblatt, D.H., Rosenblatt, A.A., and Knapp, J. A. (1987) Use of chlorine dioxide gas as a chemosterilizing agent, U.S. Patent No. 4,681,739.
13. Ridenour, G.M. and Armbruster, E.H. 1949. Bactericidal effect of chlorine dioxide. *J. Am. Water Works Assoc.,* 41:537–550.
14. Harakeh, M.S., Berg, J.D., Hoff, J.C., and Matin, A. 1985. Susceptibility of chemostat-grown *Yersinia enterocolitica* and *Klebsiella pneumoniae* to chlorine dioxide. *Appl. Environ. Microbiol.,* 49(1):69–72.
15. Berman, D. and Hoff, J.C. 1984. Inactivation of simian rotavirus SA 11 by chlorine, chlorine dioxide, and monochloramine. *Appl. Environ, Microbiol.,* 48(2):317–323.
16. Scarpino, P.V., Brigano, F.A.O., Cronier, S., and Zink, M.L. 1979. *Effect of Particulates on Disinfection of Enteroviruses in Water by Chlorine Dioxide,* EPA-600/2-79-054, Environmental Protection Technology Series, U.S. Environmental Protection Agency, Washington, D.C.
17. Hauchman, F.S., Noss, C.I., and Olivieri, V.P. 1986. Chlorine dioxide reactivity with nucleic acids. *Water Res.,* 20:357–361.
18. Farr, R.W., and Walton, C. (1993) Inactivation of human immunodeficiency virus by a medical waste disposal process using chlorine dioxide. *Infect. Control Hosp. Epidemiol.,* 14:527–529.
19. Peeters, J.E. et al. 1989. Effect of disinfection of drinking water with ozone or chlorine dioxide on survival of *Cryptosporidium parvum* oocysts. *Appl. Environ. Microbiol.,* 55:1519–1522.
20. Rosenblatt, D.H., Rosenblatt, A.A., and Knapp, J.A. (1985) Use of chlorine dioxide gas as a chemosterilizing agent, U.S. Patent No. 4,504,442.
21. U.S. EPA. 1977. *Ozone, Chlorine Dioxide, and Chloramines as Alternates to Chlorine for Disinfection of Drinking Water: State of the Art.* Water Supply Research, Office of Research and Development, U.S. Environmental Protection Agency, Cincinnati, OH.
22. Kawada, Hiroshi, Haneda, and Tadayoshi. 1995. Soil Disinfection by Using Aqueous Chlorine Dioxide Solutions. Patent application no. JP95-111095.
23. Jeng, D.K., and Woodworth, A.G. (1990) Chlorine dioxide gas sterilization under square-wave conditions. *Appl. Environ. Microbiol.,* 56:514–519.
24. Bringmann, G. 1953. Electron microscopic findings of the action of chlorine, bromine, iodine, copper, silver and hydrogen peroxide on *E. coli. Z. Hyg. Infektionskrankh,* 138:156–166.
25. Gray, P.N., and Lelah, M.D. (1997) Effects of Solid-State Chlorine Dioxide Generating Films on Corn Germination (internal technical report). Bernàrd Technologies, Inc.
26. Bull, R.J. et al. 1979. Carcinogenic activity of reaction products of alternate drinking water disinfectants, *Pharmacologist,* 21(3):218
27. Yokose, Y., Uchida, K., Nakae, D., Shiraiwa, K., Yamamoto, K., and Konishi, Y. 1987. Studies of carcinogenicity of sodium chlorite in B6C3F1 mice. *Environ. Health Perspect.,* 76:205–210.
28. Drinking Water Criteria Document on Chlorine Dioxide, Chlorite, and Chlorate [final draft], EPA contract No. 68-C2-0139, Clement International Corp., March 31, 1994.
29. Gordon, G., Kieffer, R., and Rosenblatt, D. 1972. *The Chemistry of Chlorine Dioxide: Progress in Inorganaic Chemistry,* Vol. 15. Wiley-Interscience, New York, p. 38.

30. Suh, D.H., Abdel-Rahman, M.S., and Bull, R.J. 1983. Effect of chlorine dioxide and its metabolites in drinking water on fetal development in rats. *J. Appl. Toxicol.,* 3:75–79.
31. Tuthill, R.W., Guisti, R.A., Moore, G.S., and Calabrese, F.J. 1982. Health effects among newborns after prenatal exposure to ClO_2 disinfected drinking water, *Environ. Health Perspect.,* 46:39–45.
32. Gerges, Abdel-Rahman, and Skowronski. 1985. Effects of alcide gel on fetal development on rats and mice, *J. Appl. Toxicol.,* 5(2):16
33. Haag, H.B. 1949. *The Effects on Rats of Chromic Administration of Sodium Chlorite and Chlorine Dioxide in Drinking Water,* Report to Olin Corp., Medical College of Virginia, Department of Physiology and Pharmacology, 1949.
34. Takayama, M., Sugimoto, H., Mizutani, S., and Tanni, K. 1995. Bactericidal activities of chlorine dioxide. *J. Antibacterial Antifungal Agents,* 23(7):401–406.

Incidences and Nature of Microbial Food Contamination

3 Viruses and Protozoan Parasites on Food

Dean O. Cliver

CONTENTS

I. ABSTRACT

In addition to the bacterial agents that one often reads about, viruses and parasites are important causes of foodborne and waterborne disease in the U.S. and presumably elsewhere, although incidence data from other countries are difficult to obtain. Viruses are transmitted as inert particles that are much smaller than bacteria. All of the viruses transmitted to humans via food and water (principally hepatitis A virus and the Norwalk-like viruses, occasionally a few others) are only infectious for humans and are shed in feces. These particles are inert and cannot multiply outside the body but may be inactivated (deprived of their infectivity) by cooking and a few other means. Hepatitis A virus causes a debilitating illness with an average 28-day incubation period, whereas the Norwalk-like viruses cause diarrhea and vomiting with a 1- to 2-day incubation period and a 1- to 2-day duration. Protozoa transmitted by food and water include *Giardia lamblia* and *Cryptosporidium parvum*, which are shed in human and animal feces and cause diarrhea; *Toxoplasma gondii*, which is shed in cat feces but may also infect humans via tissue cysts in undercooked meat and cause birth defects; and *Cyclospora cayetanensis*, which is shed in human feces and is rare in North America but has caused diarrhea via imported produce. Outside of North America, *Entamoeba histolytica* is still a serious threat, causing severe gastroenteritis and sometimes affecting other parts of the body. The transmission forms of protozoa shed in feces are microscopic cysts or oocysts which are larger than bacteria and cannot multiply in food or water. The infectious forms of protozoa can be killed by cooking, as can viruses. Unlike viruses, these infectious forms are typically susceptible to freezing and to irradiation. For the agents shed in feces, the first line of defense is obviously sanitation. All of these agents are more difficult than bacteria to detect in food and water. Though less "respected" than foodborne bacteria, they occasionally remind us not to take them for granted.

II. VIRUSES AND PROTOZOA

Viruses and protozoan parasites are important causes of foodborne and waterborne disease in the U.S.[2,12] Compared with diseases caused by bacteria, the illnesses these agents cause are relatively difficult to diagnose and are therefore even more under-reported than bacterial diseases. Some parasitic diseases are difficult to treat; others, as well as the viral diseases, cannot be directly treated at all. These agents, and the diseases they cause, will be discussed from the standpoints of their epidemiology, transmission, and impact on public health.

A. Viruses

In the U.S., the viruses most frequently transmitted via foods are the hepatitis A virus, which was the fourth leading cause of outbreak-associated foodborne disease in the U.S. from 1988 to 1992, and the Norwalk-like gastroenteritis viruses, which ranked number nine for that period.[2,7] Hepatitis E virus, which is often transmitted via water (and perhaps food) in Asia, Africa, and South America, evidently occurs only as imported cases in North America.[8] Foodborne viruses come from human feces, with the minor exception that the Norwalk-like viruses are also shed in vomitus.

Hepatitis A virus infects perorally and colonizes the liver.[8] After a period of multiplication, the virus elicits an immune response, and the host's body sends cytotoxic T-cells to destroy the infected liver cells. Debilitating illness, often including jaundice, results. The incubation period of the illness ranges from 15 to 50 days, with a median of 28 days. Shedding of the virus occurs approximately during the last half of the incubation period, or 10 to 14 days before onset of illness, and for a few days after onset. Natural infection results in durable immunity. Vaccination with a killed-virus vaccine was licensed in the U.S. in 1995. Passive protection of persons exposed to the virus is done with immune serum globulin, which must be administered within 2 weeks of exposure. Much transmission of hepatitis A is via person-to-person contact, but food and water vehicles are important. Food vehicles have included bivalve mollusks, any food handled by an infected person (e.g., sandwiches, salads, and other foods), and, again quite recently, frozen strawberries.[3] Prevention of hepatitis A transmission via food is based on sanitation, although cooking of produce before it is eaten will surely inactivate the virus.

The Norwalk-like viruses are sometimes called small round structured viruses, depending on whether diagnosis is done serologically or (in the latter case) by electron microscopy.[1,6] As with the hepatitis A virus, these agents infect only humans, causing gastroenteritis (often vomiting and diarrhea). The onset of illness typically occurs 24 to 48 hours after exposure, and the duration of illness is usually 24 to 48 hours. The virus sheds in vomitus, in diarrheal stool, and in normal stool for as long as a week after onset. Food vehicles are essentially the same as for hepatitis A. Infection with one of these viruses evidently does not confer long-term immunity, so there is little incentive to attempt to develop a vaccine. Laboratory diagnosis evidently is not often attempted unless a highly visible outbreak occurs.

Other viruses, the astroviruses, and (less often) rotaviruses occasionally cause foodborne and waterborne gastroenteritis. These, too, are human-specific agents shed in feces.

B. Protozoa

The principal protozoa transmitted via food and water in the U.S. are *Cryptosporidium parvum*, *Giardia lamblia*, and *Toxoplasma gondii*. *Cyclospora cayetanensis* is a recent addition to the record of foodborne disease agents in the U.S. *Entamoeba histolytica*, once a fairly common disease agent in the U.S., is fortunately now quite rare.

In 1993, in Milwaukee, WI, *Cryptosporidium parvum* caused the largest outbreak of waterborne disease in U.S. history, comprising 403,000 diarrheal illnesses.[11] Several outbreaks associated with community water supplies had been recorded earlier or have occurred since. The diarrhea caused by *C. parvum* lasts 1 to 2 weeks in normal people; there is no treatment. The transmission form is called an oocyst. It is quite small (4 to 6 μm) and almost totally chlorine resistant. Food vehicles have included apple juice (two outbreaks) and, recently, green onions.

Giardia lamblia (alias *duodenalis* or *intestinalis*) was the leading cause of recorded waterborne disease in the U.S. until 1993.[13] It can cause a prolonged diarrhea in the absence of proper diagnosis and treatment. There have been several outbreaks of giardiasis in communities whose water supplies derive from surface sources and which rely on chemical (usually chlorine) disinfection without previous filtration. There are obvious opportunities for produce contamination via water. Proper filtration removes the transmission form (cysts) from water.

Both *G. lamblia* and *C. parvum* are occasionally transmitted via food (e.g., *C. parvum* in apple juice in 1993). Another property that they share is that humans are not their only reservoir; each can infect several other species.

Cyclospora cayetanensis caused gastroenteritis outbreaks in the U.S. and Canada in 1996 and again in 1997.[4] The principal vehicle appeared to have been fresh raspberries from Guatemala, but a salad mixture of baby lettuce[4] and basil,[5] on at least one occasion, have also been implicated as vehicles. *C. cayetanensis* has not been shown to infect species other than humans. Oocysts that are shed in infected humans' feces require days or weeks in the environment to become infectious.

Entamoeba histolytica is another human-specific protozoon.[10] At one time, it caused a good deal of amebic dysentery in the U.S. but is now rarely reported. Cysts shed in feces are immediately infectious. Because chronic carriage is possible, it could represent a significant risk of contamination from infected food workers in areas where it still occurs.

Toxoplasma gondii occurs all over the world.[10] Its definitive host is the cat; infection in any feline species apparently results in fecal shedding of highly durable oocysts for a few weeks. The oocysts become infectious after 1 to 2 days and, if ingested, will produce tachyzoites that migrate into extra-intestinal tissues and produce tissue cysts. If these tissue cysts occur in a food animal and are not killed by cooking, they may cause human infections. On the other hand, oocysts from cat feces may occur in garden soil and on produce and cause human infections by that route. Although most human infections are mild, infections in pregnant women can lead to congenital infection of the fetus, producing profound abnormalities such as blindness, deafness, and hydrocephalus. There is no way of knowing with certainty what proportion of these congenital abnormalities results from transmission via foods, but it has been estimated that the approximately 2000 cases per year result in medical costs of $2.6 billion per year.

C. Contamination

Feces in soil or water may contain viruses or protozoan transmission forms (cysts or oocysts) that can contaminate food in the field or at harvest. Also, any food that is handled by an infected person may be contaminated if that person's hands were not washed carefully after defecation.

D. Prevention

Viruses are much smaller, and the protozoan transmission forms larger, than foodborne and waterborne bacteria. These agents are unable to multiply outside the host's body, which means that they cannot multiply in food or water. The inability of the agents to multiply outside the host also complicates detection, in that there are no procedures analogous to the enrichment that may be used in detecting foodborne and waterborne bacteria. Detection methods for protozoan transmission forms depend largely on skilled microscopy, assisted in some instances by serologic and genomic techniques.[9] Virus detection methods that show the greatest promise are directed to the viral coat protein, RNA, or both.[6] These methods are exacting and expensive, so they are typically applied only in the event that an outbreak has occurred. In the case of hepatitis A, food testing is further complicated by the long incubation period of the disease — appropriate samples are only rarely available as long as 4 to 6 weeks after the event that led to the outbreak. Because of the difficulties of detecting viruses in food and water, indicators of fecal — and possibly viral — contamination have been sought for years. Bacterial indicators of contamination, such as the coliform group, *Escherichia coli*, etc., are largely irrelevant; coliphages (bacteriophages that infect *E. coli*) and cytopathic human enteroviruses are under evaluation for this purpose. Even if these surrogates show correlation with the incidence of the hepatitis A and Norwalk-like viruses in food and water, they will be of no use with regard to the presence of the transmission forms of protozoa and parasites.

1. Inactivation

Heat is a good means of inactivating most of the agents under discussion. Thermal processes equivalent to milk pasteurization will apparently inactivate most of them completely. An exception is hepatitis A virus, which is somewhat more heat resistant and also withstands drying better than most of the others.[7,8] Both viruses and protozoan transmission forms are generally quite stable at refrigeration temperatures. Whereas viruses are preserved by freezing, protozoan transmission forms are generally killed. These agents tend to be more resistant than bacteria to acid and chlorine; viruses are generally less resistant than *G. lamblia*, which is far surpassed in chlorine resistance by *C. parvum*. The viruses are generally susceptible to ultraviolet light and to strong oxidizing agents, but not to quaternary ammonium compounds, in water and on exposed surfaces. Ultraviolet light and chemical disinfectants other than chlorine are being evaluated for disinfection of protozoan-contaminated food and water. Ionizing radiation is apparently effective against protozoa, but less so against viruses because of their relatively small target size.

2. Prevention via Sanitation

An important sanitation measure in growing produce is control of human and animal waste. Animal manure is probably less likely to contain pathogenic protozoan transmission forms than zoonotic bacteria; the viruses are human specific and therefore might occur in biosolids derived from wastewater treatment. Both human and animal wastes sometimes contaminate water that may be used for irrigation or for other purposes in the fields. Some foods cannot be washed at the time of harvest for fear of causing deterioration of quality. Washing food before it is consumed in the home is probably a good idea, but studies to show that this will remove viruses or protozoan cysts and oocysts seem not to have been done. General food hygiene, especially thorough hand washing, is probably the best overall recommendation for preventing transmission of these agents via foods.

III. SUMMARY

Viruses and protozoa differ most obviously from foodborne bacteria in size and in inability to multiply in food and water. Although these agents get less recognition because of difficulties of diagnosis and detection, a virus (hepatitis A) caused the largest foodborne disease outbreak and a protozoon (*C. parvum*) caused the largest waterborne disease outbreak in history. They are transmitted via food more rarely than, say, *Salmonella*; however, outbreaks such as the 1997 incident with frozen strawberries shows that they are significant hazards. They are most likely to be surface contaminants on foods and may be difficult to dislodge or kill, depending on the nature of the surface.

REFERENCES

1. Appleton, H. 1994. Norwalk virus and the small round viruses causing foodborne gastroenteritis. In: Y.H. Hui, J.R. Gorham, K.D. Murrell, and D.O. Cliver, Eds., *Foodborne Disease Handbook.* Vol. 2. *Diseases Caused by Viruses, Parasites, and Fungi.* Marcel Dekker, New York, pp. 57–79.
2. Bean, N.H., Goulding, J.S., Lao, C., and Angulo, F.J. 1996. Surveillance for foodborne-disease outbreaks: United States, 1988–1992. *Morbid. Mortal. Weekly Rep.,* 45(SS-5):1–66.

3. Centers for Disease Control and Prevention.1997. Hepatitis A associated with consumption of frozen strawberries: Michigan, March 1997. *Morbid. Mortal. Weekly Rep.,* 46:288 and 295.
4. Centers for Disease Control and Prevention. 1997. Update: outbreaks of cyclosporiasis: United States and Canada, 1997. *Morbid. Mortal. Weekly Rep.,* 46:521–523.
5. Centers for Disease Control and Prevention. 1997. Outbreak of cyclosporiasis: northern Virginia-Washington, D.C.-Baltimore, Maryland Metropolitan Area, 1997. *Morbid. Mortal. Weekly Rep.,* 46:689–691.
6. Cliver, D.O. 1995. Detection and control of foodborne viruses. *Trends Food Sci. Technol.,* 6:353–358.
7. Cliver, D.O. 1997. Foodborne viruses. In: M.P. Doyle, L.R. Beuchat, and T.J. Montville, Eds., *Food Microbiology: Fundamentals and Frontiers.* American Society for Microbiology, Washington, D.C., pp. 437–446.
8. Cromeans, T., Nainan, O.V., Fields, H.A., Favorov, M.O., and Margolis, H.S. 1994. Hepatitis A and E viruses. In: Y.H. Hui, J.R. Gorham, K.D. Murrell, and D.O. Cliver, Eds., *Foodborne Disease Handbook.* Vol. 2. *Diseases Caused by Viruses, Parasites, and Fungi.* Marcel Dekker, New York, p. 156.
9. Deng, M.Q., Cliver, D.O., and Mariam, T.W. 1997. Immunomagnetic capture PCR to detect viable *Cryptosporidium parvum* oocysts from environmental samples. *Appl. Environ. Microbiol.,* 63:3134–3138.
10. Fayer, R. 1994. Foodborne and waterborne zoonotic protozoa. In: Y.H. Hui, J.R. Gorham, K.D. Murrell, and D.O. Cliver, Eds., *Foodborne Disease Handbook.* Vol. 2. *Diseases Caused by Viruses, Parasites, and Fungi.* Marcel Dekker, New York, pp. 331–362.
11. Fayer, R., Ed. 1997. *Cryptosporidium and Cryptosporidiosis.* CRC Press, Boca Raton, FL, 251 pp.
12. Kramer, M.H., Herwaldt, B.L., Craun, G.F., Calderon, R.L., and Juranek, D.D. 1996. Surveillance for waterborne-disease outbreaks: United States, 1993–1994. *Morbid. Mortal. Weekly Rep.,* 45(SS-1):1–33.
13. Meyer, E.A., Ed. 1990. Giardiasis. In: *Human Parasitic Diseases,* Vol. 3. Elsevier, Amsterdam, 368 pp.

Human Health Effects of Microbial Food Contaminants

4 Toxicity Associated with Fumonisin-Contaminated Corn

Kenneth A. Voss, Ronald T. Riley, Ronald D. Plattner, William P. Norred, Filmore I. Meredith, James K. Porter, and Charles W. Bacon

CONTENTS

I. INTRODUCTION

Fusarium moniliforme (= *F. verticillioides*) has long been associated with outbreaks of a neurotoxic syndrome of horses called equine leukoencephalomalacia or ELEM[1–3] and a pulmonary condition of swine known as porcine pulmonary edema (PPE).[4–6] *F. moniliforme* is hepato- and nephrotoxic to various species.[5,7,8] The importance of *F. moniliforme* to human health remains unclear but continues to cause concern.[9,10] In areas of southern Africa (reviewed by Dutton[11]) and central China,[12] there is a correlation between consumption of food made from *F. moniliforme*-molded corn and unusually high rates of esophageal cancer. Also, the hepatocarcinogenicity of *F. moniliforme* has been demonstrated by feeding both naturally contaminated corn[13] and some *F. moniliforme* culture materials (molded corn prepared from single fungal isolates under controlled laboratory conditions)[14,15] to rats. Toxicity is apparently dependent upon a number of factors, and not all *F. moniliforme* isolates are toxic or carcinogenic *in vivo*.[14,16] The relationship to esophageal cancer is still unclear. Although Marasas et al.[15] found basal cell hyperplasia, they did not find esophageal cancer in rats chronically fed *F. moniliforme* culture material. Jaskiewicz et al.[14] likewise did not find esophageal cancer in rats fed culture materials from two different *F. moniliforme* isolates.

A major breakthrough occurred when Gelderblom et al.[17] and, shortly thereafter, Laurend et al.[18] independently published the chemical structure of novel water-soluble mycotoxins called fumonisins (initially named "macrofusine" by Laurend et al.), which they isolated from *F. moniliforme* (Figure 4.1). The natural occurrence of fumonisins B_1 (FB_1), the most common naturally occurring homologue, and B_2 (FB_2) in corn was soon thereafter demonstrated.[19–21] A number of other homologues and fumonisin-derived compounds have since been discovered (Figure 4.1), and it has been shown that *F. proliferatum* and perhaps other *Fusaria* also produce fumonisins.[22] Fumonisins, like the fungi, have worldwide distribution in corn and are found in corn-based products (as reviewed by Dutton[11]). It is clear that fumonisin concentrations can be quite high in corn from underdeveloped regions that is intended for human food, at times exceeding levels which are considered toxic to animals. In contrast, fumonisin concentrations in commercial corn-based food products from developed countries are generally low, averaging <1 ppm. Fumonisins, particularly FB_1, are the subject of ongoing toxicological investigations, the goals of which are to determine their organ-specific effects in various species, elucidate their modes of action, and assess their potential human health impact.

II. FUMONISIN-RELATED DISEASES IN FARM ANIMALS

A. Equine Leukoencephalomalacia

The identifying feature of equine leukoencephalomalacia (ELEM) is a unique hemorrhagic and liquifactive lesion of the white matter in which myelin appears to have "melted" away.[23] The field syndrome has a rapid onset. The affected animals present with inappetence, listlessness, ataxia, "head pressing," paralysis of facial muscles, and, as the condition progresses, recumbency and death. As in other species, there is

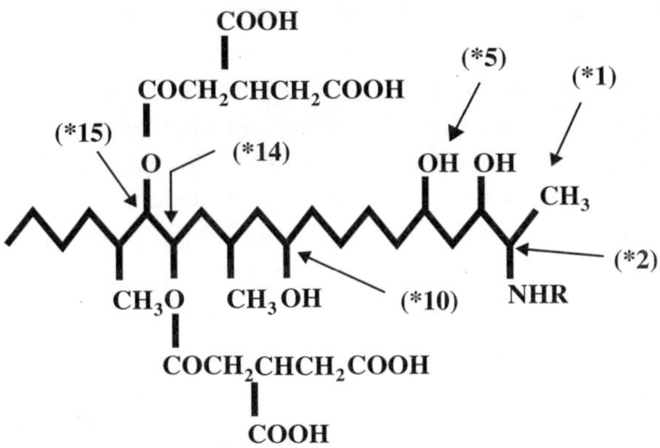

FIGURE 4.1 Structure of FB_1. Fumonisins B_2, B_3, and B_4 differ from FB_1 by the absence of one or more hydroxyl groups at carbons 5 (*5) and 10 (*10).[17,100] Replacement of the tricarballylic acid moieties at carbons 14 and 15 (*14, *15) by hydroxyl groups during nixtamalization yields hydrolyzed fumonisins,[92] and partially hydrolyzed forms having one of the tricarballylic acid groups have been identified.[101] The C series of fumonisins lacks carbon 1 (*1).[102] Ozonation of FB_1 yields a three-keto derivative.[103] The aforementioned fumonisins have a primary amino function (*2, R = H). Several fumonisins having substitutions at this position (*2) have been described: R = acetyl for fumonisins A;[17] R = a galactose or other reducing sugar[104] or N-carboxymethyl group[105] have recently been described as reaction products of FB_1 and sugar.

liver involvement ("hepatotoxic ELEM"). In advanced cases, the liver is small, pale, and firm with necrosis and fibrosis being found microscopically.[2,24] Renal lesions in horses have also been incidentally described.[24,25] That FB_1 causes ELEM was shown by Marasas et al.[26] and Kellerman et al.[25] Edema and focal hemorrhage were found in the brain of a horse given seven i.v. injections of 0.125 mg/kg BWt FB_1 over a 9-day period. In a second experiment, ELEM was produced in one horse by oral dosing of 1.25 to 4 mg/kg BWt FB_1 over a 33-day period and in another horse by giving (p.o.) 1 to 4 mg/kg BWt FB_1 over 29 days. By feeding culture materials, others[24,27,28] have since shown that FB_2, and perhaps also FB_3, are toxic to horses. Fumonisins disrupt sphingolipid metabolism in *Equidae* as in other species;[28,29] however, the relationship of sphingolipid metabolism disruption to the molecular and pathophysiological events causing brain lesions remains unknown. In 1993, the American Association of Veterinary Laboratory Diagnosticians Mycotoxin Committee suggested that, for horses, the nonroughage portion of the diet not exceed 5 ppm FB_1.

B. Porcine Pulmonary Edema

1. Clinical and Pathological Findings

Porcine pulmonary edema (PPE) is a sporadic syndrome of swine related to consumption of *F. moniliforme*- or *F. proliferatum*-molded corn.[4,5,30] Harrison et al.[31] reproduced PPE by intravenous injection of purified FB_1. PPE also has a rapid onset, but it is the lung that is the unique, species-specific target organ. Intoxicated animals become inappetent and listless, suffer respiratory distress, and ultimately succumb to respiratory failure. Enlarged, fluid-filled lungs are typically found at necropsy. Brain lesions are absent, but, as in other species, the liver may be involved and tissue sphingolipid profiles are altered.[4,32] The effect on kidneys remains unclear. Colvin et al.[33] observed single-cell necroses and hydropic degeneration in swine orally given fumonisins, while others found no morphological evidence of nephropathy at dosages causing significant disruption of renal sphingolipid profiles.[32] A guidance level of 10 ppm was recommended in 1993 by the American Association of Veterinary Laboratory Diagnosticians for swine.

2. Mechanistic Aspects of PPE

The pathogenesis of PPE remains unsolved. Ultrastructual studies[32] revealed membranous whorls in the liver (similar whorls were found in rats by Riley et al.[34]) and pulmonary alveolar macrophages. It was hypothesized that the lipid-laden macrophages cause damage to and increase permeability of the pulmonary vasculature and epithelium. However, more recent data[35,36] suggest that PPE results at least in part from primary, sphingolipid-mediated cardiotoxicity. The resulting deficits in cardiac output, contractility, stroke volume, and conductivity of the left ventricle then lead to pulmonary hypertension, vascular leakage, and edema. Left ventricular weight was decreased in fumonisin-treated swine,[35,37,38] and decreased heart weights were recently found in rats given diets high (>400 ppm) in fumonisins (provided by a *F. moniliforme* culture material) for 4 weeks.[39] That FB_1 blocked calcium current and inhibited mechanical contractility of isolated frog heart atria[40] further supports the hypothesis that pulmonary edema may be secondary to decreased cardiac function.

III. EFFECTS OF FUMONISINS IN LABORATORY ANIMALS

A. Toxic and Pathologic Effects in Rats

1. Connection Between Fumonisin and *F. moniliforme*

Subchronic studies (≤90 days)[16,17,41–44] have established the hepato- and nephrotoxicity of FB_1. Importantly, the *in vivo* kidney and liver effects of naturally contaminated corn[20] and culture materials or aqueous extracts of fumonisin-producing *F. moniliforme* isolates[17,45] are reproduced when purified FB_1 is given to rats.[17,42,43]

2. Liver Pathology: Subchronic Effects

The initial microscopic finding is scattered apoptotic hepatocytes (= single cell necrosis).[17,43,44,46] As injury progresses, loss of hepatocytes increases and cytoplasmic vacuolation, increased mitosis, cytomegaly, anisocytosis, and anisokaryosis appear. Bile duct proliferation, focal necrosis, fibrosis, oval cell proliferation, and hepatocellular hyperplasia are features that occur in more advanced lesions, giving the liver a nodular or cirrhotic appearance.[47] Cholangiofibrosis and foci of cellular alteration also occur. Serum chemical indicators of hepatotoxicity (such as alanine and aspartate aminotransferase) and alkaline phosphatase activities, as well as cholesterol, triglycerides, and bilirubin concentrations, are increased.[43,46] There is a slight sex difference in response, with females being more sensitive than males.[43,44]

3. Fumonisin B_1 and Liver Cancer

F. moniliforme culture materials[16] and fumonisins[17,48] are liver cancer promoters as shown by their ability to induce GGT-positive foci, both with and without pretreatment with the genotoxic hepatocarcinogen diethylnitrosamine (DEN). Whether or not fumonisins are liver cancer initiators is currently controversial. There is evidence that they are weak initiators *in vivo* when given at high doses,[49,50] and, like complete carcinogens, FB_1 induced GST-positive foci in liver.[52] It was also reported that FB_1 caused chromosomal breaks in rat hepatocytes *in vitro*.[51] On the other hand, evidence from studies using *Salmonella* mutagenicity and unscheduled DNA synthesis in hepatocytes as endpoints suggest that fumonisins are not genotoxic.[51,53,54]

Like *F. moniliforme*-contaminated corn[13] and culture materials,[14,15,55] chronic ingestion of purified FB_1 (50 ppm in the diet) caused cirrhosis and was hepatocarcinogenic; hepatocellular carcinomas were found after 20 to 26 months in approximately two thirds of the male BD IX rats.[47] However, the diets were marginally deficient in the lipotropes choline and methionine, a factor that may have contributed to the neoplastic response. Although this work clearly demonstrated the carcinogenic potential of FB_1, additional data on organ-specific pathology and dose response in males and females fed nutritionally sound diets are clearly needed. To this end, chronic bioassays for toxicity and carcinogenicity of FB_1 in rats and mice are nearing completion at the Food and Drug Administration's National Center for Toxicological Research (NCTR) in Jefferson, AR.

4. Kidney Pathology: Subchronic Effects

The kidney is more sensitive to fumonisins than liver in rats. The no observed effect level (NOEL; studies $\leq$90 days) for nephropathy in males is consistently and considerably lower than both the NOEL for nephropathy in females and the NOEL for hepatopathy in either sex.[42–44,56] As in liver, apoptosis is the initial microscopic finding. Apoptotic epithelial cells are scattered throughout the proximal tubules of the outer medulla (also called corticomedullary junction). At low doses, only a few such cells are present, but, with increased exposure, apoptotic cells become more numerous and mitotic figures appear. As lesions develop further, there is tubular hyperplasia with cytoplasmic basophilia, decreased epithelial height, cytoplasmic vacuolation, anisocytosis, anisokaryosis, and sloughing of apoptotic or necrotic epithelial cells into the tubular lumina. Lesions extend into the adjacent cortex or deeper into the medulla. Focal tubular atrophy and interstitial fibrosis occur in the most severe cases, but, as in liver, inflammation remains minimal. Despite the picture of simultaneous cell loss (apoptosis and tubular atrophy) and replication (mitotic figures and tubular hyperplasia), the overall balance, as indicated by decreased kidney weights routinely found in fumonisin-exposed rats, is shifted toward cell loss. The resulting imbalance between apoptosis and mitosis may have important implications for chronic toxicity and carcinogenicity.[57–59]

Clinical signs of tubular dysfunction, including increased serum creatinine and decreased serum CO_2, are also found, although these markers are not particularly sensitive.[43,45,46] Renal function in fumonisin-exposed rats has not been studied in detail. Voss et al.[60] noted transiently increased urinary output and water consumption in rats fed fumonisin provided by *F. moniliforme* (isolate M3125) culture materials. These observations supported the earlier and more extensive evidence of renal dysfunction found in rats after daily i.p. (7.5 to 10 mg/kg for 4 days) or oral administration (1 to 75 mg/kg for 11 days) of FB_1 by Bondy and colleagues.[56,61,62] Taken together, evidence for FB_1-induced renal dysfunction from these experiments included increased urine volume; decreased osmolality; increased urinary enzyme activities of GGT, lactate dehydrogenase, and N-acetyl-β-D-glucosaminidase; and, in the 4-day i.p. study, markedly decreased transport of the anion p-aminohippuric acid (75 to 80% reduction) and the cation tetraethylammonium (40% reduction) by renal cortex slices obtained from the animals at necropsy. Protenuria, consisting mainly of high molecular weight proteins, and elevated serum Mg^{++} and Ca^{++} concentrations were also found, suggesting that both glomerular and tubular function are affected by FB_1. In the 11-day gavage studies, most urinalysis findings were transient, peaked at 6 to 8 days, and returned to normal levels by study's end, suggesting that some physiological adaptation to fumonisins occurred. Microscopic lesions persisted, however, leading the authors to suggest that histopathological demonstration of apoptosis may be the most sensitive indicator of renal injury.[62]

5. Other Potential Target Organs

While it is accepted that liver and kidney are target organs in rats, there are reports of other possible targets. Esophageal epithelial (basal cell) hyperplasia,[15] thymic apoptosis,[39,61,63] decreased heart weight,[39] hypertrophy and vacuolation of the adrenal cortex (i.e., zona fasciculata[60]), and various hematological changes such as thrombocytopenia, leukocytosis, and altered differential leukocyte counts[8,60,61,63] are among occasionally reported findings in rats subchronically ($\leq$90 days) given purified fumonisins or fumonisins in *F. moniliforme* culture materials. The toxicological significance of these observations remains to be determined, as some may have been incidental, may occur only at very high dose levels, or may depend on the strain of rat. If culture materials were used as the fumonisin source, the presence of undefined confounding factors must also be considered. For example, focal seminiferous tubular hypoplasia was found in Sprague-Dawley rats fed *F. moniliforme* culture material.[8] In subsequent investigations, the same investigators did not find reproductive performance deficits, testicular pathology, or abnormalities in sperm structure or function in Sprague-Dawley rats fed a culture material of the same *F. moniliforme* isolate (MRC826).[64] Furthermore, no testicular lesions were found in F344 rats fed purified FB_1 for 90 days.[42]

Of special interest is determining the relationship, if any, between *F. moniliforme* and fumonisins to esophageal neoplastic or preneoplastic lesions. Although esophageal hyperplasia was found in rats fed an *F. moniliforme* culture material,[15] esophageal hyperplasia or neoplastic lesions have not been found in studies of purified FB_1.[42,43,47] Cell proliferation (5-bromo-2′-deoxyuridine labeling index) was transiently increased in the esophageal epithelium 3 days following a single i.v. dose of 1.25 mg/kg FB_1,[65] suggesting that FB_1 may under some conditions have a mitogenic effect on the esophageal epithelium. In contrast, the number of esophageal papillomas found in rats simultaneously given FB_1 (5 mg/kg per day for 5 weeks) and the esophageal carcinogen N-methylbenzylnitrosamine (2.5 mg/kg i.p. twice per week during the second through fourth week of fumonisin treatment) were not increased (45 to 48 weeks after treatment) compared to rats given only N-methylbenzylnitrosamine.[66] Thus, the issue remains unresolved and the possibilities that the correlation between exposure and esophageal cancer is coincidental, that the fungus produces other (esophageal) carcinogens, or, as suggested by Wild et al.,[66] that the rat is a poor experimental model cannot be discounted.

Given that sphingolipids are potentially important immunomodulators,[67,68] it can be hypothesized that fumonisins are immunotoxins; however, immunotoxicology data are limited and the results are not definitive. Immunoglobulin (IgM) response to sheep red blood cells tended to decrease in male rats given 5 to 25 mg/kg FB_1 for 14 days.[69] In males given up to 15 mg/kg FB_1, splenic clearance of *Listeria monocytogenes* was transiently decreased, and there was a trend toward slightly decreased serum IgG levels. Other endpoints including mitogen-induced proliferation of splenic mononuclear cells, intracellular Con-A-induced calcium, phagocytic activity of peripheral macrophages, or NK cell activity were unaffected. Studies in mice and poultry suggest that fumonisins inhibit macrophage function.[67,68,70–72]

B. Hepatic and Renal Effects in Other Laboratory Species

Reports indicate that apoptosis also appears early and plays a significant role in the pathogenesis of liver and kidney lesions in mice[8,42,73–75] and rabbits[7] exposed to FB_1. Although subtle differences may exist (for example, cytomegaly of hepatocytes appears more pronounced in mice), lesions in these species are similar to those in rats. Female mice are more sensitive to hepatotoxic effects than male mice or rats (either sex), and the mouse is less sensitive than the rat to the renal injury.[74,75] Information on fumonisin toxicity in nonhuman primates is limited. Once again, the liver is a target organ, as indicated by both morphological and serum chemical criteria. It has been suggested that fumonisins may, by increasing serum lipids and cholesterol, have etiological implications for atherosclerosis and other cardiovascular conditions.[5,76,77]

C. Sphingolipids and Mechanism of Toxicity

Wang et al.[78] discovered that fumonisins inhibit the enzyme ceramide synthase (sphinganine [sphingosine]-N-acyl transferase) *in vitro*, thus preventing *de novo* synthesis of ceramide (Figure 4.2). It is now well established that this occurs *in vivo*.[28,34,60,79,80] Ceramide synthase is inhibited in animals at doses that are equal to or below those causing toxicity, as determined by commonly used serum chemical, organ weight, or histopathological criteria, and a correlation between sphinganine concentrations or sphinganine (Sa)/sphingosine (So) ratio and hepatotoxic endpoints has been demonstrated repeatedly.

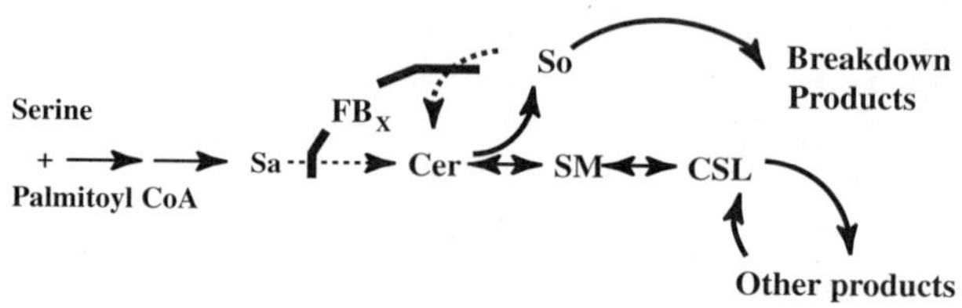

FIGURE 4.2 Simplified schematic of *de novo* synthesis of ceramide and complex sphingolipids, sphingolipid turnover, and the location of the metabolic block (inhibition of ceramide synthase) imposed by FB_1. Biochemical consequences of this block are marked accumulation of sphinganine (Sa), accumulation of minor amounts of sphingosine (So; formed by blocking the re-utilization of sphingosine liberated during complex sphingolipid turnover), increased tissue ratio of sphinganine to sphingosine (a biomarker of exposure), and loss of complex sphingolipids.

As a result of ceramide synthase inhibition, the sphingoid bases sphinganine and, to a lesser extent, sphingosine accumulate in tissues, and the Sa/So ratio in tissues, serum, or urine increases. Cellular complex sphingolipids decrease, and sphinganine and sphingosine degredation products such as sphingosine-1-phosphate may increase.[29,81,82] Taken together, these changes reflect an *overall* disruption of sphingolipid metabolism, the consequences of which may be far reaching. Sphingoid bases, ceramide, and other sphingolipids are biologically active molecules now known to play critical cell regulatory roles and to act as messengers for a variety of extracellular stimuli governing processes critical for cell growth, differentiation, apoptosis, and mitosis.[58,81] The latter are fundamental features of kidney and liver toxicity caused by fumonisins, and disruption of the critical balance between apoptosis and mitosis in tissues may be a contributing factor to carcinogenesis.[57,59] Thus, connections from ceramide synthase inhibition, to sphingolipid disruption, to apoptosis, to toxicity and carcinogenicity may *theoretically* be drawn but remain unproven insofar as fumonisins are concerned. A detailed presentation of the interrelationships between fumonisins, sphingolipids, apoptosis, and toxigenesis is beyond the scope of this chapter, and the reader is referred to several recent reviews.[58,81,83,84] Other mechanisms can certainly come into play, and there is evidence that fumonisin toxicity or carcinogenicity is mediated by or modified by oxidative damage of lipids which in turn initiates events leading to DNA damage.[85–90]

IV. APPLIED STUDIES ON FUMONISINS IN *F. MONILIFORME* CULTURE MATERIAL: EVALUATING TOXIC INTERACTIONS

Fumonisin B_1 does not exist alone on corn. Multiple fumonisin homologues (principally FB_2 and FB_3), fumonisin derivatives, and other mycotoxins are also present in corn and corn products. It is therefore important to determine the extent to which simultaneous exposure to other mycotoxins modifies FB_1 toxicity. Three recent experiments addressing this issue are summarized.

A. Comparative Toxicity of Fumonisins B_1, B_2, and B_3

Culture materials were prepared from three genetically related *F. moniliforme* isolates. The first (M3125) produced FB_1, FB_2, and FB_3 in a ratio (1.0:0.35:0.15) typically found in corn. The second (107-R-7) produced FB_2 but no detectable FB_1 or FB_3, and the third (397-R-74) produced FB_3 but no detectable FB_1 or FB_2. All three isolates produced traces of FB_4. Diets containing low, medium, or high levels of each culture material were fed to rats (n = 10/group) for 3 weeks (Table 4.1).[60] At that time, one half of the animals were killed. The remainder were given a control diet for an additional 3-week recovery period. A control group was given a diet without culture material, but was otherwise handled in the same manner. Isolates 107-R-7 and 397-R-74 produced the same body weight, serum chemical, organ weight, tissue sphingolipid, and histopathological liver and kidney effects as M3125. Thus, FB_1 is not required for *F. moniliforme* toxicity, and FB_2 and FB_3 should be considered in risk evaluation and management strategies (Table 4.1). Interestingly, dose response data suggested that FB_3 may be somewhat less potent than either FB_1 or FB_2.

Effects were reversible. Following the 3-week recovery period, no significant differences among groups in serum chemical, organ weight, or tissue sphingolipid levels remained. Likewise, no apoptosis was found in kidney or liver at that time in groups fed isolates 107-R-7 or 397-R-74. Only a few isolated apoptotic hepatocytes were seen in three of the five remaining rats fed isolate M3125, most likely reflecting the more extensive lesions seen at three weeks.

B. Nixtamalization and Toxicity

Nixtamalization is the process by which corn is converted to masa flour for tortillas. During nixtamalization, the corn is subjected to alkaline conditions sufficient to convert FB_1 to hydrolyzed FB_1 (HFB_1), also known as aminopentol or AP_1.[91,92] Because of conflicting opinions in the literature concerning HFB_1 toxicity,[48,93] the toxicity to rats (n = 10/group) of a fumonisin-containing culture material before and after nixtamalization was compared.[94] In addition, the experiment included a group fed the culture material after water extraction and a control group fed a diet without culture material (Table 4.2). No detectable FB_1 remained in the culture material after nixtamalization. The diet prepared with nixtamalized culture material contained 58 ppm HFB_1 and caused the same, but less severe, body weight, organ weight, and liver and kidney pathological effects as the untreated culture material containing 71 ppm FB_1. Water extraction of the culture material removed approximately 90% of the measurable FB_1 and more effectively detoxified the culture material than did nixtamalization.

Sphinganine levels and Sa/So ratios of liver and kidney were increased by the three test diets in this order: water-extracted culture material < nixtamalized culture material < culture material.[79] Increases were generally correlated with the degree of liver and kidney injury and were highly correlated with both body weight effects and serum chemical indications of hepatotoxicity (Table 4.2). Taken together, the results suggest HFB_1 is also toxic, although less so than FB_1. Because culture

TABLE 4.1
Effects in Rats Fed Culture Materials of Three *F. moniliforme* Isolates[a]

Diet		Findings[b]			
Isolate and Fumonisin	**FB_x ppm**	**Final BWt (g)**[c]	**Organ/Brain Weight**	**Tissue Sa/So**	**Pathology**[d]
None (controls)	—	289 [8.2]	L: 628 [93]	L: 0.19 [0.07]	L: 0/0
			K: 142 [11]	K: 0.11 [0.04]	K: 0/0
M3125	6.9	283 [13]	L: 576 [67]	L: 0.24 [0.06]	L: 0/0
(FB_1, B_2, B_3			K: 126 [7.2] ↓	K: 2.1 [1.6]	K: 5/1.2 ↑
in the ratio	53	279 [16]	L: 533 [43]	L: 0.68 [0.17]	L: 2/1.0
1:0.35:0.15)			K: 106 [4.2] ↓↓	K: 2.6 [1.0] ↑	K: 5/2.0 ↑
	303	228 [15] ↓↓	L: 478 [46]	L: 3.5 [2.4] ↑	L: 5/2.0 ↑
			K: 95 [5.7] ↓↓↓	K: 3.0 [1.2] ↑	K: 5/2.0 ↑
107-R-7 (FB_2)	4.6	288 [10]	L: 572 [25]	L: 0.25 [0.08]	L: 0/0
			K: 144 [8.1]	K: 0.39 [0.20]	K: 0/0
	32	280 [15]	L: 582 [43]	L: 0.34 [0.09]	L: 1/1.0
			K: 124 [8.5] ↓↓	K: 4.7 [4.8]	K: 5/1.4 ↑
	219	270 [12] ↓	L: 464 [17] ↓	L: 1.60 [0.48] ↑	L: 5/1.2 ↑
			K: 107 [6.6] ↓ ↓↓	K: 2.6 [2.0]	K: 5/2.0 ↑
397-R-74	6.7	289 [15]	L: 629 [101]	L: 0.25 [0.21]	L: 1/1.0
(FB_3)			K: 146 [9.2]	K: 1.1 [0.94]	K: 1/1.0
	49	285 [13]	L: 615 [26]	L: 0.24 [0.06]	L: 0/0
			K: 129 [9.4] ↓	K: 2.7 [1.7]	K: 5/1.2 ↑
	295	280 [12]	L: 555 [54]	L: 0.98 [0.26]↑	L: 4/1.0 ↑
			K: 121 [309] ↓	K: 2.1 [1.0]	K: 5/2.0 ↑

[a] Effects of feeding isolates producing FB_1, FB_2, and FB_3 (M3125); or FB_2 (107-R-7); or FB_3 (397-R-74) for 3 weeks on body weight (BWt), organ/brain weight, liver (L) and kidney (K) sphinganine:sphingosine ratio (Sa/So), and liver (L) or kidney (K) pathology findings indicative of fumonisin exposure (as described in the text).

[b] Except where indicated otherwise, values are group means with the standard deviations given in brackets; n = 5/group, except n = 10 for BWt; arrows indicate increase or decrease from controls; groups not sharing the same number of arrows differ significantly ($p < 0.05$) from each other.

[c] Body weights of all groups averaged 141 g at beginning of study.

[d] Values indicate number of animals with lesions/mean severity scores of lesions, 0 = no lesion, 1 = minimal, 2 = mild to moderate.

Source: Data summarized from Voss et al.[60]

materials were the dietary HFB_1 source, other compounds therein may have influenced the outcome, and the results should be confirmed by experiments using purified HFB_1. Nonetheless, these findings extend the observations of Norred et al.,[95] who reported that HFB_1 was a less effective ceramide synthase inhibitor than FB_1 in precision-cut rat liver slices *in vitro*.

TABLE 4.2

Effects in Rats Fed Equivalent Weights of *F. moniliforme* MRC 826 Culture Material, Water-Extracted Culture Material, or Nixtamalized Culture Material[a]

Diet		Findings[b]			
Treatment	**Fumonisin**	**Final BWt (g)[c]**	**Organ/ Brain Weight**	**Tissue Sa/So**	**Pathology[d]**
None (controls)	—	341 [9.0]	L: 640 [71] K: 142 [9.4]	L: 0.49 [0.16] K: 0.19 [0.05]	L: 0/0 K: 0/0
Water-extracted culture material	8 ppm FB_1	325 [11] ↓	L: 622 [45] K: 117 [7.2] ↓	L: 0.68 [0.11] K: 8.8 [1.2] ↑	L: 1/0.1 K: 10/1.6 ↑
Culture material	71 ppm FB_1	241 [12] ↓↓↓	L: 489 [52] ↓ K: 91 [5.4] ↓↓↓	L: 4.85 [1.1] ↑↑ K: 15.0 [0.94] ↑↑↑	L: 10/3.0 ↑ K: 10/2.5 ↑
Nixtamalized culture material	58 ppm HFB_1	309 [12] ↓↓	L: 522 [56] ↓ K: 104 [8.0] ↓↓	L: 1.51 [0.52]↑ K: 11.0 [0.52] ↑↑	L: 10/1.2 ↑ K: 10/1.9 ↑

[a] Fed for 4 weeks to determine effects on body weight (BWt), organ/brain weight of liver (L) and kidney (K), sphinganine:sphingosine ratio (Sa/So), and liver (L) or kidney (K) pathology findings indicative of fumonisin exposure (as described in the text).

[b] Except where indicated otherwise, values are group means with the standard deviations given in brackets; n = 10/group, except n = 5 for Sa/So; arrows indicate increase or decrease from controls; groups not sharing the same number of arrows differ significantly ($p < 0.05$) from each other.

[c] Body weights of all groups averaged 158 g at beginning of study.

[d] Values indicate number of animals with lesions/mean severity scores of lesions, 0 = no lesion, 1 = minimal, 2 = mild to moderate, 3 = severe.

Source: Data summarized from Voss et al.[79,94]

C. Effect of Simultaneous Exposure to Fusaric Acid

Fusaric acid is another mycotoxin produced by *F. moniliforme.* It is also common in corn[96] and is likely to occur together with fumonisins.[97] Bacon et al.[97] demonstrated a synergistic embryo-lethal effect *in ovo* by simultaneous administration of fusaric acid and fumonisin, and it has been otherwise speculated that fusaric acid may enhance fumonisin toxicity.[98,99]

Groups of male Sprague-Dawley rats (four groups, n = 5 at each fumonisin concentrations) were given diets containing 3.4, 18, or 437 ppm FB_1 provided by a culture material of *F. moniliforme* and, at each fumonisin level, supplemented with 0, 20, 100, or 400 ppm fusaric acid.[39] There were two additional control groups;

TABLE 4.3

Lack of Effect of Fusaric Acid in Rats Fed *F. moniliforme* MRC 826 Culture Material Providing 18.4 or 437 ppm Fumonisins[a]

Diet		Findings[b]			
Fusaric Acid (ppm)	$FB_1 + FB_2 + FB_3$ (ppm)	Final BWt (g)[c]	Organ/Brain Weight	Kidney Sa/So	Pathology[d]
None (controls)	None	311 [8.6]	L: 670 [113] K: 155 [10]	0.09 [0.1]	L: 0/0 K: 0/0
400	None	308 [12]	L: 643 [59] K: 149 [11]	0.12 [0.04]	L: 0/0 K: 0/0
None	18.4	305 [15]	L: 642 [60] K: 125 [5.1] ↓	5.52 [1.4] ↑	L: 0/0 K: 5/2.0 ↑
20	18.4	312 [5.6]	L: 686 [70] K: 133 [8.8] ↓	Not measured	L: 0/0 K: 5/2.0 ↑
100	18.4	314 [9.6]	L: 673 [56] K: 133 [5.3] ↓	Not measured	L: 0/0 K: 5/2.0 ↑
400	18.4	309 [14]	L: 654 [108] K: 130 [10] ↓	5.53 [0.69] ↑	L: 0/0 K: 5/2.0 ↑
None	437	234 [12] ↓	L: 473 [28] ↓ K: 91 [4.9] ↓↓	Not measured	L: 5/3.0 ↑ K: 5/3.0 ↑↑
20	437	241 [12] ↓	L: 485 [56] ↓ K: 92 [6.5] ↓	Not measured	L: 5/3.0 ↑ K: 5/3.0 ↑↑
100	437	240 [11] ↓	L: 454 [51] ↓ K: 89 [6.0] ↓↓	Not measured	L: 5/3.0 ↑ K: 5/3.0 ↑↑
400	437	243 [11] ↓	L: 487 [62] ↓ K: 90 [7.4] ↓↓	Not measured	L: 5/3.0 ↑ K: 5/3.0 ↑↑

[a] Fed for 4 weeks to determine effects on body weight (BWt), organ/brain weights of liver (L) and kidney (K), kidney sphinganine:sphingosine ratio (Sa/So) of selected groups, and liver (L) and kidney (K) pathology findings indicative of fumonisin exposure (as described in the text).

[b] Except where indicated otherwise, values are group means with standard deviations given in brackets; n = 5/group; arrows indicate increase or decrease from controls; groups not sharing the same number of arrows differ significantly ($p < 0.05$) from each other.

[c] Body weights of all groups averaged 145 to 147 g at beginning of study.

[d] Values indicate number of animals with lesions/mean severity scores of lesions, 0 = no lesion, 1 = minimal, 2 = mild to moderate, 3 = severe.

Source: Data summarized from Voss et al.[39]

one was given a diet containing 400 ppm fusaric acid, and the other was fed a diet without addition of culture material or fusaric acid (Table 4.3). Fusaric acid alone was not toxic. The body weight, serum chemical, kidney Sa/So ratio, and histopathological findings in the groups fed 3.4, 18, or 437 ppm FB_1 were consistent with fumonisin toxicity and were unaffected by simultaneous exposure to $\leq$400 ppm fusaric acid.

V. SUMMARY AND CONCLUSIONS

Since their discovery in 1988, fumonisins have been the subject of a concerted effort to determine their organ-specific effects and molecular modes of action. As a result, it is now known that some fumonisins, most notably FB_1, cause equine leukoencephalomalacia and porcine pulmonary edema, diseases long associated with the consumption of *F. moniliforme*. The relationship between fumonisins and human health, particularly their inferred relationship to esophageal cancer, remains unclear. Recent studies have shown that fumonisins cause the hepato- and nephrotoxicities in laboratory animals fed toxic *F. moniliforme* isolates. Dose-response, sphingolipid, and other *in vivo* data acquired from these investigations have provided a foundation for ongoing chronic toxicology and carcinogenicity studies. A growing body of evidence suggests that the toxic and carcinogenic effects of fumonisins are related to disruption of sphingolipid metabolism, and, because of their ability to inhibit the enzyme ceramide synthase, these compounds now serve as useful tools for researching sphingolipid metabolism and function.

NOTES

The results of 2-year chronic studies of FB_1, which were conducted by the Food and Drug Administration National Center for Toxicological Research, Jefferson, AR, were presented by P.C. Howard at the International Conference on the Toxicity of Fumonisins, Arlington, VA, May 28–30, 1999. FB_1 was hepatocarcinogenic in female B6C3F_1 mice and caused renal tubule carcinomas and adenomas in male F344 rats at dosages $\geq$ 50 ppm. Liver tumors were not found in rats.

REFERENCES

1. Haliburton, J.C. and Buck, W.B., Equine leucoencephalomalacia: an historical review, *Curr. Top. Vet. Med. Anim. Sci.*, 33, 75, 1986.
2. Haliburton, J.C., Vesonder, R.F., Lock, T.F., and Buck, W.B., Equine leucoencephalomalacia (ELEM): a study of *Fusarium moniliforme* as an etiological agent, *Vet. Hum. Toxicol.*, 21, 348, 1979.
3. Wilson, B.J. and Maronpot, R.R., Causative fungus agent of leukoencephalomalacia in equine animals, *Vet. Record*, 88, 484, 1971.
4. Diaz, G.J. and Boermans, H.J., Fumonisin toxicosis in domestic animals: a review, *Vet. Hum. Toxicol.*, 36, 548, 1994.
5. Kriek, N.P.J., Kellerman, T.S., and Marasas, W.F.O., Comparative study of the toxicity of *Fusarium verticillioides* (= *F. moniliforme*) to horses, primates, pigs, sheep and rats, *Onderstepoort J. Vet. Res.*, 48, 129, 1981.

6. Palyusik, M. and Moran, E.M., Porcine pulmonary edema with hydrothorax: a review, *J. Environ. Pathol. Toxicol. Oncol.*, 13, 63, 1994.
7. Gumprecht, L.A., Marcucci, A., Weigel, R.M., Vesonder, R.F., Riley, R.T., Showker, J.L., Beasley, V.R., and Haschek, W.M., Effects of intravenous fumonisin B_1 in rabbits: nephrotoxicity and sphingolipid alterations, *Natural Toxins*, 3, 395, 1995.
8. Voss, K.A., Norred, W.P., and Bacon, C.W., Subchronic toxicological investigations of *F. moniliforme*-contaminated corn, culture material and ammoniated culture material, *Mycopathologia,* 117, 97, 1992.
9. Marasas, W.F.O., Fumonisins: their implications for human and animal health, *Natural Toxins*, 3, 193, 1995.
10. Marasas, W.F.O., Fumonisins: history, world-wide occurrence and impact, *Adv. Exper. Med. Biol.*, 392, 1, 1996.
11. Dutton, M.F., Fumonisins, mycotoxins of increasing importance: their nature and their effects, *Pharmacol. Ther.*, 70, 137, 1996.
12. Yang, C.S., Research on esophageal cancer in China: a review, *Cancer Res.*, 40, 2633, 1980.
13. Wilson, T.M., Nelson, P.E., and Knepp, C.R., Hepatic neoplastic nodules, adenofibrosis, and cholangiocarcinomas in male Fisher 344 rats fed corn naturally contaminated with *Fusarium moniliforme*, *Carcinogenesis,* 6, 1155, 1985.
14. Jaskiewicz, K., van Rensburg, S.J., Marasas, W.F.O., and Gelderblom, W.C.A., Carcinogenicity of *Fusarium moniliforme* culture material in rats, *J. Natl. Cancer Inst.,* 78, 321, 1987.
15. Marasas, W.F.O., Kriek, N.P.J., Finchem, J.E., and van Rensburg, S.J. Primary liver cancer and oesophageal basal cell hyperplasia in rats caused by *Fusarium moniliforme*, *Int. J. Cancer*, 34, 383, 1984.
16. Gelderblom, W.C.A., Marasas, W.F.O., Jaskiewicz, K., Combrinck, S., and van Schalkwyk, D.J., Cancer promoting potential of different strains of *Fusarium moniliforme* in a short-term cancer initiation/promotion assay, *Carcinogenesis*, 9, 1405, 1988.
17. Gelderblom, W.C.A., Jaskiewicz, K., Marasas, W.F.O., Thiel, P.G., Horak, R.M., Vleggaar, R., and Kriek, N.P.J., Fumonisins: novel mycotoxins with cancer promoting activity produced by *Fusarium moniliforme*, *Appl. Environ. Microbiol.*, 54, 1806, 1988.
18. Laurent, D., Platzer, N., Kohler, F., Sauviat, M.P., and Pellegrin, F., Macrofusine et micromoniline: deux nouvelles mycotoxines isolees de mais infeste par *Fusarium moniliforme* Sheld., *Microbiologie-Aliments-Nutrition*, 7, 9, 1989.
19. Plattner, R.D., Norred, W.P., Bacon, C.W., Voss, K.A., Peterson, R., Shackleford, D.D., and Weisleder, D., A method of detection of fumonisins in corn samples associated with field cases of leukoencephalomalacia, *Mycologia*, 82, 698, 1990.
20. Voss, K.A., Norred, W.P., Plattner, R.D., and Bacon, C.W., Hepatotoxicity and renal toxicity in rats of corn samples associated with field cases of leukoencephalomalacia, *Food Chem. Toxicol.*, 27, 89, 1989.
21. Sydenham, E.W., Gelderblom, W.C.A., Thiel, P.G., and Marasas, W.F.O., Evidence for the natural occurrence of fumonisin B_1, a mycotoxin produced by *Fusarium moniliforme*, in corn, *J. Agric. Food Chem.*, 38, 285, 1990.
22. Nelson, P.E., Plattner, R.D., Shackelford, D.D., and Desjardins, A.E., Fumonisin B_1 production by *Fusarium* species other than in *F. moniliforme* section *Liseola* and by some related species, *Appl. Environ. Microbiol.*, 58, 984, 1992.
23. Buck, W.B., Haliburton, J.C., Thilsted, J.P., Lock, T.F., and Vesonder, R.F., Equine leucoencephalomalacia: comparative pathology of naturally occurring and experimental cases, *Amer. Assn. Vet. Lab. Diag. 22nd Annu. Proc.*, 239, 1979.

24. Wilson, T.M., Ross, P.F., Owens, D.L., Rice, L.G., Green, S.A., Jenkins, S.J., and Nelson, H.A., Experimental reproduction of ELEM: a study to determine the minimum toxic dose in ponies, *Mycopathologia*, 117, 115, 1992.
25. Kellerman, T.S., Marasas, W.F.O., Thiel, P.G., Gelderblom, W.C.A., Cawood, M., and Coetzer, J.A.W., Leukoencephalomalacia in two horses induced by oral dosing of fumonisin B_1, *Onderstepoort J. Vet. Res.*, 57, 269, 1990.
26. Marasas, W.F.O., Kellerman, T.S., Gelderblom, W.C.A., Coetzer, J.A.W., Thiel, P.G., and van der Lugt, J.J., Leukoencephalomalacia in a horse induced by fumonisin B_1 isolated from *Fusarium moniliforme*, *Onderstepoort J. Vet. Res.*, 55, 197, 1988.
27. Ross, P.F., Nelson, P.E., Owens, D.L., Rice, L.G., Nelson, H.A., and Wilson, T.M., Fumonisin B_2 in cultured *Fusarium proliferatum*, M-6104 causes equine leukoencephalomalacia, *J. Vet. Diagn. Invest.*, 6, 263, 1994.
28. Riley, R.T., Showker, J.L., Owens, D.L., and Ross, P.F., Disruption of sphingolipid metabolism and induction of equine leukoencephalomalacia by *F. proliferatum* culture material containing fumonisin B_2 or B_3, *Environ. Toxicol. Pharmacol.*, 3, 221, 1997.
29. Wang, E., Ross, F., Wilson, T.M., Riley, R.T., and Merrill, Jr., A.H., Increases in serum sphingosine and sphinganine and decreases in complex sphingolipids in ponies given feed containing fumonisins, mycotoxins produced by *Fusarium moniliforme*, *J. Nutrition*, 122, 1706, 1992.
30. Ross, P.F., Rice, L.G., Osweiler, G.D., Nelson, P.E., Richard, J.L., and Wilson, T.M., A review and update of animal toxicoses associated with fumonisin-contaminated feeds and production of fumonisin by *Fusarium* isolates, *Mycopathologia*, 117, 109, 1992.
31. Harrison, L.R., Colvin, B.M., Greene, J.T., Newman, L.E., and Cole, Jr., J.R., Pulmonary edema and hydrothorax in swine produced by fumonisin B_1, a toxic metabolite of *Fusarium moniliforme*, *J. Vet. Diagn. Invest.*, 2, 217, 1990.
32. Haschek, W.M., Motelin, G., Ness, D.K., Harlin, K.S., Hall, W.F., Vesonder, R.F., Peterson, R.E., and Beasley, V.R. Characterization of fumonisin toxicity in orally and intravenously dosed swine, *Mycopathologia*, 117, 83, 1992.
33. Colvin, B.M., Cooley, A.J., and Beaver, R.W., Fumonisin toxicosis in swine: clinical and pathologic findings, *J. Vet. Diagn. Invest.*, 5, 232, 1993.
34. Riley, R.T., Hinton, D.M., Chamberlain, W.J., Bacon, C.W., Wang, E., Merrill, Jr., A.H., and Voss, K.A., Dietary fumonisin B_1 induces disruption of sphingolipid metabolism in Sprague-Dawley rats: a new mechanism of nephrotoxicity, *J. Nutrition*, 124, 594, 1994.
35. Smith, G.W., Constable, P.D., Eppley, R., Tumbleson, M., Parker, H.M., and Haschek, W.M., Intravenously Administered Fumonisin B_1 Is Cardiotoxic to Swine, paper presented at the annual Society of Toxicology Meeting, Seattle, WA, March 1–5, 1998.
36. Smith, G.W., Constable, P.D., and Haschek, W.M., Cardiovascular responses to short-term fumonisin exposure in swine, *Fund. Appl. Toxicol.*, 33, 140, 1996.
37. Smith, G.W., Constable, P.D., Bacon, C.W., Meredith, F.I., and Haschek, W.M., Cardiovascular effects of fumonisins in swine, *Fund. Appl. Toxicol.*, 3111, 169, 1996.
38. Casteel, S.W., Turk, J.R., and Rottinghaus, G.E. Chronic effects of dietary fumonisin on the heart and pulmonary vasculature of swine, *Fund. Appl. Toxicol.*, 23, 518, 1994.
39. Voss, K.A., Porter, J.K., Bacon, C.W., Meredith, F.I., and Norred, W.P., Fusaric acid and modification of subchronic toxicity to rats of fumonisins in *F. moniliforme* culture material, *Food Chem. Toxicol.*, 37, 853, 1999.
40. Sauviat, M.P., Laurent, D., Kohler, F., and Pellegrin, F., Fumonisin, a toxin from the fungus *Fusarium moniliforme* Sheld., blocks both the calcium current and the mechanical activity in frog atrial muscle, *Toxicon*, 29, 1025, 1991.

41. Gelderblom, W.C.A., Snyman, S.D., Lebepe-Mazur, S., van der Westhuizen, L., Kriek, N.P.J., and Marasas, W.F.O., The cancer-promoting potential of fumonisin B_1 in rat liver using diethylnitrosamine as a cancer initiator, *Cancer Letters*, 109, 101, 1996.
42. Voss, K.A., Chamberlain, W.J., Bacon, C.W., Herbert, R.A., Walters, D.B., and Norred, W.P., Subchronic feeding study of the mycotoxin fumonisin B_1 in B6C3F_1 mice and Fischer 344 rats, *Fund. Appl. Toxicol.*, 24, 102, 1995.
43. Voss, K.A., Chamberlain, W.J., Bacon, C.W., and Norred, W.P., A preliminary investigation on renal and hepatic toxicity in rats fed purified fumonisin B_1, *Natural Toxins*, 1, 222, 1993.
44. Tolleson, W.H., Dooley, K.L., Sheldon, W.G., Thurman, J.D., Bucci, T.J., and Howard, P.C., The mycotoxin fumonisin induces apoptosis in cultured human cells and in livers and kidneys of rats, *Adv. Exper. Med. Biol.*, 392, 237, 1996.
45. Voss, K.A., Plattner, R.D., Bacon, C.W., and Norred, W.P., Comparative studies of hepatotoxicity and fumonisin B_1 and B_2 content of water and chloroform/methanol extracts of *Fusarium moniliforme* strain MRC 826 culture material, *Mycopathologia*, 112, 81, 1990.
46. Voss, K.A., Chamberlain, W.J., Bacon, C.W., Riley, R.T., and Norred, W.P., Subchronic toxicity of fumonisin B_1 to male and female rats, *Food Additiv. Contam.*, 12, 473, 1995.
47. Gelderblom, W.C.A., Kriek, N.P.J., Marasas, W.F.O., and Thiel, P.G., Toxicity and carcinogenicity of the *Fusarium moniliforme* metabolite, fumonisin B_1 in rats, *Carcinogenesis*, 12, 1247, 1991.
48. Gelderblom, W.C.A., Cawood, M.E., Snyman, S.D., Vleggaar, R., and Marasas, W.F.O., Structure-activity relationships of fumonisins in short-term carcinogenesis and cytotoxicity assays, *Food Chem. Toxicol.*, 31, 407, 1993.
49. Gelderblom, W.C.A., Cawood, M.E., Snyman, S.D., and Marasas, W.F.O., Fumonisin B_1 dosimetry in relation to cancer initiation in rat liver, *Carcinogenesis*, 15, 209, 1994.
50. Gelderblom, W.C.A., Semple, E., Marasas, W.F.O., and Farber, E., The cancer initiating potential of the fumonisin B mycotoxins, *Carcinogenesis*, 13, 433, 1992.
51. Knasmueller, S., Bresgen, N., Kassie, F., Mersch-Sundermann, V., Gelderblom, W., Zoehrer, E., and Eckl, P.M., Genotoxic effects of three *Fusarium moniliforme* mycotoxins, fumonisin B_1, moniliformin, and vomitoxin, in bacteria and in primary cultures of rat hepatocytes, *Mutation Res.*, 391, 39, 1997.
52. Mehta, R., Lok, E., Rowsell, P.R., Miller, J.D., Suzuki, C.A.M., and Bondy, G.S., Glutathione S-transferase-placental form expression and proliferation of hepatocytes in fumonisin B_1-treated male and female Sprague-Dawley rats, *Cancer Letters*, 128, 31, 1998.
53. Gelderblom, W.C.A. and Snyman, S.D. Mutagenicity of potentially carcinogenic mycotoxins produced by *Fusarium moniliforme*, *Mycotoxin Res.*, 7, 46, 1991.
54. Norred, W.P., Plattner, R.D., Vesonder, R.F., Bacon, C.W., and Voss, K.A. Effects of selected secondary metabolites of *Fusarium moniliforme* on unscheduled synthesis of DNA by rat primary hepatocytes, *Food Chem. Toxicol.*, 300, 233, 1992.
55. Kriek, N.P.J., Marasas, W.F.O., and Thiel, P.G. Hepato- and cardiotoxicity of *Fusarium verticillioides* (*F. moniliforme*) isolates from southern African maize, *Food Chem. Toxicol.*, 19, 447, 1981.
56. Suzuki, C.A.M., Hierlihy, L., Barker, M., Curran, I., Mueller, R., and Bondy, G.S., The effects of fumonisin B_1 on several markers of nephrotoxicity in rats, *Toxicol. Appl. Pharmacol.*, 133, 207, 1995.
57. Goldsworthy, T.L., Conolly, R.B., and Fransson-Steen, R., Apoptosis and cancer risk assessment, *Mutation Res.*, 365, 71, 1996.

58. Hannun, Y.A. and Obeid, L.M., Mechanisms of ceramide-mediated apoptosis, in *Eicosanoids and Other Bioactive Lipids in Cancer, Inflammation and Radiation,* Vol. 3, Honn et al., Eds., Plenum Press, New York, 1997, p. 145.
59. Majno, G. and Joris, I., Apoptosis, oncosis, and necrosis: an overview of cell death, *Am. J. Pathol.*, 146, 3, 1995.
60. Voss, K.A., Plattner, R.D., Riley, R.T., Meredith, F.I., and Norred, W.P., *In vivo* effects of fumonisin B_1-producing and fumonisin B_1-nonproducing *Fusarium moniliforme* isolates are similar: fumonisins B_2 and B_3 cause hepato- and nephrotoxicity in rats, *Mycopathologia,* 141, 45, 1998.
61. Bondy, G., Barker, M., Mueller, R., Fernie, S., Miller, J.D., Armstrong, C., Hierlihy, S.L., Rowsell, P., and Suzuki, C., Fumonisin B_1 toxicity in male Sprague-Dawley rats, *Adv. Exper. Med. Biol.*, 392, 251, 1996.
62. Bondy, G.S., Suzuki, C.A.M., Mueller, R.W., Fernie, S.M., Armstrong, C.L., and Hierlihy, S.L., Gavage Administration of the fungal toxin fumonisin $B_1$to female Sprague-Dawley rats, *J. Toxicol. Environ. Health*, Part A, 53, 135, 1998.
63. Bondy, G., Suzuki, C., Barker, M., Armstrong, C., Fernie, S., Hierlihy, L., Rowsell, P., and Meuller, R., Toxicity of fumonisin B_1 administered intraperitoneally to male Sprague-Dawley rats, *Food Chem. Toxicol.*, 33, 653, 1995.
64. Voss, K.A., Bacon, C.W., Norred, W.P., Chapin, R.E., Chamberlain, W.J., Plattner, R.D., and Meredith, F.I., Studies on the reproductive effects of *Fusarium moniliforme* culture material in rats and the biodistribution of [^{14}C] fumonisin B_1 in pregnant rats, *Natural Toxins*, 4, 24, 1996.
65. Lim, C.W., Parker, H.M., Vesonder, R.F., and Haschek, W.M., Intravenous fumonisin B_1 induces cell proliferation and apoptosis in the rat, *Natural Toxins*, 4, 34, 1996.
66. Wild, C.P., Castegnaro, M., Ohgaki, H., Garren, L., Galendo, D., and Miller, J.D., Absence of a synergistic effect between fumonisin B_1 and N-nitromethylbenzylamine in the induction of oesophageal papillomas in the rat, *Natural Toxins*, 5, 126, 1997.
67. Martinova, E.A. and Merrill, Jr., A.H., Fumonisin B_1 alters sphingolipid metabolism and immune function in BALB/c mice: immunological responses to fumonisin B_1, *Mycopathologia*, 130, 163, 1995.
68. Martinova, E.A., Solov'ev, A.S., Khrenov, A.V., Zabotina, T.N., and Alessenko, A.V., Fumonisin B_1 modulates sphingomyelin cycle product levels and the expression of CD_3 receptors in immunocompetent organs, *Biochemistry* (Moscow), 60, 461, 1995.
69. Tryphonas, H., Bondy, G., Miller, J.D., Lacroix, F., Hodgen, M., Mcguire, P., Fernie, S., Miller, D., and Hayward, S., Effects of fumonisin B_1 on the immune system of Sprague-Dawley rats following a 14-day oral (gavage) exposure, *Fund. Appl. Toxicol.*, 39, 53, 1997.
70. Dombrink-Kurtzman, M.A., Javed, T., Bennett, G.A., Richard, J.L., Cote, L.M., and Buck, W.B. Lymphocyte cytotoxicty and erythrocyte abnormalities induced in broiler chicks by fumonisins B_1 and B_2 and moniliformin from *Fusarium proliferatum*, *Mycopathologia*, 124, 47, 1994.
71. Marijanovic, D.R., Holt, P., Norred, W.P., Bacon, C.W., Voss, K.A., and Stancel, P.C., Immunosuppresive effects of *Fusarium moniliforme* corn cultures in chickens, *Poultry Sci.*, 70, 1895, 1991.
72. Qureshi, M.A. and Hagler, Jr., W.M., Effect of fumonisin B_1 exposure on chicken macrophage functions *in vitro*, *Poultry Sci.*, 71, 104, 1992.
73. Howard, P.C., Muskhelishvili, L., Dooley, K.L., Bolon, B., Warbritton, A., Voss, K.A., Lorentzen, R.J., and Bucci, T.J., Hepatic and renal apoptosis and sphingolipid changes in F344 rats and B6C3F$_1$ mice chronically fed the corn fungal toxin fumonisin B_1, *Proc. Am. Assoc. Cancer Res.,* 37, 1996, 791.

74. Sharma, R.P., Dugyala, R.R., and Voss, K.A., Demonstration of *in situ* apoptosis in mouse liver and kidney after short-term repeated exposure to fumonisin B_1, *J. Comp. Pathol.*, 117, 371, 1997.
75. Bondy, G.S., Suzuki, C.A.M., Fernie, S.M., Armstrong, C.L., Heirlihy, S.L., Savard, M.E., and Barker, M., Toxicity of fumonisin B_1 to B6C3F_1 mice: a 14-day gavage study, *Food Chem. Toxicol.*, 35, 981, 1997.
76. Fincham, J.E., Marasas, W.F.O., Taljaard, J.J.F., Kriek, N.P.J., Badenhorst, C.J., Gelderblom, W.C.A., Seier, J.V., Smuts, C.M., Faber, M., Weight, M.J., Slazus, W., Woodroof, C.W., van Wyk, M.J., Kruger, M., and Thiel, P.G., Atherogenic effects in a non-human primate of *Fusarium moniliforme* cultures added to a carbohydrate diet, *Atherosclerosis*, 94, 13, 1992.
77. Jaskiewicz, K., Marasas, W.F.O., and Taljaard, J.J.F., Hepatitis in vervet monkeys caused by *Fusarium moniliforme*, *J. Comp. Pathol.*, 97, 281, 1987.
78. Wang, E., Norred, W.P., Bacon, C.W., Riley, R.T., and Merrill, A.H., Inhibition of sphingolipid biosynthesis by fumonisins, *J. Biol. Chem.*, 266, 14486, 1991.
79. Voss, K.A., Riley, R.T., Bacon, C.W., Meredith, F.I., and Norred, W.P., Toxicity and sphinganine levels are correlated in rats fed fumonisin B_1 or hydrolyzed B_1, *Environ. Toxicol. Pharmacol.*, 5, 101, 1998.
80. Tsunoda, M., Sharma, R.P., and Riley, R.T., Early fumonisin B_1 toxicity in relation to disrupted sphingolipid metabolism in male BALB/c mice, *J. Biochem. Molec. Toxicol.*, 12, 281, 1998.
81. Merrill, Jr., A.H., Schmelz, E.-M., Dillehay, D.L., Spiegel, S., Shayman, J.A., Schroeder, J.J., Riley, R.T., and Voss, K.A., Sphingolipids — the enigmatic lipid class: biochemistry, physiology, and pathophysiology, *Toxicol. Appl. Pharmacol.*, 142, 208, 1997.
82. Yoo, H.-S., Norred, W.P., Showker, J., and Riley, R.T., Elevated sphingoid bases and complex sphingolipid depletion as contributing factors in fumonisin-induced cytotoxicity, *Toxicol. Appl. Pharmacol.*, 138, 211, 1996.
83. Riley, R.T., Voss, K.A., Yoo, H.-S., Gelderblom, W.C.A., and Merrill, Jr., A.H., Mechanism of fumonisin toxicity and carcinogenesis, *J. Food Protect.*, 57, 638, 1994.
84. Riley, R.T., Voss, K.A., Norred, W.P., Sharma, R.P., Wang, E., and Merrill, Jr., A.H., Fumonisins: mechanism of mycotoxicity, *Revue Med. Vet.*, 149, 617, 1998.
85. Gelderblom, W.C.A., Smuts, C.M., Abel, S., Snyman, S.D., Cawood, M.E., van der Westheizen, L., and Swanevelder, S., Effect of fumonisin B_1 on protein and lipid synthesis in primary rat hepatocytes, *Food Chem. Toxicol.*, 34, 361, 1996.
86. Gelderblom, W.C.A., Smuts, C.M., Abel, S., Snyman, S.D., van der Westheizen, L., Huber, W.W., and Swanevelder, S., Effect of fumonisin B_1 on the levels and fatty acid composition of selected lipids in rat liver *in vivo*, *Food Chem. Toxicol.*, 35, 647, 1997.
87. Abado-Becognee, K., Mobio, T.A., Ennamany, R., Fleurat-Lessard, F., Shier, W.T., Badria, F., and Creppy, E.E., Cytotoxicity of fumonisin B_1: implication of lipid peroxidation and inhibition of protein and DNA syntheses, *Arch Toxicol.*, 72, 233, 1998.
88. Yin, J.-J., Smith, M.J., Eppley, R.M., Page, S.W., and Sphon, J.A., Effects of fumonisin B_1 on lipid peroxidation in membranes, *Biochim. Biophys. Acta*, 1371, 134, 1998.
89. Sahu, S.C., Eppley, R.M., Page, S.W., Gray, G.C., Barton, C.N., and O'Donnell, M.W., Peroxidation of membrane lipids and oxidative DNA damage by fumonisin B_1 in isolated rat liver nuclei, *Cancer Lett.*, 125, 117, 1998.

90. Abel, S. and Gelderblom, W.C.A., Oxidative damage and fumonisin B_1-induced toxicity in primary rat hepatocytes and rat liver *in vitro*, *Toxicology*, 131, 121, 1998.
91. Bruno Doko, M., Canet, C., Brown, N., Sydenham, E.W., Mpuchane, S., and Siame, B.A., Natural co-occurrence of fumonisins and zearalenone in cereals and cereal-based foods from eastern and southern Africa, *J. Agric. Food Chem.*, 44, 3240, 1996.
92. Sydenham, E.W., Stockenstrom, S., Thiel, P.G., Shephard, G.S., Koch, K.R., and Marasas, W.F.O., Potential of alkaline hydrolysis for the removal of fumonisins from contaminated corn, *J. Agric. Food Chem.*, 43, 1198, 1995.
93. Hendrich, S., Miller, K.A., Wilson, T.M., and Murphy, P.A. Toxicity of *Fusarium proliferatum*-fermented nixtamalized corn-based diets fed to rats: effect of nutritional status, *J. Agric. Food Chem.*, 41, 1649, 1993.
94. Voss, K.A., Bacon, C.W., Meredith, F.I., and Norred, W.P. Comparative subchronic toxicity studies of nixtamalized and water-extracted *Fusarium moniliforme* culture material, *Food Chem. Toxicol.*, 34, 623, 1996.
95. Norred, W.P., Plattner, R.D., Dombrink-Kurtzman, M.A., Meredith, F.I., and Riley, R.T., Mycotoxin-induced elevation of free sphingoid bases in precision-cut rat liver slices: specificity of the response and structure-activity relationships, *Toxicol. Appl. Pharmacol.*, 147, 63, 1997.
96. Bacon, C.W., Porter, J.K., Norred, W.P., and Leslie, J.F., Production of fusaric acid by *Fusarium* species, *Appl. Environ. Microbiol.*, 62, 4039, 1996.
97. Bacon, C.W., Porter, J.K., and Norred, W.P., Toxic interaction of fumonisin B_1 and fusaric acid measured by injection into fertile chicken egg, *Mycopathologia*, 129, 29, 1995.
98. Porter, J.K., Bacon, C.W., Wray, E.M., and Hagler, Jr., W.M., Fusaric acid in *Fusarium moniliforme* cultures, corn, and feeds toxic to livestock and the neurochemical effects in the brain and pineal gland of rats, *Natural Toxins,* 3, 91, 1995.
99. Porter, J.K., Wray, E.M., Rimando, A.M., Stancel, P.C., Bacon, C.W., and Voss, K.A., Lactational passage of fusaric acid from the feed of nursing dams to the neonate rat and effects on pineal neurochemistry in the F_1 and F_2 generations at weaning, *J. Toxicol. Environ. Health*, 49, 161, 1996.
100. Poling, S.M. and Plattner, R.D., Rapid purification of fumonisins B_3 and B_4 with solid phase extraction columns, *J. Agric. Food Chem.*, 44, 2792, 1996.
101. Shephard, G.S., Thiel, P.G., Sydenham, E.W., Alberts, J.F., and Cawood, M.E., Distribution and excretion of a single dose of the mycotoxin fumonisin B_1 in a non-human primate, *Toxicon*, 32, 435, 1994.
102. Branham, B.E. and Plattner, R.D., Isolation and characterization of a new fumonisin from liquid cultures of *Fusarium moniliforme*, *J. Nat. Products*, 56, 1630, 1993.
103. McKenzie, K.S., Sarr, A.B., Mayura, K., Baily, R.H., Miller, D.R., Rogers, T.D., Norred, W.P., Voss, K.A., Plattner, R.D., Kubena, L.F., and Phillips, T.D., Oxidative degredation and detoxification of mycotoxins using a novel source of ozone, *Food Chem. Toxicol.*, 35, 807, 1997.
104. Lu, Z., Dantzer, W.R., Hopmans, E.C., Prisk, V., Cunnick, J.E., Murphy, P.A., and Hendrich, S., Reaction with fructose detoxifies fumonisin B_1 while stimulating liver-associated natural killer cell activity in rats, *J. Agric. Food Chem.*, 45, 803, 1997.
105. Howard, P.C., Churchwell, M.I., Couch, L.M., Marques, M.M., and Doerge, D.R., Formation of N-(carboxymethyl) fumonisin B_1 following the reaction of fumonisin B_1 with reducing sugars, *J. Agric. Food Chem.*, 46, 3546, 1998.

5 Mechanisms of Action of Aflatoxin B_1 at the Biochemical and Molecular Levels

Avishay-Abraham Stark

CONTENTS

I. ABSTRACT

Aflatoxin B_1 (AFB_1) is a mycotoxin that frequently contaminates human foods and animal feeds. It poses major health and economical problems in that at high doses it is an acutely acting toxin. At lower doses, it is a potent (hepato)carcinogen, mutagen, teratogen, and immunosuppressor. Numerous species of mammals, birds, and fish are susceptible to its adverse effects. The carcinogenicity, teratogenicity, and mutagenicity of AFB_1 stem from its ability to be converted to a chemically reactive metabolite that covalently binds the N^7 position of guanine residues in DNA. Mutations in oncogenes and in tumor suppressor genes in experimental hepatocellular carcinomas and lung carcinomas in AFB_1-treated animals are consistent with the mutational specificity of AFB_1. Also consistent with the mutational specificity of AFB_1 are the mutations found in codon 249 of the p53 gene in hepatocallular carcinomas in humans from areas with high exposure to AFB_1, but not in liver tumors of people from areas with low or no exposure to AFB_1. Cancers are typically diseases of the old. The triggering (initiating) events often occur decades before the onset of clinical symptoms. Thus, although many environmental causes of cancer can be identified based on population studies (epidemiology), it is more difficult to attribute a cancer-causing agent to a specific cancer on an individual basis. The discovery of specific mutations in guanine residues in p53 in liver tumors occurring in AFB_1-infested areas is an important step toward the identification of cancer-causing agents on an individual basis.

II. ACUTE TOXICITY AND CARCINOGENICITY OF AFLATOXIN TO HUMANS

Aflatoxins (AFs) are products of *Aspergillus* and *Penicillium* species. They are highly potent in the induction of acute toxicity in animals and in humans. The liver is the prime target organ for acute toxicity and carcinogenicity in all species in that the predominant route of exposure is via the alimentary tract.[1] Acute aflatoxicosis outbreaks in Uganda and Taiwan resulted in vomiting, abdominal pain, pulmonary edema, and fatty infiltration and necrosis of the liver. Consumption of AF-contaminated corn (6 to 16 ppm) in 200 Indian villages resulted in over 25% fatalities, for which the primary cause of death was gastrointestinal hemorrhage.[2–4] The symptoms of acute aflatoxicosis are indistinguishable from those of Reye's syndrome; vomiting, convulsions, coma, and death were associated with cerebral edema and fatty involvement of the liver, kidney, and heart. Aflatoxin B_1 (AFB_1) was detected in the liver, kidney, brain, bile, and gastrointestinal content of 22 out of 23 fatalities attributed to Reye's syndrome in Thailand[5] and the U.S.[6]

In addition to their prevalence in crops, foods, and feed stuff, AFs are frequently at high levels in airborne, respirable grain dust particles. They may represent an occupational and environmental hazard. For example, a significant increase in mortality from total cancer and respiratory cancer has been observed in peanut-processing workers exposed to AF-contaminated dust.[7–10] The lung is also at risk after dietary AF exposure.[11] Immunosuppression is another feature of acute AF intoxication. Exposure of lungs to aerosolic AF resulted in suppression of phagocytosis by alveolar macrophages that persisted for 2 weeks.[12] The amount of DNA adducts was proportional to the concentration of AF and to the time of exposure.[13] Although seemingly impractical, aflatoxin-containing bombs have been loaded and deployed by the Iraqis.[14]

Aflatoxin B_1 is carcinogenic to many organs. It is one of the most potent hepatocarcinogens in animals and humans. Very strong correlation exists between the daily dietary intake of AFB_1 and the incidence of primary liver cancer in humans

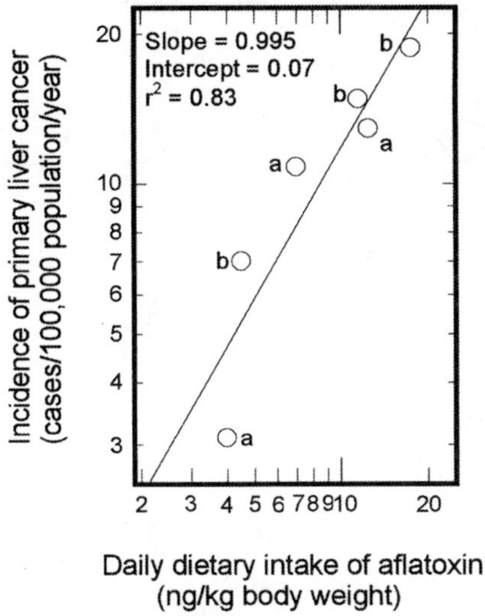

FIGURE 5.1 Correlation between ingestion of aflatoxins and the incidence of primary liver cancer in humans. The daily dietary intake of aflatoxins was calculated from the concentration of aflatoxins in foods and the daily intake of these foods (a) in Kenya,[101] and (b) in Swaziland.[102] In Mozambique, a calculated daily intake of 222 ng aflatoxins per kg body weight was associated with a primary liver cancer incidence of 35.5 cases per 100,000 population per year.[103]

(Figure 5.1). The slope of the log-log curve is close to unity, indicating that at these low doses the mutagenic AFB_1 acts predominantly as an initiator and has little effect on cell proliferation, similar to hepatocarcinogenesis induced in mice by low doses of N-acetyl-2-aminofluorene.[15]

A. Biochemical Activation of Aflatoxins

Aflatoxins must undergo oxidative metabolism in order to be converted into acutely toxic and carcinogenic reactive species. AF toxicity and carcinogenicity are exerted by covalent binding of reactive AF metabolites to cellular macromolecules: DNA, RNA, and protein.[1,16–19]

B. Biochemical Activation in the Liver

Aflatoxin metabolism has been studied most extensively in the liver. In the liver, metabolism involves the NADPH-dependent cytochrome P-450 (CYP). The levels of activation, DNA binding, mutant frequency, and sensitivity to the acute toxic effects in a certain cell type depend on the types of cytochrome P-450 and their levels in these cells.[20]

Activation of AFs toward toxicity, mutagenicity, and carcinogenicity results from oxidation at the 8,9 position that forms AF-8,9-oxide. The latter binds covalently to DNA, RNA, and protein. Hydroxylation of AF at other positions yields

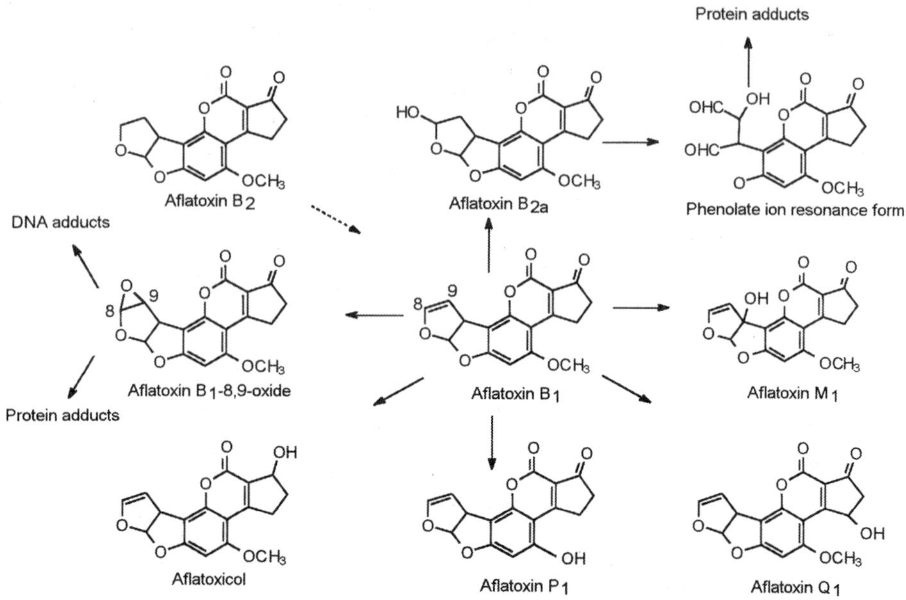

FIGURE 5.2 Metabolism of aflatoxin B_1. Oxidation by microsomal cytochrome P-450 leads to the formation of AFM_1, AFP_1, AFQ_1, and AFB_1-8,9-oxide. Unidentified oxidative reactions lead to the formation of AFL and AFB_{2a}. AFB_2 can be converted to AFB_1 by a microsomal dehydrogenase. Addition of water to AFB_1-8,9-oxide (probably by epoxide hydrolase or a similar enzyme) yields AFB_1-8,9-diol. AFB_1-8,9-diol and AFB_{2a} are converted non-enzymatically to the phenolate ion, which covalently binds proteins by formation of Schiff bases with amino moieties. AFB_1-8,9-oxide can react with nucleophilic centers in proteins, such as thiols in cysteine and methionine, and N-3 and N-7 in histidine to give other protein adducts. Addition of glutathione to AFB_1-8,9-oxide by glutathione transferase is the primary detoxification reaction of AFB_1.

AFM_1, AFP_1, AFQ_1, AFB_{2a}, and aflatoxicol (AFL) (Figure 5.2). These are less toxic and carcinogenic than AF-8,9-oxide, and, although they can undergo oxidation at the 8,9-position and bind cellular macromolecules, they are considered detoxification metabolites of AF. AF-8,9-oxide is converted by the detoxification enzyme epoxide hydrolase to the 8,9-dihydrodiol. The latter metabolite and other hydroxylated AFs are converted to their glutathione (GSH) conjugates by GSH transferase and are excreted in the urine as mercapturic acids.[21,22] AFB_2 contains a saturated 8,9 bond. Its weak mutagenicity and carcinogenicity may be due to a dehydrogenase that converts it to AFB_1 by desaturation of the 8,9 bond.[23] Ingested AF that escapes the liver is readily distributed to other organs, where it can be metabolized and exert toxicity and carcinogenicity. The resistance of a certain species to AFB_1-induced carcinogenicity is usually correlated with the capability to detoxify AFB_1, predominantly by conjugation to glutathione.

C. Biochemical Activation in the Lung

As compared to the liver, only a few studies have been conducted on the activation-inactivation metabolism in the lung. Activation of AF in the rat and rabbit lung is carried out by CYP2B4 and/or CYP4B1, and CYP1 detoxifies AF by its conversion to AFM_1.[24–26] Most of the lung metabolism occurs in the non-ciliated tracheal and bronchiolar epithelial (Clara) cells,[25,27,28] which are also the prime targets of the acute toxicity of AF.[28] Detoxification by GSH conjugation of AF, however, is sixfold lower in the rabbit lung as compared to the liver.[29] The lung activates AF to a mutagen more efficiently than benzo(a)pyrene, a known lung carcinogen.[30] Clara cells in the human lung are more rare and do not contain high concentrations of the activating enzymes as compared to those in the rodent lung, thus they are unlikely to be the center of bioactivation of AFs. Surfactant-secreting alveolar type II cells in the human lung contain high P-450 activities and are capable of bioactivation.[31,32]

In addition to cytochrome P-450, microsomal prostaglandin H synthase and cytosolic lipoxygenase are involved in xenobiotic metabolism. The process is termed co-oxidation, because metabolism depends on the oxidation of arachidonic acid. Co-oxidation is more important in extrahepatic tissues, where the overall P-450 activity is relatively low. In the Guinea pig, for example, considerable P-450-independent activation of AF occurs in the presence of indomethacin, an inhibitor of prostaglandin H synthase, but is inhibited by a lipoxygenase inhibitor, indicating that lipoxygenase is responsible for AF activation.[33–35]

The human lung is typically low in cytochrome P-450[36] and rich in prostaglandin H synthase and lipoxygenase.[37] Thus, AF was activated by whole lung microsomes to a low extent, whereas activation by the cytosol correlated with lipoxygenase activity. Separation of lung cells revealed that P-450-dependent activation was minimal or absent in all cell types, and most of the activation was contained in the lung macrophages, where co-oxidation was the predominant pathway.[32,38,39] AF activation[37] and DNA binding[40] by the human lung shows great inter-individual variability. A significant fraction of inhaled AF is translocated to the blood and the liver, resulting in binding to liver DNA.[41,42]

D. Photoactivation

In addition to microsomal metabolism, AFs are photoactivated by ultraviolet light (e.g., sunlight) to yield reactive, DNA-binding species, which form DNA adducts that are indistinguishable from those produced by microsomal metabolism. Chemically reactive (singlet) molecular oxygen, formed upon energy transfer from the photoexcited coumarin moiety of AF, oxidizes the 8,9-unsaturated bond in the furan moiety of AFB_1. AFB_2 lacks the 8,9-unsaturated bond, but its coumarin moiety readily forms singlet oxygen. Thus, photoactivated mixtures of AFB_1 and AFB_2 act synergistically in DNA binding.[43–48] Contamination of exposed skin with AFs in the sunlight may thus be hazardous.

III. MOLECULAR MECHANISM OF ACTION OF AFLATOXINS

A. Relationship Between Carcinogenicity, Mutagenicity, and Acute Toxicity and AF Binding to Cellular Macromolecules

Carcinogenicity[1] and mutagenicity[49,50] of AFs are associated with their DNA binding properties. Acute toxicity may also be associated with AF-protein binding.[51] DNA and RNA synthesis is rapidly inhibited by AF. Protein synthesis, mitochondrial electron transport, and lipid biosynthesis are also inhibited. Ingestion of AF leads to the breakdown of lysosomes in parenchymal liver cells and to the release of degradative enzymes. The latter causes acute damage to the liver (hemorrhage and necrosis).[1] The susceptibility of a certain organ or a species to AFB_1-induced carcinogenicity is correlated with the frequency of AFB_1 adducts in the DNA of the organ or species in question. For example, in Fischer rats, AFB_1 is carcinogenic to the liver but not the kidney. AFB_1 adducts are tenfold more frequent in the liver DNA as compared to kidney DNA. The Swiss mouse is resistant to AFB_1-induced hepatocarcinogenesis. The level of AFB_1-DNA adduct in mouse liver is 80-fold lower than that found in rat liver.[17]

B. Molecular Aspects of AFB_1-DNA Binding

Oxidation of AFB_1 at the 8,9-position forms AFB_1-8,9-oxide (Figure 5.2). The electrophilic C-8 sustains a nucleophilic attack by N^7 of guanine, resulting in 8,9-dihydro-8-(N^7-guanyl)-9-hydroxy AFB_1 (Figure 5.3). Excluding two reports about *in vitro* binding of chemically activated AFB_1 to cytosine[52] and formation of alkali-labile (apurinic) sites at adenine residues,[53] the reaction of metabolically activated AFB_1 *in vitro* or *in vivo*[54,55] is unique among DNA-binding carcinogens in that it exclusively binds N^7 guanine. Other carcinogens bind a variety of nucleophilic atoms in DNA. The high specificity of AFB_1 for N^7 guanine is most probably due to steric constraints of the non-covalent intercalation of the furan moiety of AFB_1 in the major groove of DNA, bringing the reactive C-8 of AFB_1-8,9-oxide into close proximity to N^7 of guanine. Dependence of AFB_1-DNA binding on the flanking sequences (see below) and inhibition of binding by covalently linked, intercalated benzofuran indicate that precise non-covalent intercalation of AFB_1-8,9-oxide must precede binding with N^7guanine.[56] Metabolites of AFB_1 that contain the 8,9-unsaturated bond are capable of DNA binding, but to a lesser extent than AFB_1.[57]

The adduct is unstable, in that modification of the N^7 position forms a positive charge in the imidazol moiety of guanine, resulting in destabilization of the glycosidic bond that leads to facile depurination of the AFB_1-N^7-guanine adduct[58] and strand scission.[59] Opening of the imidazol ring forms a stable AF-formamidopyrimidine (FAPY) adduct.[54,60] Due to their instability, N^7-guanine adducts of AFB_1 or of its metabolites are excreted in the urine and are used to monitor and measure the exposure of humans to aflatoxins.[22,61] Guanines in double-stranded DNA are preferred binding targets, as compared to those in single-stranded DNA.[62]

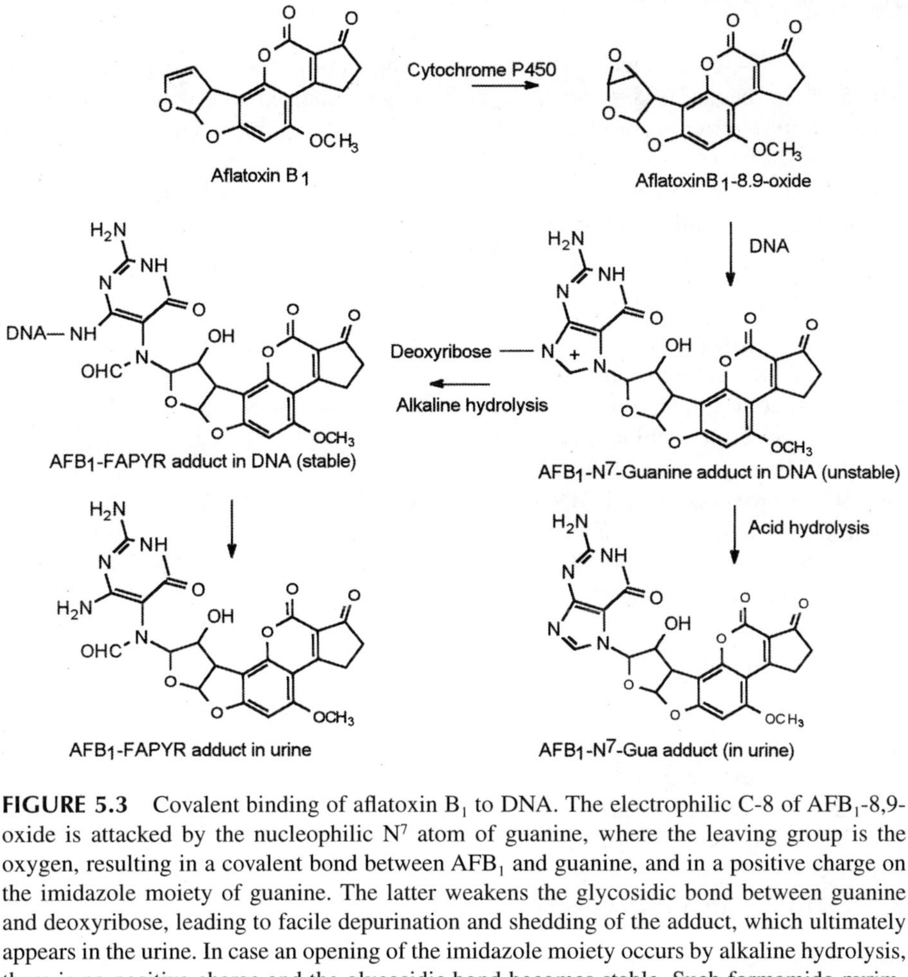

FIGURE 5.3 Covalent binding of aflatoxin B$_1$ to DNA. The electrophilic C-8 of AFB$_1$-8,9-oxide is attacked by the nucleophilic N^7 atom of guanine, where the leaving group is the oxygen, resulting in a covalent bond between AFB$_1$ and guanine, and in a positive charge on the imidazole moiety of guanine. The latter weakens the glycosidic bond between guanine and deoxyribose, leading to facile depurination and shedding of the adduct, which ultimately appears in the urine. In case an opening of the imidazole moiety occurs by alkaline hydrolysis, there is no positive charge and the glycosidic bond becomes stable. Such formamido pyrimidine (FAPYR)-AFB$_1$ adducts are also removed and appear in the urine. The appearance of N^7-guanine and FAPYR adducts are indicative of exposure to aflatoxins, and their quantity reflects the severity of the exposure.

Theoretical models show that GGGG sequences are major targets for covalent binding. Within this sequence, the third G residue was calculated to be the most favorable to bind AFB$_1$.[63] In double-stranded DNA, the flanking nucleotides greatly affect the efficiency of AFB$_1$ binding. For example, the Gs in AGA, TGA, and GCC are poor targets; in TGT, CGC, GGG, and GGC, moderate targets; and in CCG, GGG, CCGG, CCGCC, and CGGCG, strong targets. Reaction rates of AFB$_1$ with moderate and strong targets are 2 to 3 and 6 to 10 times faster than with poor ones, respectively.[64] *In vivo*, not all regions of DNA are equally susceptible to binding; for example, genes coding for ribosomal RNA in the nucleolus[65] and inter-nucleosomal DNA[66] preferentially bind AFB$_1$, because they are less protected by chromosomal proteins.

C. Genetic Consequences of AFB_1-DNA Binding

The AFB_1-N^7-Gua adduct, FAPY adducts, and apurinic sites are pre-mutagenic lesions.[67] In bacteria (*Salmonella typhimurium*), approximately 60 adducts per genome are required to produce a lethal event, and 40 adducts per genome induce a frequency of 10^{-4} reversions (point mutations) to histidine prototrophy.[43] Only 0.3 adducts per genome are required to induce the same frequency of forward mutations in guanine-hypoxanthine phosphoribosyl transferase.[49]

1. Mutagenesis

The mutations in AFB_1-treated cells are best explained as being due to the AFB_1-N^7-Gua adducts rather than to apurinic sites.[58] An apurinic site is a pre-mutagenic lesion in that an error-prone polymerase induced under SOS conditions has a preference to place an adenine against the abasic site, leading to a GC→TA transversion.[68,69] The most prevalent mutation is a GC→TA transversion. Less common are GC→CG transversions and GC→AT transitions,[67] and a minor fraction consists of frameshifts.[70,71] AFB_1 mutagenesis depends on several error-prone DNA repair systems in bacteria and is an inducer of the SOS response.[58,67,72–74] AFB_1 induces an SOS-like system in mammalian cells.[75]

2. Immunosuppression

Aflatoxin B_1 is an immunosuppressor in mice, rats, and rabbits[12,76–78] and is immunotoxic to lectin-stimulated human peripheral blood lymphocytes *in vitro*[79] due to its capability to impair DNA, RNA, and protein synthesis. Actively dividing cells such as those of the immune system are thus sensitive to these effects more than non-dividing ones.

3. Teratogenicity of Aflatoxin B_1

Aflatoxin B_1 causes malformations in numerous organs in embryos of various species of mammals, birds, amphibians, and fish.[80–82] The exact mechanism by which AFB_1 acts as a teratogen is not clear. However, because teratogenicity is associated with mutations and gross alteration in the structure of chromosomes (e.g., sister chromatid exchanges, chromosomal aberrations, induction of micronuclei in humans[83]), the DNA binding property of AFB_1 may be the key reaction. AFB_1 ingested by pregnant females is readily transferred via the placenta to animal[84] and human[85] embryos.

4. Relationship Between Primary Liver Cancer and the Specificity of AFB_1 in DNA Binding and Mutagenesis

The tumor suppressor gene p53 is the most frequently mutated oncogene in diverse human tumors. Its activation involves primarily point mutations that lead to amino acid substitutions. Sequence analysis of the mutations reveals that there is an association between the types of the altered DNA bases, the mutagen, and carcinogenesis in a specific tissue or organ. For example, altered p53 in malignant skin cancers in humans contained CC→TT mutations and C→T substitutions exclusively at dipyrimidine sites. These alterations are known to be caused only by ultraviolet light. Alterations in p53 in internal cancers did not contain these ultraviolet-specific mutations.[86,87]

c-Ki-*ras*, c-Ha-*ras*, and c-N-*ras* protooncogenes are frequently mutated in experimental liver tumors in rats and mice. In AFB_1-treated mice, 75 of 76 liver tumors contained mutations in codon 12 (GGT), where all of the mutations were located in guanine residues,[88] strongly indicating that the primary lesion was the binding of AFB_1 to guanine residues at this codon. Similar predominance of GGT→GAT, GGT→TGT, and GGT→GTT mutations in codon 12 of c-Ki-*ras* or c-N-*ras* occurs in liver tumors in rats.[89,90] Although frequently mutated in many human cancers, *ras* activation is not associated with hepatocellular carcinoma,[91] even in regions where dietary exposure to AFB_1 is a risk factor.[92] But, the p53 tumor suppressor is frequently mutated in many cancers in humans, including hepatocellular carcinoma. Surveys compiling numerous (>1000) individual cases of hepatocellular carcinoma in humans revealed that high levels of AFB_1 in the environment and in the diet in West Africa, Mozambique, Southern Africa, Qidong and Tongan in China, and Mexico correlated with a high prevalence of hepatocellular carcinoma and with a high prevalence of mutations involving guanines in codon 249 of p53.[93–97] Such guanine residues are strong or moderate targets for AFB_1 binding.[64] In Alaska, Europe, or Japan, where exposure to AF is low or does not occur, hepatocellular carcinomas show very low frequency of mutations in codon 249.[98–100] Thus, the G→T transversions in codon 249 of p53 are consistent with the mutagenic specificity of AFB_1 observed *in vitro* and *in vivo*. The identification of specific AFB_1-induced mutations in tumors of humans living in areas known to be contaminated with AFB_1 makes it possible to more accurately identify the causing agent. The AFB_1 story is one of the best examples for molecular epidemiology in action.

REFERENCES

1. Busby, Jr., W.F. and Wogan, G.N., Aflatoxins, in *Chemical Carcinogens,* Searle, C.E., Ed., ACS Monograph 182, American Chemical Society, Washington, D.C., 1984, pp. 946–1135.
2. Shank, R.C., *Mycotoxins and N-Nitroso Compounds: Environmental Risks,* Vol. I, Shank, R.C., Ed., CRC Press, Boca Raton, FL, 1981, pp. 107–140.
3. Shank, R.C., Epidemiology of aflatoxin carcinogenesis, in *Environmental Cancer,* Kraybill, H.F. and Mehlman, M.A., Eds., John Wiley & Sons, New York, 1977, pp. 291–318.
4. van Rensburg, S.J., Role of epidemiology in causation of mycotoxin health risk, in *Mycotoxins in Human and Animal Health,* Rodricks, J.V., Hesseltine, C.W., and Mehlman, M.A., Eds., Pathotox, Park Forest South, IL, 1977, pp. 699–711.
5. Shank, R.C., Burgeois, C.H., Kechamars, N., and Chandavimol, P., Aflatoxins in autopsy of specimens from Thai children with an acute disease of unknown etiology. *Food Cosmet. Toxicol.,* 9:501–507, 1971.
6. Nelson, D.B. et al., Aflatoxins and Reye's syndrome: a case control study. *Pediatrics,* 66:865–869, 1980.

7. van Nieuwenhuize, J.P. et al., Aflatoxinen: epidemiologish onderzoek naar carcinogeniteit bij langdurige "low level" exposite van eon fabriekspopulatie. *T. Soc. Geneesk,* 51:754–760, 1973.
8. Sorenson, W.G. et al., Aflatoxin in respirable corn dust particles. *J. Toxicol. Environ. Health,* 7:669–672, 1981.
9. Hayes, R.B. et al., Aflatoxin exposures in the industrial setting: an epidemiological study of mortality. *Food Chem. Toxicol.,* 22:39–43, 1984.
10. Dvorackova, I., Aflatoxin inhalation and alveolar cell carcinoma. *Br. Med. J.,* 1:691, 1986
11. Harrison, J.C. and Garner, R.C., Immunological and HPLC detection of aflatoxin adducts in human tissues after an acute poisoning incident in S.E. Asia. *Carcinogenesis,* 12:741–743, 1991.
12. Jakab, G.J. et al. Respiratory aflatoxicosis: suppression of pulmonary and systemic host defenses in rats and mice. *Toxicol. Appl. Pharomacol.,* 125:198–205, 1994.
13. Zarba, A. et al., Aflatoxin B_1-DNA adduct formation in the rat liver following exposure by aerosol inhalation. *Carcinogenesis,* 13:1031–1033, 1992.
14. Zilinskas, R.A., Iraq's biological weapons. The past as future?, *J. Am. Med. Assoc.,* 278:418–424, 1997.
15. Cohen, S.M. and Ellwein, L.B., Cell proliferation in carcinogenesis. *Science,* 249:1007–1011, 1990.
16. Stark, A.A. et al., Aflatoxin B_1 mutagenesis and adduct formation in *Salmonella typhimurium. Proc. Natl. Acad. Sci. USA,* 76:1343–1347, 1979.
17. Wogan, G.N., Essigmann, J.M., Croy, R.G., Busby, W.F., Jr., Groopman, J.D., and Stark, A.A., Macromolecular binding of aflatoxin B_1 and sterigmatocystin: relationship of adduct patterns to carcinogenesis and mutagenesis, in *Naturally Occurring Carcinogensmutagens and Modulators of Carcinogenesis,* Miller, E.C., Miller, J.A., Hirono, I., Sugimura, T., and Takayama, A., Eds., University Park Press, Baltimore, MD, 1979, pp. 1933.
18. Stark, A.A., Molecular aspects of aflatoxin B_1 mutagenesis and carcinogenesis, in *Mycotoxins and Phycotoxins,* Steyn, P.S. and Vleggaar, R., Eds., Elsevier Science, Amsterdam, 1986, pp. 435–445.
19. Crespi, C.L., Penman, B.W., Leaky, J.A.E., Arlotto, M.P., Stark, A.A., Parkinson, A., Turner, T., Steimel, D.T., Rudo, K., Davis, R.L., and Langenbach, R., Human cytochrome P-450A3: sequence, role of the enzyme in the metabolic activation of promutagens, comparison to nitrosamine activation by human cytochrome P-450IIE1. *Carcinogenesis,* 11:1293–1300, 1990.
20. Mace, K. et al., Aflatoxin B_1-induced DNA adduct formation and p53 mutations in CYP-450-expressing human liver cell lines. *Carcinogenesis,* 18:1291–1297, 1997.
21. Moss, E.J., Neal, G.E., and Judah, D.J., The mercapturic acid pathway metabolites of a glutathione conjugate of aflatoxin B_1. *Chem. Biol. Interact.,* 55:139–155, 1985.
22. Scholl, P.F., Musser, S.M., and Groopman, J.D., Synthesis and characterization of aflatoxin B_1 mercapturic acids and their identification in rat urine. *Chem. Res. Toxicol.,* 10:1144–1151, 1997.
23. Roebuck, B.D., Siegel, W.G., and Wogan, G.N., *In vitro* metabolism of aflatoxin B_2 by animal and human liver. *Cancer Res.,* 38:999–1002, 1978.
24. Massey, T.E., Cellular and molecular targets in pulmonary chemical carcinogenesis: studies with aflatoxin B_1. *Can. J. Physiol. Pharmacol.,* 74:621-628, 1996.
25. Daniels, J.K. et al., Biotranasformation of aflatoxin B_1 in rabbit lung and liver microsomes. *Carcinogenesis,* 11:823–827, 1990.
26. Daniels, J.M. et al., DNA binding and mutagenicity of aflatoxin B_1 catalyzed by isolated rabbit lung cells. *Carcinogenesis,* 14:1429–1434, 1993.

27. Coulombe, R.A. et al., Metabolism of AFB_1 in the upper airways of the rabbit: role of the nonciliated tracheal epithelial cells. *Cancer Res.,* 46:4091–4096, 1986.
28. Stewart, R.K. and Massey, T.E., Kinetic studies of glutathione-S-transferase (GST)-catalyzed conjugation of aflatoxin B_1-8,9-epoxide in rabbit lung and liver. *Toxicologist,* 13:63, 1993.
29. Ball, R.W. et al., Comparative activation of aflatoxin B_1 by mammalian pulmonary tissue. *Toxicol. Lett.,* 75:119–125, 1995.
30. Devereux, T.R. et al., Xenobiotic metabolism in human alveolar type II cells isolated by centrifugal elutriation and density gradient centrifugation. *Cancer Res.,* 46:5438–5443, 1986.
31. Anttila, S. et al., Smoking and peripheral types of cancer are related to high level of pulmonary cytochrome P-450IA in lung cancer patients. *Int. J. Cancer,* 47:681–685, 1991.
32. Liu, L. and Massey, T.E., Bioactivation of AFB_1 by lipoxygenases, prostaglandin H synthase and cytochrome P-450 monooxygenase in Guinea-pig tissues. *Carcinogenesis,* 13:533–539, 1992.
33. Liu, L. et al., *In vitro* prostaglandin H synthase- and monooxygenase-mediated binding of AFB_1 to DNA in Guinea-pig tissue microsomes. *Carcinogenesis,* 11:1915–1919, 1990.
34. Liu, L. et al., *In vitro* cytochrome P-450 monooxygenase and prostaglandin H synthase-mediated AFB_1 biotransformation in Guinea-pig tissues: effect of β-naphthoflavone treatment. *Arch. Toxicol.,* 67:379–385, 1993.
35. Kelly, J.D. et al., Aflatoxin B_1 activation in the human lung. *Toxicol. Appl. Pharmacol.,* 144:88–95, 1997.
36. Donnelly, P.J. et al., Biotransformation of aflatoxin B_1 in the human lung. *Carcinogenesis,* 17:2487–2494, 1996.
37. Donnelly, P.J. et al., Arachidonic acid dependent bioactivation of AFB_1 in human lung. *Toxicologist,* 14:336, 1994.
38. Donnelly, P.J. et al., Bioactivation of AFB_1 in human alveolar type II cells and macrophages. *Proc. Am. Assoc. Cancer Res.,* 36:153, 1995.
39. Harrison, J.C. and Garner, R.C., Immunological and HPLC detection of aflatoxin adducts in human tissues after an acute poisoning incident in S.E. Africa. *Carcinogenesis,* 12:741–743, 1991.
40. Biswas, G. et al., Comparative kinetic studies of aflatoxin B_1 binding to pulmonary and hepatic DNA of rat and hamster receiving the carcinogen intratracheally. *Teratol. Carcinog. Mutagen.,* 13:259–268, 1993.
41. Coulombe, R.A. et al., Pharmacokinetics of intratracheally administered aflatoxin B_1. *Toxicol. Appl. Pharmacol.,* 109:19–206, 1991.
42. Imaoka, S. et al., Rat pulmonary microsomal cytochrome P-450 enzymes involved in the activation of procarcinogens. *Mut. Res.,* 284:233–241, 1992.
43. IsraelKalinsky, H., Tuch, J., Roitelman, J., and Stark, A.A., Photoactivated aflatoxins are mutagens to *Salmonella typhimurium* and bind covalently to DNA *in vitro. Carcinogenesis,* 3:423–429, 1982.
44. IsraelKalinsky, H., MalcaMor, L., and Stark, A.A., Comparative aflatoxin B_1 mutagenesis in *Salmonella typhimurium* TA100 in metabolic and photoactivation systems. *Cancer Res.,* 44:1831–1839, 1984.
45. Stark, A.A., MalcaMor, L., Herman, Y., and Liberman, D.F., DNA strand scission and apurinic sites induced by photoactivated aflatoxins. *Cancer Res.,* 48:3070–3076, 1988.
46. Shaulsky, G., Johnson, R.L., Shockor, J.P., Taylor, L.C.E., and Stark, A.A., Properties of aflatoxin DNA adducts formed by photoactivation and characterization of the major photoadduct as aflatoxinN^7 guanine. *Carcinogenesis,* 11:519–527, 1990.

47. Stark, A.A., Gal, Y., and Shaulsky, G., Involvement of singlet oxygen in the photoactivation of aflatoxins B_1 and B_2 to DNA binding forms *in vitro*. *Carcinogenesis,* 11:529–534, 1990.
48. Stark, A.A. and Liberman, D.F., Synergism between aflatoxins in covalent binding to DNA and in mutagenesis in the photoactivation system. *Mutation Res.,* 247:77–86, 1991.
49. Stark, A.A. et al., Aflatoxin B_1 mutagenesis, DNA binding, and adduct formation in *Salmonella typhimurium*. *Proc. Natl. Acad. Sci. USA,* 76:1343–1347, 1979.
50. Stark, A.A. and Giroux, C.N., Mutagenesis in *Streptococcus pneumoniae* (*pneumococcus*) by transformation with DNA modified by the carcinogen-mutagen, aflatoxin B_1. *Mut. Res.,* 107:23–32, 1983
51. McLean, M. and Dutton, M.F., Cellular interactions and metabolism of aflatoxin: an update. *Pharmacol. Ther.,* 65:163–92, 1995.
52. Yu, F.L. et al., Evidence for the covalent binding of aflatoxin B_1-dichloride to cytosine in DNA. *Carcinogenesis,* 12:997–1002, 1991.
53. D'Andrea, A.D. and Haseltine, W.A., Modification of DNA by aflatoxin B_1 creates alkali-labile lesions in DNA at positions of guanine and adenine. *Proc. Natl. Acad. Sci. USA,* 75:4120–4124, 1978.
54. Essigmann, J.M. et al., Structural identification of the major DNA adduct formed by aflatoxin B_1 *in vitro*. *Proc. Natl. Acad. Sci. USA,* 74:1870–1874, 1977.
55. Croy, R.G. et al., Identification of the principal aflatoxin B_1-DNA adduct formed *in vivo* in rat liver. *Proc. Natl. Acad. Sci. USA,* 75:1745–1749, 1978.
56. Kobertz, W.R. et al., An intercalation inhibitor altering the target selectivity of DNA damaging agents: synthesis of site-specific aflatoxin B_1 adducts in a p53 mutational hotspot. *Proc. Natl. Acad. Sci. USA,* 94:9579–584, 1997.
57. Heinonen, J.T. et al., Determination of aflatoxin B_1 biotransformation and binding to hepatic macromolecules in human precision liver slices. *Toxicol. Appl. Pharmacol.,* 136:1–7, 1996.
58. Bailey, E.A. et al., Mutational properties of the primary aflatoxin B_1-DNA adduct. *Proc. Natl. Acad. Sci. USA,* 93:1535–1539, 1996.
59. Stark, A.A. et al., DNA strand scission and apurinic sites induced by photoactivated aflatoxins. *Cancer Res.,* 48:3070–3076, 1988.
60. Mao, H. et al., An intercalated and thermally stable FAPY adduct of aflatoxin B_1 in a DNA duplex: structural refinement from 1H NMR. *Biochemistry,* 37:4374–4387, 1998.
61. Groopman, J.D. et al., Molecular epidemiology of aflatoxin exposures: validation of aflatoxin-N^7-guanine levels in urine as a biomarker in experimental rat models and humans. *Environ. Health Perspect.,* 99:107–113, 1993.
62. Jacobsen, J.S. et al., DNA replication-blocking properties of adducts formed by aflatoxin B_1-2,3-dichloride and aflatoxin B_1-2,3-oxide. *Mut. Res.,* 179:89–101, 1987.
63. Bonnett, M. and Taylor, E.R., The structure of aflatoxin B_1-DNA adduct at N^7 of guanine. Theoretical intercalation and covalent adduct models. *J. Biochem. Structure Dynamics*, 7:127–149, 1989.
64. Muench, K.F. et al., Sequence specificity in aflatoxin B_1-DNA interactions. *Proc. Natl. Acad. Sci. USA,* 80:6–10, 1983.
65. Irvin, T.R. and Wogan, G.N., Quantitation of aflatoxin B_1 adduction within the ribosomal RNA gene sequences of rat liver DNA. *Proc. Natl. Acad. Sci. USA,* 81:664–668, 1984.

66. Bailey, G.S. et al., Carcinogen AFB_1 is located preferentially in internucleosomal deoxyribonucleic acid following exposure *in vivo* in rainbow trout. *Biochemistry,* 19:5836–5842, 1980.
67. Prieto-Alamo, M.J. et al., Mutational specificity of aflatoxin B_1: comparison of *in vivo* host-mediated assay with *in vitro* S9 metabolic activation. *Carcinogenesis,* 17:1997–2002, 1996.
68. Schapper, R.M., Kunkel, T.A., and Loeb, L.A., Infidelity of DNA synthesis associated with bypass of apurinic sites. *Proc. Natl. Acad. Sci. USA,* 80:487–491, 1984.
69. Tessman, I. and Kennedy, M.A., DNA polymerase II of *Escherichia coli* in the bypass of abasic sites in *vivo. Genetics,* 136:439–448, 1994.
70. Zhu, Y. et al., Aflatoxin B_1, 2-aminoanthracene, and 7,12-dimethylbenz[a]anthracene-induced frameshift mutations in human APRT. *Environ. Molec. Mutagen.,* 26:234–239, 1995.
71. Bennett, C.B., Luo, X., and Humayun, M.Z., Frameshifts: genetic requirements for frameshift reversion induced by bulky DNA adducts in M13 DNA. *Mut. Res.,* 249:19–27, 1991.
72. Stark, A.A. and Giroux, C.N., Mutagenicity and cytotoxicity of the carcinogen-mutagen aflatoxin B_1 in *Streptococcus pneumoniae* (*Pneumococcus*) and *Salmonella typhimurium*: dependence on DNA repair functions. *Mut. Res.,* 106:195–208, 1982.
73. Sahasrabudhe, S. et al., Mutagenesis by aflatoxin in M13 DNA: base-substitution mechanisms and the origin of strand bias. *Mol. Gen. Genet.,* 217:20–25, 1989.
74. Foster, P.L., Groopman, J.D., and Eisenstadt, E., Induction of base substitution mutations by aflatoxin B_1 is mucAB dependent in *Escherichia coli. J. Bacteriol.,* 170:3415–3420, 1988.
75. Sarasin, A.R. and Hanawalt, P.C., Carcinogens enhance survival of UV-irradiated simian virus 40 in treated monkey kidney cells: induction of a recovery pathway?, *Proc. Natl. Acad. Sci. USA,* 75:346–350, 1978.
76. Venturini, M.C. et al., Mycotoxin T-2 and aflatoxin B_1 as immunosuppressors in mice chronically infected with *Toxoplasma gondii. J. Comp. Pathol.,* 115:229–237, 1996.
77. Dimitri, R.A. and Gabal, M.A., Immunosuppressant activity of aflatoxin ingestion in rabbits measured by response to *Mycobacterium bovis* antigen. I. Cell-mediated immune response measured by skin test reaction. *Vet. Hum. Toxicol.,* 38:333–336, 1996.
78. Raisuddin, S. et al., Immunostimulating effects of protein A in immunosuppressed aflatoxin-intoxicated rats. *Int. J. Immunopharmacol.,* 16:977–984, 1994.
79. Methenitou, G. et al., Immunomodulative effects of aflatoxins and selenium on human peripheral blood lymphocytes. *Vet. Hum. Toxicol.,* 38:274–277, 1996.
80. Roll, R., Matthiaschk, G., and Korte, A., Embryotoxicity and mutagenicity of mycotoxins. *J. Environ. Pathol. Toxicol. Oncol.,* 10:1–7, 1990.
81. Bloom, S.E., Sister chromatid exchange studies in the chick embryo and neonate: actions of mutagens in a developing system. *Basic Life Sci.,* 29:509–533, 1984.
82. Geissler, F. and Faustman, E.M., Developmental toxicity of aflatoxin B_1 in the rodent embryo *in vitro*: contribution of exogenous biotransformation systems to toxicity. *Teratology,* 37:101–111, 1988.
83. Miele, M. et al., Aflatoxin exposure and cytogenetic alterations in individuals from Gambia, West Africa. *Mut. Res.,* 349:209–217, 1996.
84. Autrup, H.. Transplacental transfer of genotoxins and transplacental carcinogenesis. *Environ. Health Perspect.,* 101:33–38, 1993.

80. Denning, D.W. et al., Transplacental transfer of aflatoxin in humans. *Carcinogenesis,* 11:1033–1035, 1990.
86. Ziegler, A. et al., Tumor suppressor gene mutations and photocarcinogenesis. *Photochem. Photobiol.,* 63:432–435, 1996.
87. Brash, D.E. et al., A role for sunlight in skin cancer: UV-induced p53 mutations in squamous cell carcinoma. *Proc. Natl. Acad. Sci. USA,* 88:10124–10128, 1991.
88. Donnelly, P.J. et al., Activation of K-ras in aflatoxin B_1-induced lung tumors from AC3F1 (A/J x C3H/HeJ) mice. *Carcinogenesis,* 17:1735–1740, 1996.
89. McMahon, G. et al., Characterization of c-Ki-ras and N-ras oncogenes in aflatoxin B_1-induced rat liver tumors. *Proc. Natl. Acad. Sci. USA,* 87:1104–1108, 1990.
90. Soman, N.R. and Wogan, G.N., Activation of the c-Ki-ras oncogene in aflatoxin B_1-induced hepatocellular carcinoma and adenoma in the rat: detection by denaturing gradient gel electrophoresis. *Proc. Natl. Acad. Sci. USA,* 90:2045–2099, 1993.
91. Tada, M. et al., Analysis of ras gene mutations in human hepatic malignant tumors by polymerase chain reaction and direct sequencing. *Cancer Res.,* 50:1121–1124, 1990.
92. Leon, M. and Kew, M.C., Analysis of ras gene mutations in hepatocellular carcinoma in southern African blacks. *Anticancer Res.,* 15:859–861, 1995.
93. Montesano, R., Hainaut, P., and Wild, C.P., Hepatocellular carcinoma: from gene to public health. *J. Natl. Cancer Inst.,* 89:1844–1851, 1997.
94. Shen, H.M. and Ong, C.N., Mutations of the p53 tumor suppressor gene and ras oncogenes in aflatoxin hepatocarcinogenesis. *Mut. Res.,* 366:23–44, 1996.
95. Yang, M. et al., Mutations at codon 249 of p53 gene in human hepatocellular carcinomas from Tongan, *China Mut. Res.,* 381:25–29, 1997.
96. Soini, Y. et al., An aflatoxin-associated mutational hotspot at codon 249 in the p53 tumor suppressor gene occurs in hepatocellular carcinomas from Mexico. *Carcinogenesis,* 17:1007–1012, 1996.
97. Aguilar, F. et al., Geographic variation of p53 mutational profile in nonmalignant human liver. *Science,* 264:1317–1319, 1994.
98. De-Benedetti, V.M. et al., p53 is not mutated in hepatocellular carcinomas from Alaska natives. *Cancer Epidemiol. Biomarkers Prev.,* 4:79–82, 1995.
99. Volkmann, M. et al., p53 overexpression is frequent in European hepatocellular carcinoma and largely independent of the codon 249 hot spot mutation. *Oncogene,* 9:195–204, 1994.
100. Hayashi, H. et al., The mutation of codon 249 in the p53 gene is not specific in Japanese hepatocellular carcinoma. *Liver,* 13:279–181, 1993.
101. Peers, F.G., Gilman, G.A., and Linsel, C.A., Dietary aflatoxins and human liver cancer: a study in Swaziland. *Int. J. Cancer,* 17:167–176, 1976.
102. Peers, F.G. and Linsel, C.A., Dietary aflatoxins and liver cancer — a population-based study in Kenya. *Br. J. Cancer,* 27:473–484, 1973.
103. van Rensburg, S.J. et al., Primary liver cancer rate and aflatoxin intake in a high cancer area. *South Afr. Med. J.,* 48:2508a–2508d, 1974.

Detection and Monitoring of Microbial Food Contaminants

6 Rapid Methods of Microbiological Analysis: Update

Daniel Y.C. Fung

CONTENTS

I. INTRODUCTION

Rapid methods and automation in microbiology are dynamic fields of study that address the utilization of microbiological, chemical, biochemical, biophysical, immunological, and serological methods for the study of improving isolation, early detection, characterization, and enumeration of microorganisms and their products in clinical, food, industrial, and environmental samples. In the past 10 years, food microbiologists have started to adapt rapid and automated methods in their laboratories (Figure 6.1). Conventional methods of detection, enumeration identification, and characterization of microbes are described in reference books such as *Compendium of Methods for the Microbiological Examination of Foods* (Vanderzant and Splittstoesser, 1992), *Official Methods of Analysis of the AOAC* (AOAC, 1995), *Bacteriological Analytical Manual* (FDA, 1998), *Standard Methods for the Examination of Dairy Products* (APHA, 1992), and *Modern Food Microbiology* (Jay, 1996).

Important publications on the subject of rapid methods for medical specimens, water, food, industrial, and environmental samples are in a series of papers by Fung and colleagues (Fung, 1991, 1992, 1994, 1995; Fung et al., 1989), and books such as *Mechanizing Microbiology* (Sharpe and Clark, 1978), *Foodborne Microorganisms and Their Toxins: Developing Methodology* (Pierson and Stern, 1986), *Rapid Methods*

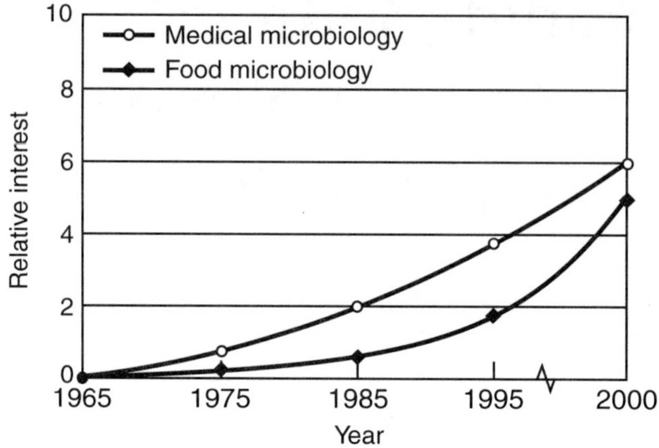

FIGURE 6.1 Relative interest in rapid methods among medical microbiologists and food microbiologists. (From Fung, D.Y.C., *Food Technol.,* 44(6):64-67, 1995. With permission of Institute of Food Technologists.)

in Food Microbiology (Adams and Hope, 1989), *Instrumental Methods for Quality Assurance in Foods* (Fung and Matthews, 1991), and *Rapid Analysis Technique in Food Microbiology* (Patel, 1994). Hartman et al. (1992) had an excellent chapter on rapid methods and automation in *Compendium of Methods for the Microbiological Examination of Foods* (Vanderzant and Splittstoesser, 1992). Swaminathan and Feng (1994) also provided updated materials in rapid methods.

The purpose of this paper is to review the basic principles and practical applications of a variety of instruments and procedures directly and indirectly related to improved methods for microbiology in quality assurance and research in food microbiology.

II. IMPROVEMENTS IN SAMPLING AND SAMPLE PREPARATION

One of the most useful instruments developed for sample preparation is the Stomacher (Tekmar; Cincinnati, OH). This instrument is designed to massage food samples in a sterile bag. The food sample is first placed in the sterile disposable plastic bag, and appropriate sterile diluents are added. The bag with the food is placed in the open chamber. After the chamber is closed, the bag is then massaged by two paddles for a suitable time period, usually from 1 to 5 minutes. No contact occurs between the instrument and the sample. During massaging, microorganisms are dislodged into the diluent for further microbiological manipulation. Massaged slurries are then used for microbiological analysis. A similar system called the Masticator Homogenizer is marketed by IUL Instruments in Germany. Recently, Sharpe developed a new instrument called the Pulsifier, which can dislodge bacteria from food by pulsification in a bag. Fung et al. (1998) evaluated the Pulsifier and determined that it is as accurate as the Stomacher but provided much clearer liquid samples due to minimal food debris.

An instrument for sample preparation is the Gravimetric Diluter (Spiral Biotech; Bethesda, MD). One of the routine procedures in food microbiological work is to aseptically measure a sample of food (e.g., 1 g of meat) and then aseptically add an exact amount of sterile diluent (e.g., 9 mL) to make a desired dilution (1:10). With the Gravimetric Diluter, the analyst needs only to aseptically place an amount of food (e.g., 1.2 g), into the Stomacher bag, set a desired dilution (1:10), and set the instrument to deliver the appropriate amount of sterile diluent (e.g., 9.8 mL). Thus, the dilution operation can be done automatically. A new version of this instrument is called the Diluflo and has been in use satisfactorily in the author's laboratory since 1992.

III. ALTERNATIVE METHODS FOR VIABLE CELL COUNT PROCEDURE

The conventional viable cell count or standard plate count method is time consuming, both in terms of operation and collection of data. Several methods have been explored to improve the efficiency of operation of the viable cell count procedure. The spiral plating method is an automated system for obtaining viable cell counts (Spiral Biotech; Bethesda, MD). The instrument can spread a liquid sample on the surface of agar contained in a Petri dish in a spiral shape (the Archimedes spiral) with a concentration gradient starting from the center and decreasing as the spiral progresses outward on the rotating plate. The volume of liquid deposited at any segment of the agar plate is known. After the liquid containing microorganisms is spread, the agar plate is incubated overnight at an appropriate temperature for the colonies to develop. The colonies appearing along the spiral pathway can be counted either manually or electronically.

New versions of the spiral plater have recently been introduced: the Autoplater (Spiral Biotech; Bethesda, MD) and the Whitley Automatic Spiral Plater (Bioscience International, Inc.; Rockville, MD). With these instruments, an analyst need only present the liquid sample, and the instrument completely and automatically processes the sample, including re-sterilizing the unit for the next sample. A newer version of this system has been introduced by IUL in which the stylus is automatically sliced after each use, thus bypassing sterilization of the stylus by disinfectants as is done by the original spiral platers.

Another alternative method for viable cell count is the Isogrid system (QA Laboratories, Ltd.; San Diego, CA). This system consists of a square filter with hydrophobic grids printed on the filter to form 1600 squares for each filter. Food samples are weighed, blended, and enzyme treated before passage through the membrane filter containing the hydrophobic grids by vacuum. The filter is then placed on agar containing a suitable nutrient for growth of the bacteria, yeasts, or molds. The hydrophobic grids prevent colonies from growing farther than the square grids; thus, all colonies have a square shape. This facilitates counting of the colonies both manually and electronically. This method has been successfully used to make viable cell counts for a variety of foods, including milk, meat, black pepper, flour, peanut butter, mushrooms, rice, fish, shrimp, and oysters. Other applications of the Isogrid system include determination of total coliforms, fecal coliforms, *Escherichia coli, Salmonella* spp., etc. (Sharpe, 1991).

Rehydratable nutrients are imbedded into a series of films in the Petrifilm system (3M Co.; St. Paul, MN). The top layer of the protective cover is lifted, 1 mL of liquid sample is introduced to the center of the unit, and the cover is then replaced. A plastic template is placed on the cover to make a round mold. The rehydrated medium with nutrient will support the growth of microorganisms after suitable incubation time and temperature. The colonies can be counted directly in the unit. The unit is about the size and thickness of a plastic credit card, thus providing great savings of space in storage and incubation. Petrifilm units have been developed for total bacterial count, coliform count, fecal coliform count, yeast and mold counts, and a hemorrhagic *E. coli* O157:H7 kit.

The Redigel system (RCR Scientific, Inc.; Goshen, IN) is another convenient viable cell count system. This system consists of sterile nutrients with a pectin gel in a tube. The tube is ready to be used any time and no heat is needed to "melt" the agar. A 1-mL food sample is first pipetted into the tube. After mixing, the entire content is poured into a special Petri dish previously coated with calcium. When liquid comes in contact with the calcium a Ca-pectate gel is formed and the complex swells to resemble conventional agar. After an appropriate incubation time and temperature, the colonies can be counted exactly like conventional standard plate count method. Besides total count, Redigel also has systems for coliform, fecal coliform, yeast and molds, lactic acid bacteria, *Staphylococcus aureus,* and *Salmonella*. Fung and Chain (1991) compared 17 different foods (pasteurized milk, raw milk, cheddar cheese, chocolate chips, rice, wheat germ, corn meal, whole wheat, flour, peanuts, pecans, ground beef, chicken, ground black pepper, thyme, broccoli, mushrooms, and turkey pot pie; 20 samples each) and obtained a correlation of 0.964.

The four aforementioned methods have potential as alternatives to the conventional agar pour plate method. Chain and Fung (1991) made a comprehensive analysis of all four methods compared to the conventional method on seven different foods (skinless chicken breast, fresh ground beef, fresh ground pork, packaged whole shelled pecans, raw milk, thyme, and whole wheat flour; 20 samples each) and showed that the new systems and the conventional method were highly comparable and exhibited a high degree of accuracy and agreement ($r = 0.95+$). It should be noted that these newer methods require some training and experience before satisfactory results can be obtained consistently. They are good methods if operated carefully.

In the direct epifluorescent filter technique (DEFT) method, the liquified sample is first passed through a filter that retains the microorganisms, the filter is then stained with acridine orange, and the slide is observed under ultraviolet microscopy. Live cells usually stain orange-red, orange-yellow, or orange-brown, whereas dead cells fluoresce green. The slides can be read by the eye or by a semi-automated counting system marketed by Bio-Foss. A viable cell count can be made in less than an hour. With the use of an image analyzer, an operator can count 50 DEFT slides per hour

(Pettipher, 1986). Tortorello and Gendel (1993) further developed this procedure by using fluorescent antibodies in conjunction with DEFT to enumerate *E. coli* O157:H7 in milk and juice.

IV. INSTRUMENTS FOR ESTIMATION OF MICROBIAL POPULATIONS AND BIOMASS

Many methods have been developed in recent years to estimate the total number of microorganisms by parameters other than the viable colony count, as described in the previous section. For a new method be acceptable, it should have some direct correlation with the total viable cell count. Thus, standard curves correlating parameters such as adenosine triphosphate (ATP) level, detection time of electrical impedance or conductance, generation of heat, radioactive CO_2, etc. with viable cell counts of the same sample series must be made. In general, the larger the number of viable cells in the sample, the shorter the detection time of these systems. A scattergram is then plotted and used for further comparison of unknown samples. The assumption is that as the number of microorganisms increases in the sample, these physical, biophysical, and biochemical events will also increase accordingly.

Theoretically, these methods can detect as few as one viable cell in the sample if the incubation period is long enough (days or weeks). On the practical side, usually the limit is 10^{3-4} cells/mL. When a sample has 10^5 to 10^6 organisms per /mL detection, can be achieved in about 4 to 6 hours.

All living things utilize ATP. In the presence of a firefly enzyme system (luciferase and luciferin system), oxygen, and magnesium ions, ATP will facilitate the reaction to generate light. The amount of light generated by this reaction is proportional to the amount of ATP in the sample; thus, the light units can be used to estimate the biomass of cells in a sample. The light emitted by this process can be monitored by a variety of fluorimeters. These procedures can be automated for handling of large numbers of samples. Some of the instruments can detect as little as 10^2 to 10^3 fg. The amount of ATP in one colony-forming unit has been reported as 0.47 fg with a range of 0.22 to 1.03 fg.

Lumac (Landgraaf; The Netherlands) markets several models of ATP instruments and provides customers with test kits with all necessary reagents, such as a fruit juice kit, hygiene monitoring kit, etc. The reagents are injected into the instrument automatically and readout is reported as relative light units (RLUs). By knowing the number of microorganisms responsible for generating known RLUs, one can estimate the number of microorganisms in the food sample. In some food systems, such as wine, the occurrence of any living matter is undesirable; thus, monitoring of ATP can be a useful tool for quality assurance in the winery. Recently, much interest has been expressed in using ATP estimation not for total viable numbers but as a sanitation check by companies such as Lumac, BioTrace (Plainsboro, NJ), Lightning (IDEXX; Westbrook, ME), Hy-Lite (Glengarry, Biotech; Cornwall, Canada), Charm 4000 Luminometer (Charm Sciences; Malden, MA), and others.

As microorganisms grow and metabolize nutrients, large molecules change to smaller molecules in a liquid system and cause a change in electrical conductivity and resistance in the liquid, as well as at the interphase of electrodes. By measuring the changes in electrical impedance, capacitance, and conductance, scientists can estimate the number of microorganisms in the liquid because the larger the number of microorganisms in the fluid, the faster the change in these parameters which can be measured by sensitive instruments. A detailed analysis on the subject of impedance, capacitance, and conductance in relation to food microbiology has been made by Eden and Eden (1984).

The Bactometer (bioMerieur Vitek, Inc.; Hazelwood, MO) is an instrument designed to measure impedance changes in foods. Samples are placed in the wells of a 16-well module. After the module is completely or partially filled, it is plugged into the incubator unit to start the monitoring sequence. At first, there is a stabilization period for the instrument to adjust to the module, then a baseline is established. As the microorganisms metabolize the substrates and reach a critical number (10^{5-6} cells/mL), the change in impedance increases sharply, and the monitor screen shows a slope similar to the log phase of a growth curve. The point at which the change in impedance begins is the detection time, and this is measured in hours from the start of the experiment. The detection time is inversely proportional to the number of microorganisms in the sample. By knowing the number of microorganisms per milliliter in a series of liquid samples and the detection time of each sample, one can establish a standard curve. From the curve, one can decide the cutoff points to monitor certain specifications of the food products.

Impedance methods have been used to estimate bacteria in milk, dairy products, meats, and other foods (Bishop and White, 1985; Eden and Eden, 1984; Waes and Bossuyt, 1984; Zindulis, 1984). Of particular interest is the application of this method for determining the shelf-life potential of pasteurized whole milk by Bishop et al. (1984).

The Malthus system (Crawley, U.K.) works by measuring the conductance of the fluid as organisms grow in the system. It also generates a conductance curve similar to the impedance curve of the Bactometer, and it also uses detection time to monitor the density of the microorganisms in the food.

The Malthus system has been used for microbial monitoring of brewing liquids (Evans, 1988), milk (Visser and deGroote, 1984), and hygiene monitoring (McMurdo and Whyward, 1984). Gibson and colleagues (Gibson et al., 1987, 1980; Ogden, 1986) have done considerable work using the Malthus system to study seafood microbiology.

Besides estimating viable cells in foods, both the Bactometer and Malthus systems can detect specific organisms by the use of selective and differential liquid media. New developments of these two systems are constantly being made. For example, the Malthus system has developed a tube system to detect CO_2 production by yeast using indirect conductance measurements. They also introduced disposable units in the system. An automatic instrument for measuring direct and indirect impedance has been developed in Europe and is called the RABIT (Rapid Automated Bacterial Impedance Technique).

The Omnispec bioactivity monitor system (Wescor, Inc.; Logan, UT) is a tristimulus reflectance colorimeter that monitors dye pigmentation changes mediated by microbial activity. Dyes can be used that produce color changes as a result of pH changes, changes in the redox potential of the medium, or the presence of compounds with free amino groups. Samples are placed in microtiter wells or other types of containers and scanned by an automated light source with computer interface during the growth stages (0 to 24 hours). The change of color or hue (a*, from green to red; b*, from blue to yellow; L*, from black to white) can be monitored similar to the impedance curve and conductance curve. Manninen and Fung (1992) evaluated this system in a study of pure cultures of *Listeria monocytogenes* and food samples and found high correlation coefficients (*r*) of 0.90 to 0.99 for pure bacterial cultures and 0.82 for minced beef between the colony counts predicted by the colorimetric technique and the results of the traditional plate count method. They also showed that detection times for bacterial cultures such as *Enterobacter aerogenes, E. coli, Hafnia alvei,* and several strains of *L. monocytogenes* were substantially shorter (2 to 24 hours) using the instrument than using the traditional method, and they concluded that the colorimetric detection technique employed by the Omnispec system simplifies the analyses, saves labor and materials, and provides a high sampling capacity. Tuitemwong (1993) completed an extensive study using the Omnispec 4000 to monitor growth responses of food pathogens in the presence or absence of membrane-bound enzymes. This instrument is highly efficient in large-scale studies of microbial interaction with different compounds in liquid and food.

V. MINIATURIZED MICROBIOLOGICAL TECHNIQUES

Identification of microorganisms is an important part of quality assurance and control programs in the food industry. The author has developed many miniaturized methods to reduce the volume of reagents and media (from 5–10 mL to about 0.2 mL) for microbiological testing in microtiter plates. The basic components of the miniaturized system are the microtiter plates for test cultures, a multiple inoculation device, and containers to house solid media (large Petri dishes) and liquid media (another series of microtiter plates). The procedure involves placing liquid cultures (pure cultures) to be studied into sterile wells of a microtiter plate to form a master plate. Each microtiter plate can hold up to 96 different cultures, 48 duplicate cultures, or various combinations as desired. The cultures are then transferred by a sterile multipoint inoculator (96 needles protruding from a template) to solid or liquid media. Sterilization of the inoculator is by alcohol flaming. Each transfer represents 96 separate inoculations in the conventional method. After incubation at an appropriate temperature, the growth of cultures on solid media or liquid media can be observed and recorded, and the data can be analyzed. These miniaturized procedures save considerable time in operation, effort in manipulation, materials, labor, and space. These methods are ideal for studying large numbers of isolates or for research involving challenging large numbers of microbes against a host of test compounds.

The miniaturized methods have been used to study large numbers of isolates from foods (Fung and Hartman, 1975; Lee et al., 1982, 1985) and to develop bacteriological media and procedures (Chein and Fung, 1991). Miniaturized methods for studying food yeast were also developed in the author's laboratory (Fung and Liang, 1989; Lin and Fung, 1985, 1987). Currently, these miniaturized methods are being used to study food mycology (Hart and Fung, 1990; Hart et al., 1991). Many useful microbiological media were discovered through this line of research. For example, an aniline blue *Candida albicans* medium was developed and marketed by Difco under the name Candida Isolation Agar. The sensitivity and specificity were 98.0 and 99.5%, respectively, with a predictive value of 99.1% (Goldschmidt et al., 1991).

On the commercial side, many diagnostic kits to identify microorganisms have been developed and marketed since the 1970s. Currently, API, Enterotube, R/B, Minitek, MicroID, and IDS are available. Most of these systems were first developed for the identification of enterics (*E. coli, Salmonella, Shigella, Proteus, Enterobacter* spp., etc.). Later, many of the companies expanded the capacity to identify nonfermentors, anaerobes, gram-positive organisms, and even yeasts and molds. Most of the early comparative analyses centered around evaluation of these kits for clinical specimens. Cox et al. (1987) and Fung and Cox (1981) studied these systems from the standpoint of food microbiology and concluded that these systems generally provide 90 to 95% accuracy when compared with conventional methods. Comparative analysis of diagnostic kits and selection criteria for miniaturized systems were made by Fung et al. (1984) and Cox et al. (1984). They concluded that these miniaturized systems are accurate, efficient, labor saving, space saving, and less expensive than the conventional procedure. Their usefulness in clinical and food microbiological laboratories will continue to be important.

VI. NOVEL TECHNIQUES

Many sophisticated instruments have been developed to identify isolates from clinical specimens, such as the Sensititre (Radiometer Amer; Westlake, OH) and Biolog (Hayward, CA). One of the most automated systems for the identification of isolates (clinical and foods) is the Vitek system. The system depends on the growth of target organisms in specially designed media housed in tiny chambers in a plastic "card." The card is then inserted into the incubation chamber. The instrument periodically scans the wells of the cards and sends information to the computer, which then compares this information to the database and identifies the unknown cultures in the cards. The system is entirely automated and computerized and provides hard copies for recordkeeping. Most evaluations of the usefulness of the Vitek system had used clinical specimens, but Bailey et al. (1985) were successful in utilizing the Vitek system to identify Enterobacteriaceae from foods. The system is capable of identifying enterics, yeast, *Bacillus*, selected Gram-positive pathogens, and other organisms.

Concepts and applications of miniaturized kits, immunoassays, and DNA hybridization for recognition and identification of foodborne bacteria are reviewed by Cox et al. (1987). The DNA probe (Genetrak; Farmingham, MA) is a sensitive method to detect pathogens such as *Campylobacter, Salmonella, Listeria,* and *E. coli*. At first, the system utilized radioactive compounds for assay. The second generation of probes uses enzymatic reactions to detect the presence of pathogens. Another major change in this area is the development of probes to detect target RNA. In a cell, there is only one copy of DNA; however, there may be 1000 to 10,000 copies of ribosomal RNA. Thus, the new generation of probes is designed to probe target RNA. After enrichment of cells (either in liquid or colonies on solid agar), DNA or RNA of target cells can be extracted and released into the liquid and then detected by the appropriate DNA and/or RNA probes. These methods are in the process of being automated. Currently, kits are available for *Salmonella, Listeria, Campylobacter, Yersinia*, etc. As the need arises, more organisms will be added to the list.

Polymerase chain reaction (PCR) systems are the latest development in DNA amplification technology, and recently their use in food microbiology has attracted much attention. Normally, the procedures are highly complicated and an extremely clean environment is necessary to perform the test. Recently, much research has been directed toward simplifying the procedure for laboratory analysts. DuPont recently commercialized a system called BAX® for PCR. According to Scott Fritschel of Qualicon LLC (a unit subsidiary of DuPont):

> "The BAX® screening system is designed to work from overnight enrichment broths. In the case of the BAX® system for *Salmonella*, the food is enriched overnight in any standard non-selective broth (lactose broth, buffered peptone water, etc.). A 1:10 dilution into BHI is performed followed by a 3-hour grow-back incubation. An aliquot is taken into a tube and treated with lytic enzymes and heat. An aliquot of the cell lysate is used to rehydrate the PCR reagents housed in a tube, and a control tablet and the sample is placed into the thermal cycler along with the corresponding control tube. Thermal cycling, UV visualization of the gel, and photography to document the results are then performed to detect PCR products.
>
> "This type of development will help to transfer PCR technology to the food laboratories in the near future. Some obstacles to PCR technology include the occurrence of inhibitors in the food samples and the fact that DNA from dead cells can also be amplified, thus giving a false-positive result for a food that may be safe to eat. For pure cultures, this technology is very useful. Dilution of samples (1:10) will reduce the inhibitors, and enrichment of the target culture will ensure amplification of live cells, even in the presence of other organisms."

Another important development also by DuPont is the RiboPrinter® Characterization System. Again, according to Scott Fritschel:

> "The RiboPrinter is designed to accept isolated colonies of bacteria as the sample. A sample is prepared as follows:
>
> "1. A colony of bacteria is picked from an agar plate using a sterile plastic stick (provided in the sample kit).
>
> "2. Cells from the stick are suspended in a buffer solution by mechanical agitation.
>
> "3. An aliquot of the cell suspension is loaded into the sample carrier to be placed into the instrument. That's it! Each sample carrier has space for eight individual colony picks.
>
> "In both products, we [Qualicon] have invested a lot of research in allowing the user to work from samples that microbiologists are comfortable handling. We do not want to require that the user be familiar with the techniques of molecular biology in order to benefit from the power that these techniques can bring to food microbiology."

Modification of the basic PCR procedure includes reverse transcriptase PCR, nested/multiplex PCR, randomly amplified polymorphic DNA (RAPD), fluorescent probes in PCR, etc. When these methods will find their way to the food microbiology laboratory is not certain. Recently, a system named Probelia was introduced by Pasteur Institute. It utilizes uracil instead of thymine in the base pairing and also includes an enzyme, uracil dehydrolase, to cleave contaminants in the PCR environment. An effective internal control is also included in the assay system.

The enzyme-linked immunosorbent assay (ELISA) method commercialized by Organon Teknika (Durham, NC) utilizes two monoclonal antibodies specific for *Salmonella* detection. In a comparative study involving 1289 samples, Eckner et al. (1987) found that there was no significant difference between the conventional method and the ELISA method for food samples, except cake mix and raw shrimp. Another ELISA system, the Tecra system (International BioProducts; Redmond, WA), was developed in Australia and uses polyclonal antibodies to detect *Salmonella*. These methods have also been used to detect *Listeria* and *E. coli*. Many companies are providing a host of monoclonal and polyclonal antibodies for a variety of diagnostic tests, some including food pathogens (Fung et al., 1988).

In the Vitek ImmunoDiagnostic Assay System (VIDAS), all intermediate steps are automated. It is a multiparametric immunoanalysis system that utilizes the enzyme-linked fluorescent immunoassay (ELFA) method. According to the manufacturer, "The end result of the test protocol is a fluorescent product and the VIDAS reader utilizes a special optical scanner that measures the degree of fluorescence. From the moment the solid phase receptacles and the reagent strips are placed in the instrument, the VIDAS is fully automated." This is a revolutionary development because one of the drawbacks of the ELISA test is the many steps necessary for adding reagents and washing test samples. Many automated ELISA instruments are on the market now and include TECRA OPUS (International BioProducts; Redmond, MD), Bio-Tek Instruments (Highland Park, VT), Automated EIA Processor (BioControl; Bethell, WA), etc.

In a related development, several self-contained small units (REVEAL, VIP, etc.) have been marketed recently. After enrichment (with or without oxyrase type of stimulation) an analyst applies a small aliquot (boiled or unboiled) to the kit. Reaction occurs in a few minutes. These kits have been used to rapidly screen *E. coli* O157:H7 and *Salmonella* in ground beef.

The UNIQUE system for *Salmonella* is another way of using immunocapture technology. A dip stick with antibody against *Salmonella* is applied to the preenriched liquid. The antibody captures the *Salmonella* if present. This charged dip stick is then placed in a fresh enrichment broth and the cells are allowed to multiply for a few hours. After the second enrichment step, the dip stick with a much larger population of *Salmonella* attached to it will be subject to further ELISA procedures. The entire test is housed in a convenient plastic self-contained unit. This type of method is very useful for a small laboratory with low-volume testing of pathogens.

Immunomagnetic capture methods have attracted much attention lately. Vicam and Dynal have developed magnetic beads coated with a variety of antibodies to capture target cells or cellular components in foods. After the magnetic beads have a chance to interact with potential target cells in a tumbling apparatus for an hour, the beads are physically separated from the food or liquid by a powerful magnet applied to the side of the test tube. Further microbiological procedures such as direct plating or ELISA tests can be made on these charged beads.

Motility enrichment is a very useful concept in rapid isolation and identification of food pathogens. A commercial system called the *Salmonella* 1-2 test (BioControl; Bothell, WA) utilizes motility as a form of selection. The food sample is first pre-enriched for 24 hours in lactose broth, and then 0.1 mL is inoculated into one of the chambers in an L-shaped system. The chamber contains selective enrichment liquid medium. There is also a small hole connecting the liquid chamber with the rest of the system, which has a soft agar through which salmonellae can migrate. An opening on the top of the second chamber allows the analyst to deposit a drop of polyvalent anti-H antibody. If the sample contains salmonellae from the lower side of the L-shaped unit, the salmonellae will migrate through the hole and up the agar column. Simultaneously, the antibody against flagella of salmonellae will move downward by gravity. When the antibody meets the salmonellae they will form a visible "immunoband." The presence of an immunoband in this system is a positive test for *Salmonella* spp. The system is easy to use and has gained popularity because of its simplicity.

Oxyrase (Mansfield, OH), a membrane fraction of *E. coli*, was found in the author's laboratory to stimulate the growth of a large number of important facultative anaerobic food pathogens. In the presence of a hydrogen donor such as lactate, oxyrase can convert O_2 to H_2O, thereby reducing the oxygen tension of the medium and creating anaerobic conditions that favor the growth of facultative anaerobic organisms. In a medium containing 0.1 units/mL of the enzyme, the growth of *L. monocytogenes, E. coli* O157:H7, *Salmonella typhimurium, Streptococcus faecalis,* and *Proteus vulgaris* was greatly enhanced; colony counts were greater by 1 to 2 log units (depending on the initial count and the strain studied) after incubation in the presence of the enzyme for 5 to 8 hours at 35 to 42°C compared with control without oxyrase.

By combining the oxyrase enzyme and a unique U-shaped tube, Yu and Fung (1991a,b; 1992) developed an effective method to detect *L. monocytogenes* and *Listeria* spp. from laboratory cultures and meat systems. Niroomand and Fung (1994) studied the effects of oxyrase in stimulation of growth of *Campylobacter* from foods. Tuitemwong et al. (1994) also found that these membrane fragments and those obtained from *Acetobacter* and *Gluconobacter* can stimulate growth of starter cultures in food fermentation.

There are many other systems that involve modern biochemistry, chemistry, and immunology. For example, one can use protein profiles for microbial "fingerprinting" (AMBIS; San Diego, CA), or cell composition as a way to identify bacterial cultures (Hewlett-Packard; Palo Alto, CA).

VII. CONCLUSIONS

This chapter describes a variety of methods that are designed to improve current methods, explore new ideas, and develop new concepts and technologies for the improvement of applied microbiology. Although many of these methods were first

TABLE 6.1.
Attributes for an Ideal Automated Microbiology Assay System

1. Accuracy for the intended purposes — sensitivity, minimal detectable limits, specificity of test system, versatility, potential applications, comparison to referenced methods
2. Speed — in productivity, in obtaining results, in number of samples processed per run or per day
3. Cost — minimum initial costs, costs per test, reagent costs, other
4. Acceptability — by scientific community, by regulatory agencies
5. Simplicity of operation — sample operation, operation of test equipment, computer versatility
6. Training — on-site, reasonable length, quality of training personnel
7. Reagents — preparation, stability, availability, consistency
8. Company reputation
9. Technical service — speed and availability, cost and scope of technical background
10. Utility and space requirement

developed for clinical microbiology, they are now being used for food microbiology. After being involved in the field of rapid methods and automation in microbiology for 25 years, this author has summarized his ideas concerning the attributes for an ideal automated microbiology assay system in Table 1. The *Journal of Rapid Methods and Automation* was started in 1992 by the author to encourage rapid dissemination of information concerning current developments in rapid methods and automation. This field will certainly grow, and many food microbiologists will find these new methods very useful in their routine work in the immediate future. Many methods described here are already being used by applied microbiologists nationally and internationally.

ACKNOWLEDGMENT

This material is based on work supported by the Cooperative State Research Service, U.S. Department of Agriculture, under agreement No. 8890341874511. Contribution No. 99270-A Kansas Agricultural Experimental Station.

REFERENCES

Adams, M.R. and Hope, C.F.A., 1989. *Rapid Methods in Food Microbiology.* Elsevier, Amsterdam.

AOAC, 1995. *Official Methods of Analysis of the AOAC,* Vols. I and II. Association of Official Analytical Chemists, Arlington, VA.

APHA, 1992. *Standard Methods for the Examination of Dairy Products.* American Public Health Association, Washington, D.C.

Bailey, J.S., Cox, N.A., Thomson, J.E., and Fung, D.Y.C., 1985. Identification of Enterobacteriaceae in foods with the Automicrobic system. *J. Food Protect.,* 48:147.

Bishop, J.R. and White, C.H., 1985. Estimation of potential shelf-life to cottage cheese utilizing bacterial numbers and metabolites. *J. Food Protect.,* 48:663.

Bishop, J.R., White, C.R., and Firstenberg-Eden, R., 1984. A rapid impedimetric method for determining the potential shelf-life of pasteurized whole milk. *J. Food Protect.,* 47:471.

Chain, V.S. and Fung, D.Y.C., 1991. Comparison of Redigel, Petrifilm, spiral plate system, ISOGRID, and standard plate count for the aerobic count on selected foods. *J. Food Protect.,* 54:208.

Chein, S.P. and Fung, D.Y.C., 1991. Acriflavin violet red bile agar for the isolation and enumeration of *Klebsiella pneumoniae*. *Food Microbiol.,* 7:73.

Cox, N.A., McHan, F., and Fung, D.Y.C., 1977. Commercially available mini-kits for the identification of Enterobacteriaceae: a review. *J. Food Protect.,* 40:866.

Cox, N.A., Fung, D.Y.C., Goldschmidt, M.S., Bailey, J.S., and Thomson, J.E., 1984. Selecting a miniaturized system for identification of Enterobacteriaceae. *J. Food Protect.,* 47:74.

Cox, N.A., Fung, D.Y.C., Bailey, J.S., Hartman, P.A., and Vasavada, P.C., 1987. Miniaturized kits, immunoassays and DNA hybridization for recognition and identification of food-borne bacteria. *Dairy Food Sanitat.,* 7:628.

Eckner, K.F., FLowers, R.S., Robinson, B.J., Mattingly, J.A., Gabis, D.A., and Silliker, J.A., 1987. Comparison of *Salmonella* Bio-EnzaBead immunoassay method and conventional culture procedure for detection of *Salmonella* in foods. *J. Food Protect.,* 50:379.

Eden, R. and Eden, G., 1984. *Impedance Microbiology.* Research Studies Press, Letchworth, U.K.

Evans, H.A.V., 1985. A note on the use of conductivity in brewery microbiology control. *Food Microbiol.,* 2:19.

FDA, 1998. *Bacteriological Analytical Manual,* 6th ed. Association of Official Analytical Chemists, Food and Drug Administration, Arlington, VA.

Fung, D.Y.C., 1991. Rapid methods and automation for food microbiology. In: *Instrumental Methods for Quality Assurance in Foods*, D.Y.C. Fung and R.F. Matthews, Eds., Marcel Dekker, New York.

Fung, D.Y.C., 1992. Historical development of rapid methods and automation in microbiology. *J. Rapid Methods Automation Microbiol.,* 1(1):1.

Fung, D.Y.C., 1994. Rapid methods and automation in food microbiology: a review. *Food Rev. Int.,* 10(3):357–375.

Fung, D.Y.C., 1995. What's needed in rapid detection of food-borne pathogens? *Food Technol.,* 44(6):64–67.

Fung, D.Y.C. and Chain, V.S., 1991. Comparative analysis of Redigel and aerobic plate count methods for viable cell count of selected foods. *Food Microbiol.,* 8:299.

Fung, D.Y.C. and Cox, N.A., 1981. Rapid identification systems in the food industry: present and future. *J. Food Protect.,* 44:877.

Fung, D.Y.C. and Hartman, P.A., 1975. Miniaturized microbiology techniques for rapid characterization of bacteria. In: *New Approaches to the Identification of Microorganisms,* C.G. Heden and T. Illeni, Eds., Wiley, New York, pp. 347–370.

Fung, D.Y.C. and Liang, C., 1989. A new fluorescent agar for the isolation of *Candida albicans*. *Bulletin d'information des Laboraores des Service Veterinaries (France).* No. 29/30, pp. 1–2.

Fung, D.Y.C. and Matthews, R.F., Eds., 1991. *Instrumental Methods for Quality Assurance in Foods,* Marcel Dekker, New York.

Fung, D.Y.C., Goldschmidt, M.C., and Cox, N.A., 1984. Evaluation of bacterial diagnostic kits and systems at an instructional workshop. *J. Food Protect.,* 47:68.

Fung, D.Y.C., Sharpe, A.N., Hart, B.C., and Lin Y., 1998. The Pulsifier: a new instrument for preparing food suspensions for microbiological analyses. *J. Rapid Methods Automation Microbiol.,* 6(1):43–50.

Fung, D.Y.C., Bennett, R., and Lehleitner, G.C., 1988. Rapid diagnosis in bacteriology: contribution of polyclonal and monoclonal antibodies. In: *Biologie Prospective, le Colloque de Pont-e-Mousson,* M.M. Gaeteau, J. Henry, and G. Siest, Eds., John Libbey, London, pp. 21–26.

Fung, D.Y.C., Cox, N.A., Goldschmidt, M.C., and Bailey, J.S., 1989. Rapid methods and automation: a survey of professional microbiologists. *J. Food Protect.,* 52:65.

Gibson, D.M. and Hobbs, G., 1987. Some recent developments in microbiological methods in seafood quality. In: *Seafood Quality Determination,* D.F. Kramer and J. Listen, Eds., Elsevier, Amsterdam, pp. 283–298.

Gibson, D.M. and Ogden, I.D., 1980. Assessing bacterial quality of fish by conductance measurement. *J. Appl. Bacteriol.,* 49:12.

Goldschmidt, M.C., Fung, D.Y.C., Grant, R., White, J., and Brown, T., 1991. New aniline blue dye medium for rapid identification and isolation of *Candida albicans. J. Clinical Microbiol.,* 29(6):1098–1099.

Hart, R.A. and Fung, D.Y.C., 1990. Evaluation of dye media selective for *Aspergillus* and/or *Penicillium.* In: *Proc. Annual Meeting of the American Society of Microbiology,* Anaheim, CA, May 13–17, p. 279.

Hart, R.A., Mo, O., Borius, F., and Fung, D.Y.C., 1991. Comparative analysis of Tryptan blue agar and Congo red agar for the enumeration of yeast and mold using HGMF system. *J. Food Safety,* 11:227.

Hartman, P.A. et al., 1992. Rapid methods and automation. In: *Compendium of Methods for the Examinations of Foods,* C. Vanderzant and D. Splittstoesser, Eds., American Public Health Association, Washington, D.C.

Jay, J.M., 1996. *Modern Food Microbiology,* 4th ed., Van Nostrand-Reinhold, New York.

Lee, C.Y., Fung, D.Y.C., and Kastner, C.L., 1982. Computer-assisted identification of bacteria on hot-boned and conventionally processed beef. *J. Food Sci.,* 47:363.

Lee, C.Y., Fung, D.Y.C., and Kastner, C.L., 1985. Computer-assisted identification of microflora on hot-boned and conventionally processed beef: effect of moderate and slow initial chilling rate. *J. Food Sci.,* 50:553.

Lin, C.C.S. and Fung, D.Y.C., 1985. Effect of dyes on growth of food yeast. *J. Food Sci.,* 47:770.

Lin, C.C.S. and Fung, D.Y.C., 1987. Critical review of conventional and rapid methods for yeast identification. *CRC Crit. Rev. Microbiol.,* 14(4):273.

Manninen, M.T. and Fung, D.Y.C., 1992. Estimation of microbial numbers from pure bacterial cultures and from minced beef samples by reflectance colorimetry with Omnispec 4000. *J. Rapid Methods Automation Microbiol.,* 1(1):41.

McMurdo, I.H. and Whyward, S., 1984. Suitability of rapid microbiological methods for the hygiene management of spray drier plant. *J. Soc. Dairy Technol.,* 34(1):4.

Niroomand, F. and Fung, D.Y.C., 1994. Effect of oxygen reducing membrane fragments on growth of *Campylobacter, J. Rapid Methods Automation Microbiol.,* 2:247.

Ogden, I.D., 1986. Use of conductance methods to predict bacterial counts in fish. *J. Appl. Bacteriol.,* 61:36.

Patel, P.D., 1994. *Rapid Analysis Techniques in Food Microbiology,* Chapman & Hall, New York.

Pettipher, G.L., 1986. Review: the direct epifluorescent filter technique. *J. Food Technol.,* 21:535.

Pierson, M.D. and Stern, N.J., 1986. *Foodborne Microorganisms and Their Toxins: Developing Methodology,* Marcel Dekker, New York.

Sharpe, A.W., 1991. Rapid methods: consideration to adoption by regulatory agencies. In: *Proc. 105 AOAC Annual International Meeting,* Aug. 12–15, Phoenix, AZ.

Sharpe, A.W. and Clark, D.S., Eds., 1978. *Mechanizing Microbiology.* Charles C Thomas, Springfield, IL.

Swaminathan, B. and Feng, P., 1994. Rapid detection of food-borne pathogenic bacteria. *Rev. Microbiol.,* 48:401.

Tortorello, M. and Gendel, S.M., 1993. Fluorescent antibodies applied to direct epifluorescent filter techniques for microscopic enumeration of *Escherichia coli* O157:H7 in milk and juice. *J. Food Protect.,* 56:672.

Tuitemwong, K., 1993. Characteristics of Food Grade Membrane Bound Enzymes and Applications in Food Microbiology and Food Safety, Ph.D. dissertation, Kansas State University, Manhattan, KS.

Tuitemwong, K., Fung, D.Y.C., and Tuitemwong, P., 1994. Acceleration of yogurt fermentation by bacterial membrane fraction biocatalysis. *J. Rapid Methods Automation Microbiol.,* 3:127.

Vanderzant, C. and Splittstoesser, D., Eds., 1992. *Compedium of Methods for the Examination of Foods.* American Public Health Association, Washington, D.C.

Visser, I.J.R. and deGroote, J., 1984. Prospects for the use of conductivity as an aid in the bacteriological monitoring of pasteurized milk. *Antonin van Leewenhoek,* 50:202.

Waes, G.M. and Bossuyt, R.G., 1984. Impedance measurements to detect bacteriophage problems in cheddar cheese. *J. Food Protect.,* 47:349.

Yu, L.S.L. and Fung, D.Y.C., 1991a. Oxyrase enzyme and motility enrichment Fung-Yu tube for rapid detection of *Listeria monocytogenes* and *Listeria* spp. *J. Food Safety,* 11:149.

Yu, L.S.L. and Fung, D.Y.C., 1991b. Effect of oxyrase enzyme in *Listeria monocytogenes* and other facultative anaerobes. *J. Food Safety,* 11:163.

Yu, L.S.L. and Fung, D.Y.C., 1992. Growth kinetics of *Listeria* in the presence of oxyrase enzyme in a broth model system. *J. Rapid Methods Automation Microbiol.,* 1(1):15.

Zindulis, J., 1984. A medium for the impedimetric detection of yeasts in foods. *Food Microbiol.,* 1:159.

7 Biosensors for On-Line Monitoring

J. Rishpon, T. Neufeld, V. Sacks, and I. Eshkenazi

CONTENTS

I. ABSTRACT

Direct on-line monitoring of such factors as nutrition value, chemical residues, antibiotics, and bacterial/pesticide contamination will improve both the quality and safety of food. To achieve this goal, we have developed simple, fast, reliable, and disposable biosensors for on-the-spot identification and quantification of food contaminates, as well as for examining food quality. Three enzyme-based biosensors are presented: (1) a biosensor for determining the presence of parathion by directly measuring enzymatic activity, (2) a biosensor designed for detecting toxic organophosphates by inhibiting enzymatic activity, and (3) an enzyme biosensor constructed of a three-enzyme cascade reaction for detecting the presence of lactose in cow's milk. We have also developed a novel disposable, amperometric, enzyme-channeling immunosensor for a quantitative, rapid, separation-free enzyme immunoassay using model systems with anti-peroxidase antibody and bacteria (*Staphylococcus aureus, Salmonella*). Enzyme biosensors and immunosensors are ideal tools for the quantitative characterization of food ingredients and food contaminates. Such sensors are simple to operate, and the procedures are rapid, accurate, and inexpensive, requiring neither special skills nor complicated training. The use of a micro-flow cell ensures the continuous flux of new substrate and prevents the accumulation or adsorption of products near the working electrode. The miniature size of the electrochemical cell, combined with screen-print electrodes, offers the advantage of working with low volumes of compounds and reagents, which is especially important when dealing with hazardous materials.

II. INTRODUCTION

Foodborne infection is a major public health problem worldwide, resulting in millions of food-poisoning cases each year. Various microorganisms and toxic agents, such as pesticides, can be transmitted to humans through nutritive products and cause severe health damage. Moreover, certain food ingredients considered innocent can be hazardous when they cause serious problems for individuals who are either allergic to such specific components or cannot digest them properly.

Bacteria are among the most common food contaminants. Many kinds of bacteria can damage food quality, thereby causing infection, especially, but not exclusively, of the digestion system. Because not every edible compound can be pasteurized, certifying the purity of certain products before their arrival at the market is important.

Another group of substances known as food contaminants consists of organophosphate (OP) compounds. In modern agriculture, a wide variety of such compounds have been and are still used today as insecticides, despite their known toxicity. Insecticide methods include a combination of fumigation and such biochemicals as OP. The acute toxicity of OP compounds to the human nervous system results from its strong binding to the neuro-enzyme acetylcholine esterase (AChE) which forms a very stable complex that inhibits enzymatic activity. Detecting organopollutants that threaten living organisms is a crucial civil issue, not to mention a military concern, because such compounds make up an active component in chemical warfare.[1]

Inferior food quality is not necessarily related only to the presence of traces of toxic materials. To some extent, naturally occurring ingredients such as lactose can be dangerous, as well. Lactose is a disaccharide containing two different monosaccharide units: D-glucose and D-galactose. In the food industry, the major source of lactose is cow's milk. The normal concentration of lactose in milk is around 4 to 5% (40 to 50 mg/mL); only small amounts of non-lactose carbohydrates have been found. Lactose is widely used in baking ingredients, in ready-to-eat products, and in commercial milk formulas for infants. Nevertheless, for 5 to 15% of the population, the consumption of lactose is forbidden because such individuals lack one of the enzymes (mainly lactase) along the digestive cascade of this sugar. For such persons, consuming lactose is harmful; therefore, finding an inexpensive and simple method to determine the lactose concentration in various edible components is essential.

Increased public awareness of such crucial problems has motivated many industries to improve the quality of their products. Direct on-line monitoring of factors such as nutritional value, chemical residues, antibiotics, and bacterial contamination

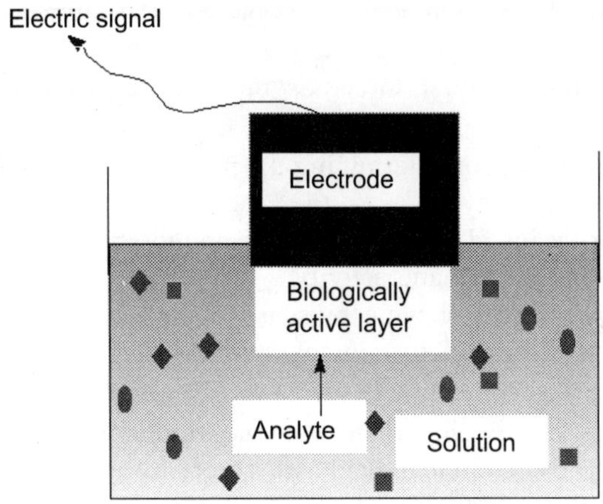

FIGURE 7.1 A bioelectrochemical sensor.

will result in improved food quality and safety, as well as greater efficiency of food production. To achieve these goals, analytical devices allowing rapid, on-the-spot detection must be used. Such devices must not only provide fast and sensitive measurements but also be simple to operate and inexpensive. In this respect, a bioelectrochemical sensor is an attractive choice.

A bioelectrochemical sensor is an analytical device that uses biologically sensitive material for the direct detection of chemical or biochemical species. Biosensors generally consist of a biologically sensitive component, such as an enzyme, antibody, or receptor which is attached to an electrode, and a suitable substrate, as illustrated in Figure 7.1. During the biochemical reaction, ions or electroactive substances may be produced.[2] By using well-established electrochemical techniques, the biochemical response is converted into an amplified, quantifiable, electrical signal. Amperometric and potentiometric enzyme electrodes, which are well known in the characterization of bioactive surfaces, have been commercialized, and their use in diagnostic devices has been an active research area during the past few years. Bioelectrochemical sensors combine the selectivity of biological recognition with the high sensitivity and relative simplicity of modern electroanalytical techniques. Normally, such sensors are used for applications in which electronic signals provide quantitative information on the concentration of selected substances in an analyzed medium.

Biological recognition events *in vivo* are generally accompanied by enzyme cascade reactions that are used to amplify the recognition events and to control the biochemical conditions that are in close proximity. Since the introduction by Clark in 1962 of the first enzyme-based sensor for glucose,[3] enzyme-based biosensors involving a wide variety of substrates have been used extensively. Most enzyme systems used involve the catalysis of redox reactions, in which either the substrate or the product is electrically charged.[4,5]

We will discuss the development and application of simple, fast, reliable, and disposable biosensors for identifying and quantifying food contaminates, as well as for examining food quality. First, three enzyme-based biosensors are described:

1. A simple biosensor for determining the presence of parathion by measuring enzymatic activity
2. A biosensor designed for detecting organophosphates which is based on the inhibition of enzymatic activity
3. A more complicated enzyme biosensor constructed of a three-enzyme cascade for detecting lactose in cow's milk

Enzyme biosensors, in general, and enzyme immunoassays (EIAs), in particular, play important roles in clinical diagnostics, veterinary medicine, environmental control, and bioprocess analysis. Because of its high selectivity and sensitivity, EIA enables the detection of a broad spectrum of analytes in complex samples. Decentralizing quantitative immunoassays from hospital laboratories, where complex instrumentation and highly qualified technical staff are required, can be achieved by combining immunoenzymatic systems with electrodes to create immunoelectrochemical sensors that provide immediate, on-the-spot results. To detect viral antigens and bacteria in food samples, we combined EIA with an enzyme biosensor to produce a novel enzyme-channeling immunosensor comprising two enzymes and two antibodies.

III. EXPERIMENTAL

A. General

The amperometric biosensors are based on a current measurement system with three electrodes: (1) a working electrode, (2) a reference electrode, and (3) a counter electrode. The biosensing elements were attached to the working electrode via an activated nylon membrane (Immunodyne ABC 5-µm and 3-µm cutoff; Pall, U.S.) or via cross-linkage with polyethylene imine (PEI).[9] The electrodes were connected to a computer-controlled potentiostat. The potentiostat and the software were from B.A.S. Bioanalytical Systems. An electrochemical cell is shown in Figure 7.2.

B. Enzyme Biosensor

Three types of enzyme sensors were developed:

1. The first directly detects parathion by its reaction with the adsorbed enzyme parathion hydrolase.
2. The second is based on measuring the degree of inhibition of the enzyme acetylcholine esterase by organophosphate substrates.
3. The third detects lactose by a three-enzyme cascade: ß-galactosidase, glucose oxidase, and horseradish peroxidase.

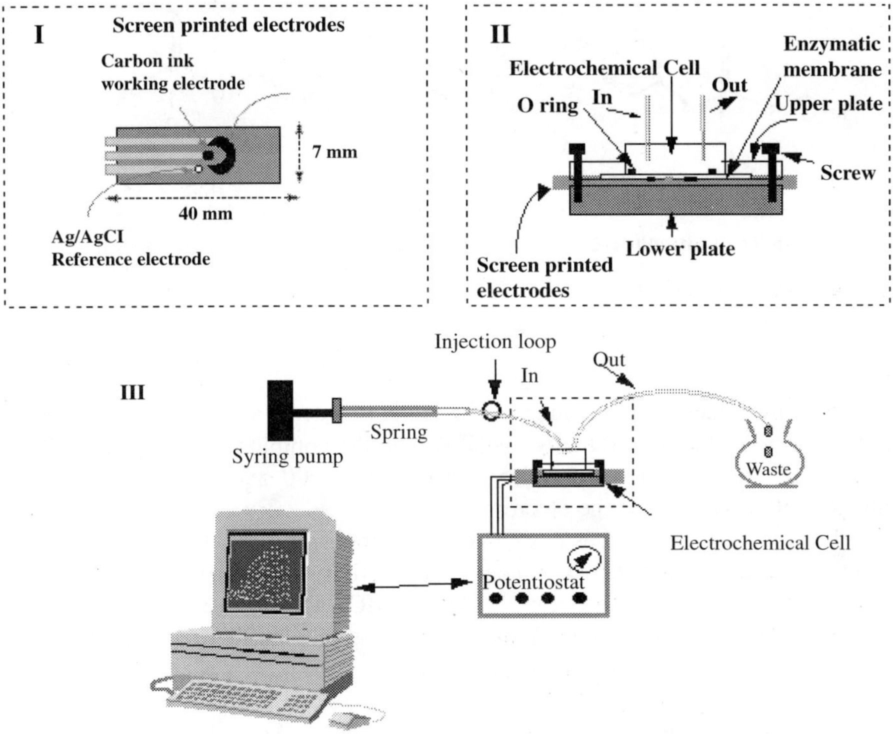

FIGURE 7.2 The electrochemical cell apparatus: (I) Screen-printed electrodes. (II) Micro flow-cell volume is 30 mL; the substrate reacts with the adsorbed enzyme and then is washed away. (III) The entire measuring system includes the flow system and the micro electrochemical flow cell based on screen-printed electrodes and attached to a computer-controlled potentiostat.

The three biosensors consisted of screen-printed electrodes[6] containing all three electrodes (working, counter, and reference) and were printed on a Milar sheet (EMTEK; Misgav Dov, Israel). A nylon membrane (Immunodyne ABC; Pall, U.S.) was placed on the electrodes, which were then placed in a homemade micro flow cell (cell volume 30 μl). The cell was washed with buffer solution using a 1-mL syringe. The substrate solution was injected into the cell from another syringe and through a 5-μL injection loop, which ensures a constant volume of the examined sample. Using a flow-cell system allowed the repetitive measurements of the enzymatic reaction as the product was removed and the substrate reintroduced. The flow rate was optimized for each system. A schematic diagram of the screen-printed electrodes and the micro flow-cell apparatus is presented in Figure 7.2.

In the first biosensor, the enzyme parathion hydrolase, isolated from *Pseudomonas* spp.,[7] was immobilized on the nylon membrane (the enzyme and parathion were obtained from Migal; Kiryat Shmona, Israel). After parathion was injected into the flow-cell system and reached the immobilized enzyme, the following hydrolysis occurred:

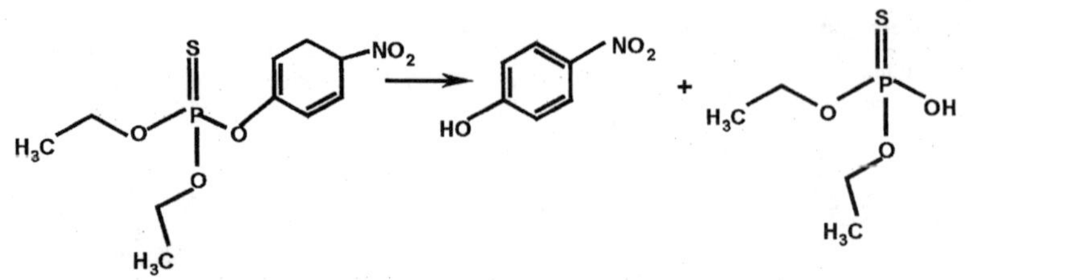

O,O'-diethyl O-4-nitrophenyl phosphorothioate (parathion) — **p-nitrophenol (PNP)** — **di-ethylthiophosphoricacid (DEPA)**

The p-nitrophenol thus produced immediately reacted at the working electrode, and the current peak was directly visualized on the computer screen. The potential between the working and reference electrode was 900 mV. At this potential, p-nitrophenol is oxidized at the electrode.

The second biosensor was designed to measure an initial activity of the enzyme AChE and a subsequent decrease in enzymatic activity after exposure to OP.[8] AChE-type V-S from electric eel (Sigma; St. Louis, MO) was immobilized on the nylon membrane. Buffer solution containing 0.1 *M* phosphate buffer, 0.1 *M* KCl, and 0.1 *M* $K_3[Fe(CN)_6]$ was injected into the cell through the flow system. The flow rate was 100 μL/min; a voltage of 300 mV was maintained between the working and reference electrode. Following the initial signal (the background current), different concentrations of the substrate acetylthiocholine (ATCh) were injected into the flow cell, and a signal of the initial activity of the enzyme was recorded by monitoring the degradation product thiocholine. The net reaction is as follows:

$$(CH_3)_3N^+CH_2CH_2S - OCCH_3 + H_2O \xrightarrow{\text{AChE}} (CH_3)_3N^+CH_2CH_2SH + CH_3COOH$$

$$2(CH_3)_3N^+CH_2CH_2SH + 2[FeCN_6]^{-3} \longrightarrow (CH_3)_3N^+CH_2CH_2S - SCH_2CH_2N^+(CH_3)_3$$

$$+ 2[FeCN_6]^{-4}$$

The peaks were visualized on the computer screen, and the peak areas reflected the actual activity of the enzyme. Thereafter, the enzyme was exposed to 5-μL samples of the inhibitor solution for different incubation times. The inhibitor chosen was the OP dimethyl-2,2-dichlorovinyl phosphate (DDVP) (Machteshim, Chemical Industries, Ltd., Israel). Because the current obtained reflects the decrease in enzyme activity, the degree of inhibition is directly inferred from the ratio between enzyme activity before and after exposure to the inhibitor. The enzymatic activity was restored either by adding a high concentration of the substrate or by adding 0.1 m*M* 2-pyridine aldoxime (PAM), purchased from Sigma, which abrogates the inhibition by causing interference between the enzyme and the DDVP.

The third system developed was an amperometric biosensor for direct lactose measurement in raw cow's milk. Three enzymes were immobilized on the electrode surface: (1) ß-galactosidase (ß-GAL) from *Escherichia coli*, type VIII; (2) glucose oxidase (GOD) from *Aspergillus niger*, type VII; and (3) horseradish peroxidase (POD), type VI, all purchased from Fluka. The reaction solution contained 5-aminosalicilic acid (5-ASA) (Sigma) as mediator.

The assay procedure included the following: Standard solutions of lactose or milk samples (diluted 1:10) were added to the electrochemical cell together with 15 mL of 0.1-*M* phosphate buffer (pH 6) and 2-m*M* Mg^{+2}. A peak denoting reduction current was obtained after adding either the lactose or the milk samples. The amount of lactose was calculated by the peak area using the standard addition method.

The biochemical reaction began with the cleavage of lactose by ß-GAL, producing glucose, which then reacted with GOD, releasing H_2O_2. The cleavage of peroxide by POD was accompanied by the oxidation of the mediator 5-ASA, which is present in the solution. The oxidized form of the mediator was measured by the printed carbon electrode polarized at 0 mV vs. a silver/silver chloride. The reduction of the mediator on the electrode creates a current peak visualized on the computer screen which is proportional to the primary lactose concentration. The entire reaction is represented in the following scheme:

$$\text{Lactose} \xrightarrow{\text{ß-GAL}} \text{Glucose} \xrightarrow{\text{GOD}} H_2O_2 + (\text{5-ASA}) \xrightarrow{\text{POD}}$$

$$(\text{5-ASA})^+ + e^- \longrightarrow (\text{5-ASA}) + \text{Signal}$$

C. Amperometric Enzyme-Channeling Immunosensor

To detect bacteria in a pure culture, we developed a novel, rapid, one-step, separation-free immunosensor that is based on an enzyme-channeling immunoassay (ECIA).[10,11] The principle of the method is shown in Figure 7.3. According to this method, the product of one enzyme reaction is the substrate for a second enzyme reaction. The binding interaction brings the two enzymes into close proximity and catalyzes the conversion of the initial substrate to a final detectable electroactive product. Cyclic regeneration of the substrate, accumulation of a redox mediator, and control of the hydrodynamic conditions at the sensor/solution interface enable the direct preferential measurement of a surface-bound enzyme label with high sensitivity. The method does not require a separation step to remove excess free label from the bulk solution. This type of biosensor not only measures separate components but also determines the concentration of bacteria.

We used a sandwich immunoassay. Protein A, the major cell wall component of *Staphylococcus aureus*, can be detected by binding the bacteria to an electrode surface that has been modified by immunoglobulin G from rabbit (RbIgG). Graphite electrodes (pencil leads) were coated by a polyethylenimine (50%) film with the co-immobilized capture antibody, RbIgG, and GOD (all from Sigma). The remaining unbound sites were blocked with 0.1-*M* glycine (Fluka). Standard solutions of *S. aureus* (10^2–10^6 cells/mL) (Sigma) were prepared in phosphate-buffered saline. The electrode was placed in a buffer solutions, an aliquot (10 µL) of *S. aureus* standard solution was added, and the RbIgG-GOD electrode was rotated for 10 minutes at 1000 r.p.m. at room temperature to enhance the kinetics of immunobinding between

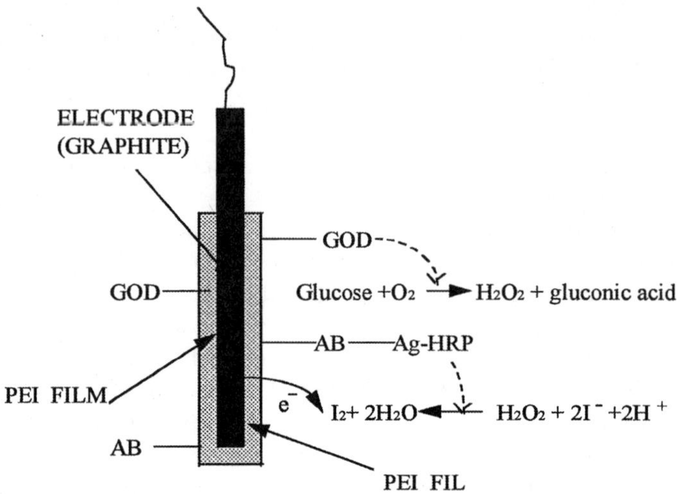

FIGURE 7.3 Enzyme-channeling immunoassay. Schematic illustration of the separation-free, amperometric, enzyme-channeling immunoassay with immobilized antibody and glucose oxidase on the PEI-modified electrode surface.

the cells and the antibody. Then, the electrode was washed and transferred to the electrochemical cell containing 0.1-*M* acetate buffer, 0.1-*M* NaCl, 3-m*M* KI, and 0.01% BSA. Aliquot (10 μL) of RbIgG-HRP (4 μg/mL) was added to the cell, thus placing the bacterial cells between two antibodies. After 10 minutes of incubation, an aliquot (10 μL) of 0.1-*M* glucose was added to the cell, and the response to glucose was recorded. On the electrode surface, the GOD had catalyzed the oxidation of glucose to H_2O_2, and the iodine produced in the peroxidase-catalyzed H_2O_2/iodide redox system was monitored amperometrically by the electrochemical reduction of iodine back to iodide. As the GOD and the horseradish peroxidase (HRP) were in close proximity, one could follow the labeled peroxidase activity, which was related to the GOD activity.

We have used a similar assay for the detection of *Salmonella.* Heat-killed *Salmonella typhimurium* cells (positive control), affinity-purified antibodies to *Salmonella*, and peroxidase-labeled affinity purified antibody to *Salmonella* were obtained from Kirkegaard & Perry Lab, Inc. (Gaithersburg, MD).

Electron microscopy of *S. aureus* cells on the electrode surface was performed with a JEOL-840a scanning electron microscope with an acceleration voltage of 25 kV, a magnification of 10,000, and an objective aperture of 1 mm.

IV. RESULTS AND DISCUSSION

A. Quantitative Determination of Parathion

p-Nitrophenol is the product of parathion degradation by parathion hydrolase. For tracing parathion, the electrochemical activity of p-nitrophenol enables the use of

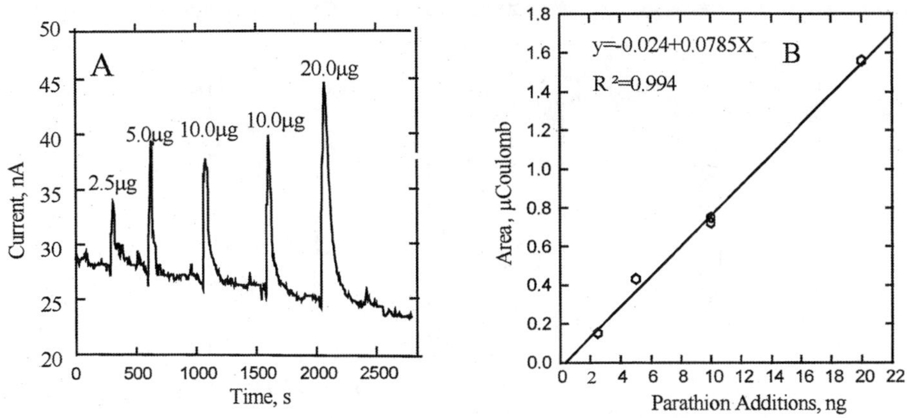

FIGURE 7.4 (A) Time-dependent modified electrode response to parathion. Different amounts of parathion were injected directly into the system by the flow system. The flow rate was 0.5 mL/min. The response current was immediately developed and decreased after the products were washed by the buffer solution. (B) Calibration curve of parathion hydrolase biosensor response to parathion.

electrochemical detection methods, in general, and of biosensors, in particular. By following enzymatic activity, the biosensor described here can detect and measure a chosen substrate directly rather than monitor its degradation products. When added to the reaction solution, parathion reached the immobilized enzyme on the electrode and hydrolysis occurred. The p-nitrophenol thus produced immediately reached the working electrode, donated its electrons, and created a signal. Various concentrations of p-nitrophenol were added to an electrolyte buffer solution.

Different amounts of parathion solution were added to the electrochemical cell. Figure 7.4a displays the electrode response after adding the parathion. To obtain a fast and repetitive response, we first exposed the enzyme to the substrate for a certain time period and then washed the product that had adsorbed to the electrode. Using a cell-flow system allowed the continuous flow of buffer solution during the measurement, so the product was removed from the electrode, allowing a subsequent injection of the parathion. Furthermore, immobilizing the enzyme on the electrode reduced the diffusion distance of the enzyme-substrate system, providing higher local concentration of the active product when compared to that found in bulk. The strong correlation between the parathion concentration and the current response is evident. The areas under each peak, obtained by integration of the signals, was linearly related to the parathion concentration, thus providing the calibration curve presented in Figure 7.4b.

In conclusion, we have described a biosensor that can trace small amounts of a desired substrate by using a naturally occurring parathion hydrolysis that produces an electrochemically active, measurable product. When dealing with hazardous environmental pollutants, the ability to detect low concentrations is extremely important. Moreover, the high accuracy, the small size of the biosensor, and the low price make it a useful as well as a convenient candidate for an outdoor, on-the-spot detector.

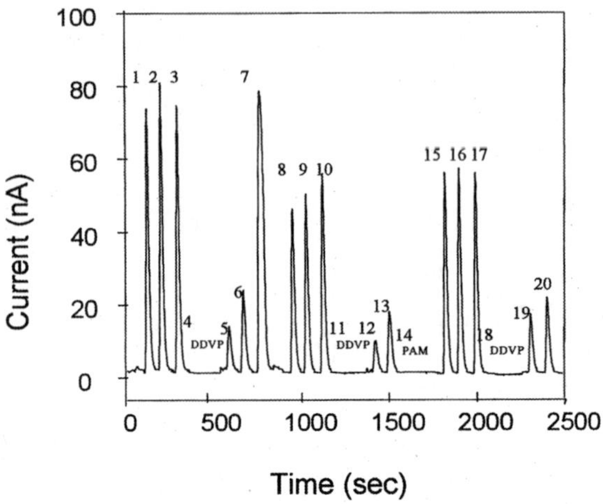

FIGURE 7.5 Biosensor response to DDVP. Signals 1–3, 8–10, 15–17 = addition of 1 m*M* ATCh; signals 4, 11, and18 = addition of 50 ng of DDVP; signals 5–6, 12–13, 19–20 = ATCh addition after exposure to DDVP; signal 7 = addition of 10 m*M* ATCh; signal 14 = PAM addition.

B. Measurement of OP Compounds

The use of biosensors that are based on inhibiting the activity of the neuro-enzyme AChE for the detection of OP compounds has been described in the literature.[12] Such sensors measure the decrease in enzyme activity in the presence of OP compounds. AChE activity can be measured by potentiometric pH or by amperometric detection of the product of acetylcholine degradation, thiocholine, or the H_2O_2 that is formed after choline oxidation by choline oxidase. Because many kinds of OP are powerful inhibitors of AChE activity, we used the degree of enzyme inhibition to quantify OP concentration in a desired sample.

Constant concentrations of the substrate acetylthiocholine were added, and the electrode response was monitored; thereafter, the enzyme was exposed to samples of DDVP for various incubation times. The enzyme electrode response to ATCh and DDVP is shown in Figure 7.5. The signals obtained were very sharp and reproducible. The optimal substrate concentration obtained from the calibration curve was 1 m*M*. The degree of inhibition depended on two main factors: quantity of the inhibitor and time allowed for the enzyme and the inhibitor to interact. After the addition of 50 ng DDVP, a significant decrease in enzymatic activity occurred.

We tested different quantities of DDVP, varying between 1 and 100 ng, with different exposure times to the enzyme, and the enzyme activity was calculated. The results are presented in Figure 7.6. The enzyme activity directly depended on both the applied inhibitor concentration and the exposure time. As the inhibitor concentration increased, the degree of activity of the enzyme declined. Similarly, as the incubation time was extended, enzymatic activity decreased. The optimal flow rate of the inhibitor when reaching the cell was 1 μL/min. As the cell volume was 30

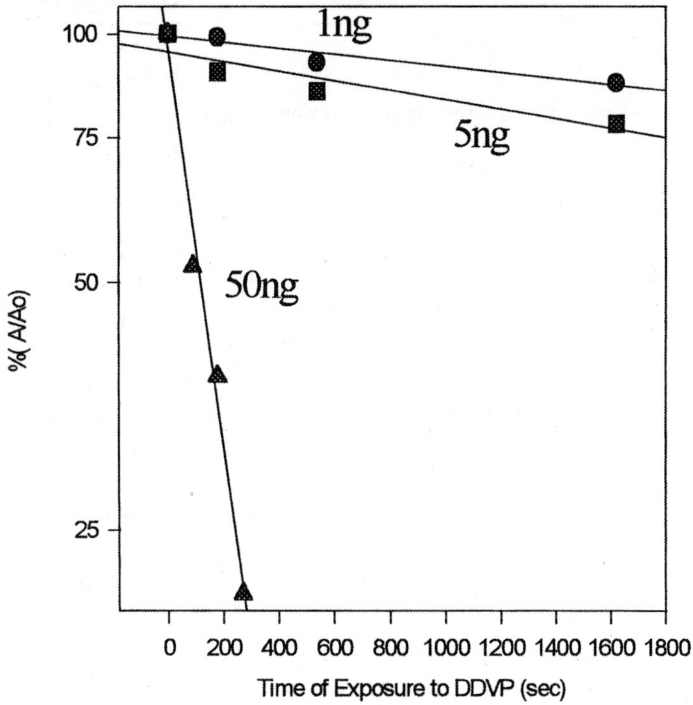

FIGURE 7.6 Enzyme activity in response to DDVP. Logarithmic dependence of enzyme activity on the amount and exposure time to DDVP.

μL, this flow rate allowed a maximum exposure of the inhibitor to the enzyme because each inhibitor molecule was delayed in the electrochemical cell.

The results presented here are also very important for determining an unknown quantity of DDVP. By changing the incubation time and monitoring the enzyme activity, we could calculate the concentration of the OP compounds. In summary, we have developed a disposable amperometric flow injection sensor that is applicable to identifying and quantifying OP compounds in the environment. The sensor is sensitive, rapid, small, and inexpensive. The miniature flow-cell system, combined with screen-printed electrodes, provides the advantage of working with low volumes of samples containing hazardous OP compounds.

C. Quantitative Determination of Lactose in Cow's Milk

Several procedures are currently available for lactose determination. Among the physical methods are gas, liquid, and high-pressure liquid chromatography;[13] polarimetry;[14] and gravimetric analysis.[15] The major disadvantages of these methods are excessive time consumption, relative insensitivity, rather complex sample preparation, and high cost. Enzymatic methods involving ß-galactosidase and glucose oxidase are also well studied. Such methods are highly specific, using immobilized

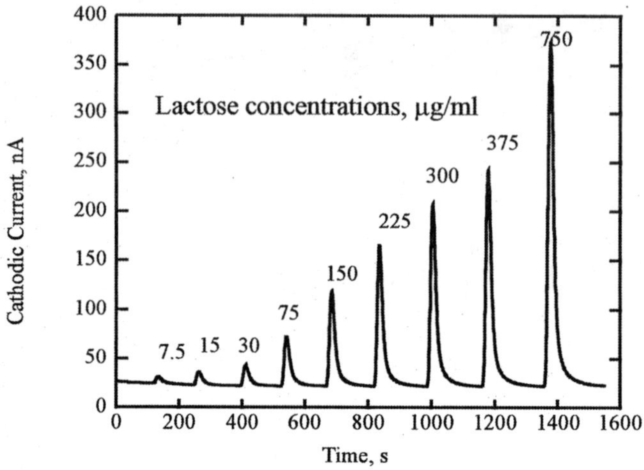

FIGURE 7.7 Correlation between lactose concentration and electric response of the biosensor.

rather than free enzymes together with colorimetric methods and chromogenic agents[16] for response detection.

Extensive research has been directed toward development of electrochemical enzymatic biosensors for lactose determination. A device such as that presented here is highly specific and sensitive, simple to operate, and inexpensive. When adding lactose to the biosensor, we obtained an electric response in the nano-amperes scale. The linear region of the response lay between 10 μg/mL and 340 μg/mL lactose concentration. The amount of lactose in the milk samples was about 15 to 20 μg/mL after final dilution. Figure 7.7 shows the peaks observed after injection of different lactose concentration. The optimal pH for this system was 6.2. Figure 7.8 shows the estimation of lactose concentration in a milk sample.

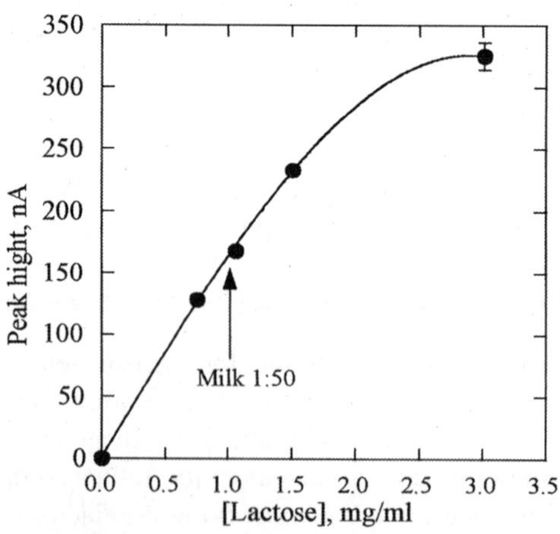

FIGURE 7.8 Calibration curve and detection of raw milk by the triple enzyme sensor.

Finally, we examined the reliability of the biosensor and its shelf-life. Lactose concentration values in raw cow's milk measured in our laboratory were confirmed by measurement of the same samples in a government laboratory. The results deviated by 3% at the most. The shelf-life of the biosensor was at least 5 months. Although the activity of the electrodes decreases with time, calibration of the electrode before using can overcome this problem.

In conclusion, we have developed a very sensitive system that can trace a few μM of lactose (about 6000 times diluted when compared with the concentration in pure milk). Moreover, the biosensor is reproducible, is stable for a relatively long time (at least 6 months when stored in solution), has a short time response, and above all, is based on a simple procedure that can be used on the spot in dairy farms.

D. Quantitative Determination of Bacteria

Figure 7.9 shows an electron micrograph of two graphite electrodes, one with immobilized captured antibody and the other without antibody, which were used as a control experiment. The two electrodes were placed in a solution containing 10^6 cells of *S. aureus*. Clearly, the electrode without antibody did not bind bacteria, confirming the absence of nonspecific adsorption of bacteria to the electrodes.

The responses of the electrodes after incubation with different concentrations of *S. aureus* are shown in Figure 7.10. Current changes caused by the HRP activity were related to variations in cell concentrations, ranging from 1.0×10^3 to 1.0×10^6 cells/mL. In another set of measurements we had added the bacteria directly to the electrochemical cell, and after incubation of 10 minutes we added the second antibody and glucose. The results obtained showed that bacteria at concentrations lower than 10^3 cell/mL can be detected; however, saturation occurred at concentrations higher than 5000 cells/mL. It should be noted that polyclonal antibodies were used in the *S. aureus* assay, enabling the RbIgG-HRP in the solution to compete with the RbIgG on the electrode. Presumably, the saturation effects that we observed were anticipated by the binding of all the RbIgG-HRP to the bacteria in the solution. We believe that the use of monoclonal antibodies will increase the dynamic range of the measurements. Introduction of one washing step after the incubation period with the bacteria resulted in a wide detection range, between 10^2 and 10^6 cells/mL (Figure 7.11). Similar results were obtained with *Salmonella*.

The immunoassay presented here was carried out with an amperometric immunosensor, using a disposable cell and electrodes. The sensor enables the preferential measurement of surface-bound conjugate relative to the excess, enzyme-labeled reagent in the bulk sample solution.

V. CONCLUSIONS

To summarize, both enzyme biosensors and immunosensors may serve as ideal tools for the quantitative characterization of food ingredients, as well as food contaminates. The procedures are not only rapid, accurate, and inexpensive but also require no particular skill or complicated training. Moreover, the sensors are simple to operate and can be used equally well indoors and outdoors. The use of a flow cell

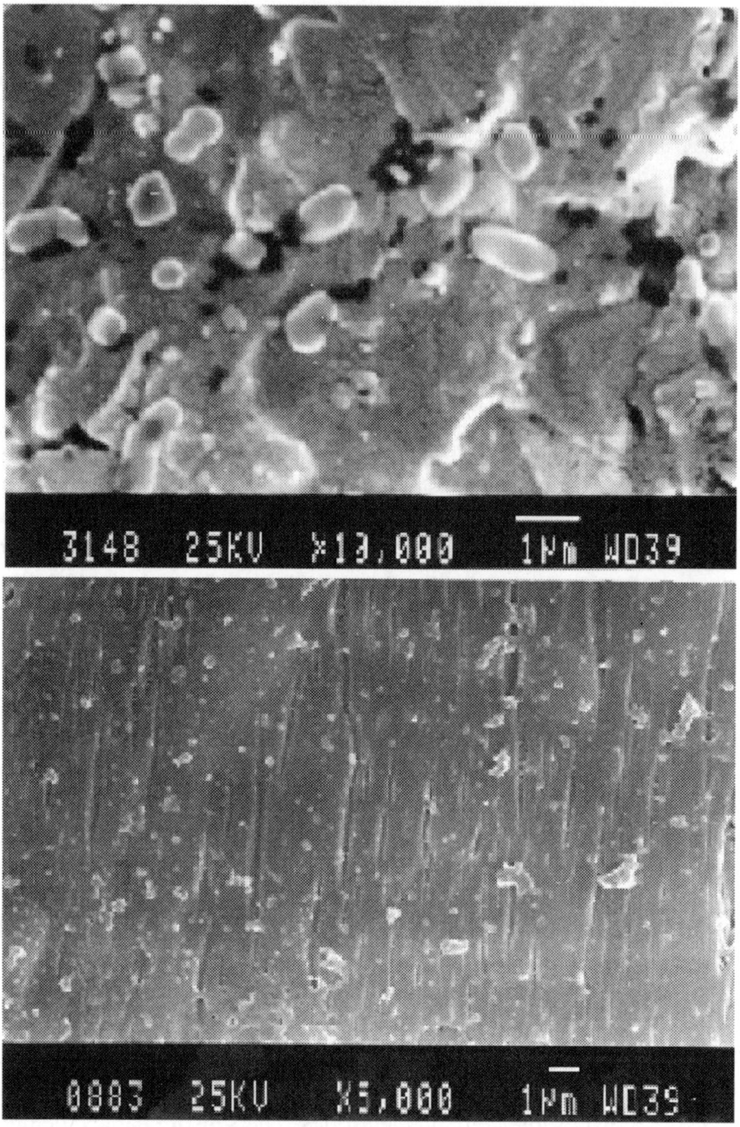

FIGURE 7.9 Electron microscopy of bacteria modified electrode. (Top) Electrode bearing an immobilized antibody on its surface. (Bottom) Electrode without antibody captured on its surface.

ensures the continuous flux of new substrate, thus preventing accumulation or adsorption of products near the working electrode. The miniature size of the electrochemical cell, combined with the screen-printed electrodes, offers the advantage of working with low volumes of compounds and reagents which is especially

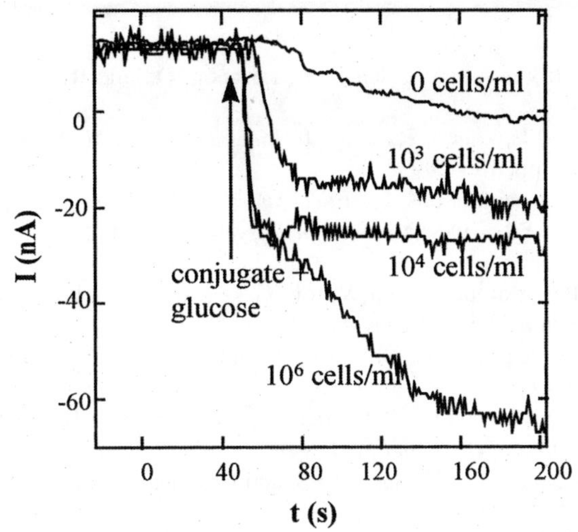

FIGURE 7.10 The response curves generated by amperometric immunosensor specific for protein A in the presence of various quantities of *S. aureus* cells.

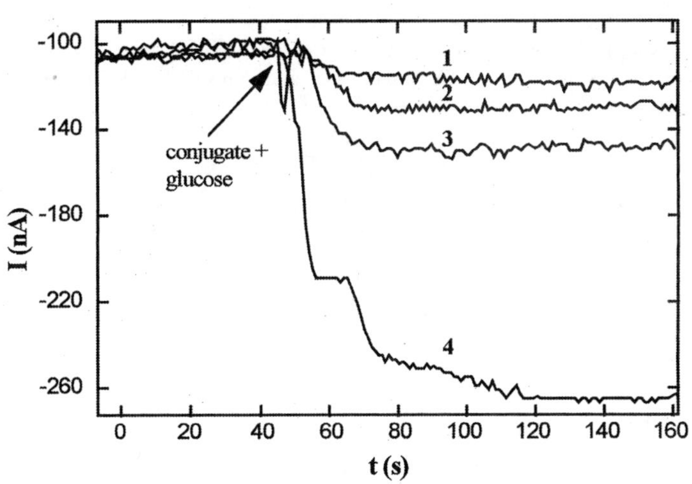

FIGURE 7.11 The response curves generated by amperometric immunosensor specific for *Salmonella* in the presence of various quantities of *Salmonella* cells: 1 = 0 cells/mL; 2 = 10^3 cells/mL; 3 = 10^4 cells/mL; 4 = 10^6 cells/mL.

important when dealing with hazardous materials. Such biosensors could be useful for improving food quality by allowing a critical examination of edible materials before their distribution to markets or by checking individual food components before their use in commercial products.

REFERENCES

1. Ashani, Y., Rothschild, N., Segall, Y., Levanon, D., and Raveh, L., 1991. *Life Sci.,* 49:367–374.
2. Bard, A.J. and Faikner, L.R., 1980. In *Electrochemical Methods: Fundamentals and Applications*, pp. 136–249.
3. Clark, L.C., Jr., and Lyons, C., 1962. *Ann. N.Y. Acad. Sci.,* 102:29.
4. Mosbach, K., Ed., 1988. *Methods in Enzymology,* Vol. 137. Academic Press, New York.
5. Turner, A.P.F., Karube, I., and Wilson, G.S., Eds., 1987. *Biosensors: Fundamentals and Applications.* Oxford University Press, New York.
6. Wang, J., Cai, X., Rivas, G., Shiraishi, H., and Dontha, N. 1997. *Biosens. Bioelectron.,* 12(7):587–599.
7. Rothschild, N., Levanon, D., Gertler, A., Sakal, E., and Henis, Y., 1990. In: *Proc. Annual Meeting of the Israel Society of Microbiology,* p. 23.
8. Huang, Y., Qiao, C., Williamson, M.S. and Devonshire, A.L., 1997. *Chin. J. Biotechnol.,* 13(3):177–183.
9. Guilbault, G.G., Hock, B., and Schmidt, R., 1992. *Biosens. Bioelectron.,* 7:411–419.
10. Rishpon, J. and Ivnitski, D., 1996. *Ann. N.Y. Acad. Sci.,* 799:508–513.
11. Rishpon, J. and Ivnitski, D., 1997. *Biosens. Bioelectron.,* 12(3):195–204.
12. Ivnitskii, D.M. and Rishpon, J., 1993. *Analyt. Chim. Acta,* 282:517–525.
13. Yang, M.T., Milligan, L.P., and Mathison, G.W., 1981. *Chromatography,* 209:316–322.
14. Garrison, E.R., 1939. *J. Assoc. Off. Agric. Chem.,* 33:489–495.
15. Dyer, N., Stoute, V.A., Gershwin, D., and Ngo, T.T., 1991. *Analyt. Biochem.,* 194:16–24.

8 Applications of the Polymerase Chain Reaction for Detection, Identification, and Typing of Foodborne Microorganisms

Pina M. Fratamico

CONTENTS

I. DETECTION OF FOODBORNE PATHOGENS

Public awareness of microorganisms transmitted by food that pose a severe threat to human health has increased dramatically in recent years. Foodborne pathogenic microorganisms include bacterial pathogens such as species of *Salmonella* and *Campylobacter, Listeria monocytogenes,* and *Escherichia coli* O157:H7; protozoan parasites such as *Giardia, Cryptosporidium*, and *Cyclospora*; and enteric viruses including hepatitis A and Norwalk. It is estimated that the annual incidence of microbial foodborne disease cases in the U.S. ranges from 6.5 to 33 million, with as many as 9000 deaths annually.[1] Thus, there is an urgent need for rapid, sensitive, and reliable assays to accurately identify and detect the presence of pathogenic organisms in foods.

A. Traditional Methods

Conventional methods for detection, isolation, and identification of foodborne bacteria require enrichment culturing of food samples for various lengths of time (depending on the target organism) followed by subculturing on a medium that contains selective agents and/or allows for differentiation of targeted bacterial colonies from among the background microflora. Presumptive colonies are ultimately identified by performing a series of biochemical and serological tests. Additionally, tests for the presence of toxins produced by the target organism, as well as for other virulence characteristics, may also be performed. Although traditional cultural methods can be sensitive, such methods are often laborious and time consuming, requiring 4 to 7 days or longer before definitive results can be obtained.

B. Rapid Methods

During the past 10 to 15 years, great strides have been made in the development of rapid methods for detection, identification, and enumeration of foodborne pathogens. In addition to providing results in a shorter length of time than conventional methods, rapid assays can also be more sensitive, specific, and accurate than classical methods. Two types of rapid methods are immunoassays and nucleic acid-based assays.

1. Immunoassays

Immunologic methods depend on the binding specificity of an antibody to an antigen usually displayed on the surface of a microorganism or to a secreted toxin. Immunoassays, such as enzyme-linked immunosorbent assays (ELISA), latex agglutination, or immunodiffusion assays, which employ polyclonal and/or monoclonal antibodies directed against specific antigens of microorganisms, have found wide application in food testing. A variety of immunoassay kits are available commercially. Limitations of immunoassays include: (1) cross-reactivity of the antibody with antigenically similar organisms giving rise to false positive results; (2) quantitative differences in gene expression that may be related to growth parameters; (3) enzyme inhibitors present in foods; and (4) limited availability of antibodies against specific pathogens or specificity only at the species level or below. In general, immunoassays yield presumptive results which must be confirmed by biochemical and serological testing of isolates.

2. Genetic Methods

Nucleic acid hybridization techniques employing labeled gene probes have been used for the detection and identification of specific microorganisms. Generally, probes are labeled nucleotide sequences, ranging in length from 15 to 20 to several thousand base pairs, which under specific conditions hybridize to complementary target sequences. The resulting nucleic acid hybrids can then be detected in a number of ways. The sensitivity of gene probe hybridization assays are in the range of 10^4 to10^6 target gene copies. Applications of gene probes for detection of various foodborne pathogens have been described.[2,3] Commercially prepared colorimetric gene probe kits employing specific DNA probes in a dipstick hybridization format are available from GENE-TRAK® (Hopkinton, MA) for detection of *Listeria, E. coli, Salmonella, Staphylococcus, Campylobacter,* and *Yersinia enterocolitica,* and a gene probe assay for detection of *Listeria monocytogenes* is available from Gen-Probe® (San Diego, CA). Sensitivity of the assays is about 10^6 colony forming units (CFU)/mL.

The polymerase chain reaction (PCR) is a powerful technique that has revolutionized the molecular biology laboratory. It has found wide application and simplified and enhanced procedures such as DNA cloning, sequencing, and mutating and genetic fingerprinting. In recent years, an increasing number of reports have demonstrated that PCR can also be employed as a diagnostic tool to rapidly and reliably detect pathogenic microorganisms both in the clinical setting and in food and environmental microbiology laboratories. PCR is a technique allowing *in vitro* amplification of a specific sequence of DNA using a pair of oligonucleotide primers (used to initiate replication) that hybridize to opposite strands of the targeted DNA and "bracket" the section of the DNA that is to be repeatedly copied. Commonly, a target gene sequence unique to the species of interest is selected, and it is frequently a virulence gene. Following PCR, the amplified product (also referred to as an amplicon), which is of defined length and contains the oligonucleotide primer sequences at its ends, can be detected in a number of ways (see below). The significant feature of PCR is that a unique template sequence can be repeatedly replicated against a background of a large number of contaminating DNA sequences.

II. PERFORMING PCR

To perform PCR, a master mixture is usually prepared (to ensure uniformity of reagent concentrations among the PCR reactions) consisting of PCR buffer containing Tris-HCl (pH 8.3), KCl, and $MgCl_2$; the four deoxynucleotide triphosphates (dNTPs); a pair of DNA primers (typically 15 to 30 bases long); *Taq* DNA polymerase; template DNA (i.e., sequence that is to be amplified by the PCR); and water. The simplest method for preparation of template DNA from bacterial cells is to boil intact cells in a water bath or to lyse them using a thermal cycler set at 99°C for 10 min. The master mixture and template DNA released from lysed bacteria are mixed in individual PCR reaction tubes. Gelatin, bovine serum albumin, or non-ionic detergents such as Triton X-100 are often included in PCR reactions to help stabilize *Taq* DNA polymerase and prevent the formation of secondary structures. The inclusion of low concentrations of non-ionic detergents also may reverse the inhibitory effects of ionic detergents such as SDS which may be used in sample processing, and they also help to inhibit the activity of proteolytic enzymes that may be present. The sequence and concentration of the primers are very important for overall assay success. The primer/template ratio influences the specificity of the PCR, thus it should be optimized to prevent formation of non-specific products and to obtain good product yield.

For total reaction volumes of 50 mL (or 100 mL volumes can be used), 1 to10 mL of template DNA from the material to be analyzed is added to 49 to 40 mL of reaction mixture, respectively (i.e., master mixture), in 0.2-mL volume reaction tubes. The tubes are placed in a thermal cycling instrument, which is programmed to go through cycles of high-temperature DNA denaturation to separate the DNA strands, followed by lower temperature primer annealing and primer extension (i.e., template copying). The amount of template DNA approximately doubles with each cycle, thus theoretically, after 30 cycles, the DNA can be amplified 10^9-fold, assuming 100% efficiency during each cycle. However, in practice, because the amount and activity of the enzyme become limiting and because of other factors, approximately 10^6-fold amplification of the target sequence is achieved. Typically, after an initial high-temperature denaturation step, the PCR cycling parameters are 94 to 95°C for 1 min, 50 to 65°C annealing temperature for 1 min (depending on melting temperature [T_m] of primers), and 72°C for 1 min for primer extension. To promote completion of partially extended products and annealing of single-stranded complementary products, a final extension step at 72°C for 5 to 15 min is frequently included after the last cycle. Unwanted, non-specific sequences may be amplified if the annealing temperature is too low, and expected amplicons may not be detectable if the temperature is too high. Several annealing temperatures should be tested, starting at 5°C below the calculated T_m of the primers. The various reagents, kits, and equipment required to perform the PCR are available from many commercial suppliers, and oligonucleotide primers can be synthesized in the laboratory or custom synthesized by a number of commercial manufacturers.

III. ADVANTAGES OF PCR-BASED METHODS

There are a number of advantages of PCR-based methods over other types of assays for detection of foodborne pathogens. The PCR is a rapid technique, with both high sensitivity and specificity. With the choice of appropriate oligonucleotide primers targeting unique genetic markers, PCR assays can be designed to be very specific for the target organism. This contrasts with cross-reactions of antisera that can occur with immunoassays. Due to the relatively high sensitivity of PCR assays, pre-enrichment and enrichment steps could potentially be eliminated completely or shortened considerably. Instead of plating and performing biochemical and serological tests, the PCR may be performed directly on food samples after a short enrichment step.[4]

Because of the lack of rapid and sensitive assays for detection of foodborne viruses and parasites, the extent of cases and outbreaks of diseases caused by these microorganisms may be grossly underestimated.[1] Viruses and parasites are more difficult to recover from foods, and enrichment culturing in liquid medium, as is performed to increase levels of bacteria, is not possible with foodborne viruses and parasites. Nucleic acid-based amplification systems are a promising alternative to cumbersome, time-consuming, and less sensitive traditional methods used for detection and identification of viruses and parasites.

The PCR is also a functional tool for detection and identification of bacteria that cannot be identified by culture techniques. For example, a PCR-based 16S rRNA technique was used to identify *Tropheryma whippelii*, a non-culturable, possibly foodborne bacterium responsible for Whipple's disease.[5] As another example, there is no appropriate cultural method that can distinguish strains of enterotoxigenic *E. coli* (ETEC) from non-pathogenic strains. Several PCR-based assays have been developed for detection of these organisms in foods, targeting genes involved in virulence of ETEC and allowing identification of pathogenic strains among a majority of non-pathogenic isolates.[6,7] The PCR also has the ability to detect bacteria that are viable but nonculturable by traditional procedures. An additional advantage is that PCR-based methods are amenable to automation, leading potentially to more cost-effective, high-throughput testing of foods for the presence of undesirable microorganisms.

IV. CONCERNS WITH THE USE OF PCR-BASED METHODS

To an unaccustomed laboratory analyst, the successful implementation of PCR for detection of pathogens in foods and other complex samples such as blood or fecal specimens can be fraught with some problems and setbacks. The usefulness of PCR for detection of microorganisms in these types of samples is limited by the presence of substances such as bilirubin, bile salts, hemoglobin degradation products, polyphenolic compounds, proteinases, complex polysaccharides, and fat, which inhibit the DNA polymerase, bind magnesium, or denature the DNA.[8,9] Sensitivity is dramatically decreased if the PCR is performed on crude samples containing inhibitors; therefore, sample preparation steps and/or DNA extraction are often required prior to performing PCR. Appropriate controls should be performed in each assay to assess whether or not negative results are due to the presence of amplification inhibitors in the sample. Also, because there may be very low numbers of the target organisms in foods, either a concentration technique or enrichment culturing is often applied to the samples to increase the relative number of target microorganisms and achieve adequate PCR sensitivity.

The PCR detects DNA from non-viable as well as from live target cells. With enrichment culturing, however, only live cells multiply, thus diluting out the presence of dead cells. Nevertheless, the PCR may not be suitable for testing pasteurized, cooked, or irradiated samples, which may contain considerable numbers of dead microorganisms.[10,11] Yet, it may be important to know if organisms such as *Staphylococcus aureus* or *Clostridium botulinum* were present in such samples, as this signals the possibility that toxins produced by these organisms that can survive irradiation and heat processing steps may be present.

Various parameters determine the overall efficiency, reproducibility, and specificity of PCR assays, and these require optimization. Several important parameters affecting the PCR include the quality of the DNA template, magnesium ion concentration, primer design and concentration, pH, PCR temperature cycling profile (e.g., annealing, extension, and denaturation temperatures and duration times, number of cycles), dNTP concentration, and size of amplicons. With each new PCR assay for different target organisms and for different foods, protocol optimization is required.

Carryover of DNA from one reaction that contaminates subsequent reactions and/or cross contamination between samples can cause false positive results; however, implementation of adequate quality control measures can help to minimize amplicon carryover and contamination in PCR assays. Plugged pipette tips or positive displacement pipettes should be used to avoid sample and reagent contamination. It is advisable to have separate workstations for each of these four tasks: reagent mixture preparation, sample preparation, template addition, and post-PCR detection. A bench-top hood equipped with ultraviolet lights is useful for decontamination. At each workstation, there should be dedicated equipment and supplies; in addition, 10 to 20% bleach solutions should be used to regularly clean workbenches, pipettors, and other equipment and supplies. To control carryover contamination, an enzymatic method can be used based on uracil *N*-glycosylase (UNG) which preferentially degrades uracil-containing DNA synthesized in the presence of dUTP.[12] Another method involves the addition of isopsoralen derivatives to the PCR mixture; following amplification the reaction tube is irradiated with ultraviolet light, resulting in the formation of cyclobutane monoadducts with pyrimidine bases in the DNA.[13] The modified DNA cannot serve as template in subsequent PCR reactions. Other chemical inactivation techniques, such as use of hydroxylamine for post-amplification inactivation of PCR products, have also been described.[14]

V. STRATEGIES TO OVERCOME INTERFERENCE PROBLEMS

A number of strategies have been employed to overcome PCR problems associated with interference from food or other types of complex samples.[15,16] These include centrifugation of samples followed by washing of the target bacteria to remove residual inhibiting substances prior to the PCR,[17] recovery of bacteria by filtration,[18,19] purification of crude DNA extracts by gel chromatography[9,20] and by DNA affinity columns,[21] and removal of inhibitors and concentration of bacteria in food samples by buoyant density centrifugation.[22] Food and other complex samples can be diluted to decrease the concentration of inhibitors; however, sensitivity of PCR detection is reduced correspondingly with the dilution factor.

Most commonly, sample preparation methods consist of cell lysis to release the DNA, followed by purification and concentration of the DNA. A method of extraction using phenol-chloroform followed by precipitation of the DNA using ethanol has commonly been used; however, also available are numerous commercially prepared template purification kits and reagents that allow release of DNA from the cells and separation from PCR inhibitory components. Some of the protocols may involve lysis of cells, followed by binding to solid-phase matrices and elution of the DNA, which is then suitable for PCR. Other protocols consist of concentration of bacteria by centrifugation, cell lysis, and use of commercially available extraction reagents to sequester PCR inhibitors from food or other complex matrices.

Another technique, referred to as immunomagnetic separation (IMS), involves the use of magnetic beads coated with antibodies reactive against the target organism to sequester the bacteria from a large part of the contaminating microflora and from interfering food components. The IMS technique also results in concentration of target bacteria in the sample, thus assay sensitivity is enhanced. This technique has been employed to remove inhibiting components from enrichment samples of ground beef and bovine feces prior to performing the PCR.[23]

VI. DETECTION AND IDENTIFICATION OF MICROORGANISMS BY PCR

A. Bacteria

Numerous reports have appeared on the use of the PCR for detection of bacterial pathogens either found as natural contaminants or seeded in foods. Some of these include detection of *Yersinia* and Shiga toxin-producing *E. coli* in culture and in artificially inoculated foods;[24] *Campylobacter* in water, sewage, and food samples;[25] stressed *Salmonella* in dairy and egg products;[26] and *L. monocytogenes* in food products (see Table 8.1).[27,28]

Multiplex PCR assays employ multiple sets of primers to amplify more than one target sequence simultaneously in a single reaction tube. Multiplex PCR assays have been used to detect and/or identify one organism by amplification of more than one gene, or multiple organisms can be detected simultaneously by targeting unique sequences from each organism.[23,29,30] For maximal amplification efficiency and specificity, multiplex assays require optimization of temperature cycling conditions and reagent concentrations, and primers with nearly identical annealing temperatures and which do not display homology either internally or to one another should be used. A multiplex PCR method was developed for *E. coli* O157:H7 by employing primers specific for the *eaeA* gene, conserved sequences of Shiga toxin 1 and 2 genes (stx_1, stx_2), and for a hemolysin gene found on a 60-MDa plasmid. Amplification products of 1087 (*eaeA*), 227/224 (stx_1/stx_2), and 166 (hemolysin) base pairs were successfully amplified simultaneously in a single reaction.[29] This multiplex PCR method was used to specifically identify enterohemorrhagic *E. coli* (EHEC) displaying serotype O157. Because *E. coli* O157:H7 and *Salmonella* can contaminate similar foods and other types of samples, a multiplex PCR was designed to allow simultaneous detection of both *E. coli* O157:H7 and *Salmonella* spp. in sample enrichment cultures.[23] Apple cider, beef carcass wash water, ground beef, and bovine feces were inoculated with approximately less than 1 to 250 CFU/g or mL of both *E. coli* O157:H7 and *S. typhimurium*. Following enrichment culturing, the samples were subjected to a DNA extraction technique or to immunomagnetic separation and then tested by the multiplex PCR assay. Four pairs of primers were employed in the PCR: three primers for amplification of *E. coli* O157:H7 *eaeA*, both stx_1 and stx_2, and plasmid sequences; and one primer for the amplification of a portion of

TABLE 8.1
PCR-Based Methods for Detection of Foodborne Microorganisms

Microorganisms	Target Gene/s	Type of Sample	Sensitivity	Ref.
		Bacteria		
Salmonella spp., *E. coli* O157:H7	*invA*, *eaeA*, stx_1, stx_2, hemolysin	Apple cider, beef carcass wash water, ground beef, bovine feces	≤1 CFU/g or mL	23
Yersinia, Shiga toxin-producing *E. coli*	*virF, yadA*, stx_1, stx_2	Tofu, chocolate milk	Not given	24
C. jejuni, C. coli	Intergenic sequence between *flaA* and *flaB*	Water, sewage, beef, chicken, pork	≤3 CFU/g	25
Salmonella spp.	Random genomic fragment	Eggs, ice cream, cheese	5.9 cells/25 g	26
L. monocytogenes	*prfA*	Salmon	94 CFU/g	27
L. monocytogenes	*iap*	Catfish fillets, milk	1–2 CFU/g, 20 CFU/mL	28
Y. enterocolitica	*virF, ail*	Pork, cheese, zucchini	0.5 CFU/cm^2	30
		Yeasts and Molds		
Dekkera-Brettanomyces strains	*Dekkera* genomic DNA fragment	Sherry	Fewer than 10 cells	32
Bacteria, yeasts, molds	Elongation factor	Milk	10 cells/mL	33
Aflatoxigenic molds	ver-1, omt-1, apa-2	Corn	10^2 spores/g	34
		Viruses		
Hepatitis A virus	C-terminus capsid protein VP3 and A-terminus protein VP1	Oysters	10 PFU/20g	35
Enteroviruses	Conserved 5′ non-coding region	Sewage effluent, groundwater	0.028 PFU/mL	36
Hepatitis A virus	C-terminus capsid protein VP3 and A-terminus protein VP1	Liquid waste, shellfish	4 virus particles	37
Small round structured viruses, enteroviruses	5′ UTR, VP7 gene, 3D, capsid, RNA polymerase	Human stool, seafood	3–30 $TCID_{50}$/1.25 g	38
		Parasites		
C. parvum	18S rRNA gene	Water	1–10 oocytes	39
C. parvum	CpR1, *C. parvum*-specific genomic fragment	Raw milk	1–10 oocytes/20 mL	40
T. gondii	P30 gene	Cured meats	5×10^3 trophozoites/g	41
Cyclospora sp., *Eimeria* spp.	18S rRNA gene	Raspberry wash sediments	10–19 oocytes/PCR	42

the *Salmonella invA* gene. Four fragments of the expected sizes were amplified in a single reaction and visualized following agarose gel electrophoresis in all of the samples inoculated with ≤1 CFU/g or mL. Results could be obtained in approximately 30 hours, which included sample enrichment.

B. Yeasts and Molds

Polymerase chain reaction has also found application for the rapid detection of molds and yeasts involved in food spoilage. By a multiplex PCR assay, Pearson and McKee[31] were able to simultaneously detect three strains of yeast that cause food spoilage. Other reports described PCR-based assays to detect spoilage yeasts in red wine;[32] viable bacteria, molds, and yeasts in milk;[33] and specific aflatoxigenic molds in grains such as corn.[34]

C. Viruses

In the past decade, the expanding information on the sequences of viral genomes has allowed the development of PCR-based assays to ascertain the presence of viruses in food, environmental, and other samples. Several techniques for virus detection include magnetic immunoseparation PCR,[35] an integrated cell culture-reverse transcriptase PCR (RT-PCR) method for detection of enteroviruses,[36] antigen capture PCR for detection of hepatitis A virus in environmental samples,[37] and a semi-nested RT-PCR for viruses in seafoods.[38] A concern with PCR methods for detection of enteric viruses in foods is the inability to differentiate infectious from non-infectious (inactivated) viruses. This limitation can be overcome if the PCR is preceded by propagation of infectious virus particles in cell or tissue culture.

D. Parasites

A number of PCR assays have been designed for detection of protozoan parasites in food and environmental samples and in fecal specimens. PCR is rapid, sensitive, and specific and is an attractive alternative to tedious techniques involving microscopy which are less amenable to batch processing of samples. Johnson et al.[39] found that the sensitivity of PCR was similar to that obtained by immunofluorescence for the detection of *Cryptosporidium* oocysts in wastewater concentrates. PCR combined with digoxigenin-labeled probe hybridization was used for detection of *C. parvum* in raw milk,[40] for detection of *Toxoplasma gondii* in cured meats,[41] and for detection and identification of *Cyclospora* spp. and *Eimeria* spp. in raspberries.[42] Detailed reviews on the use of PCR for detection of foodborne pathogens have been published by Hill[43] and Olsen et al.[3]

VII. DETECTION OF PCR PRODUCTS

A number of techniques can be employed to detect and confirm the identity of PCR amplification products. The most commonly used procedure is agarose gel electrophoresis with ethidium bromide staining, which allows visualization of products due to the ability of the stain to intercalate into the DNA. Accurate size determination of the amplicons is accomplished by comparison to a set of size markers that are applied to adjacent wells of the same gel. Other nucleic acid stains such as SYBR® Green (Molecular Probes, Inc.; Eugene, OR), which may be more sensitive than ethidium bromide, are commercially available. Amplicons can also be stained by a silver-staining technique that does not require ultraviolet trans-illumination for visualization of the DNA. Hybridization techniques in which products are transferred to membranes followed by detection using radioactive or non-radioactively labeled specific probes are more sensitive than gel electrophoresis, allowing a 10^2- to 10^3-fold increase in detection sensitivity.[44] Alternatively, specific oligonucleotides can be covalently bound to membranes, then hybridized to amplification products that were biotinylated during the PCR. Streptavidin-horseradish peroxidase binds to the biotinylated DNA, which then permits colorimetric detection.[45]

Numerous ELISA formats have been developed for detection of amplicons. For example, PCR amplification can be performed with a digoxigenin-labeled primer and a biotinylated probe. The biotin-digoxigenin hybrids can then be quantified by an ELISA through binding to streptavidin-coated microtiter plates followed by detection using a peroxidase-labeled anti-digoxigenin antibody.[46] Detection of a *L. monocytogenes*-specific *hlyA* gene product was achieved using an antibody directed to RNA-DNA hybrids formed upon hybridization with a probe bound to a microtiter plate.[47]

With electrochemiluminescence detection systems, PCR products are labeled with biotin and ruthenium (II) trisbipyridal (Ru(bpy)3(2+)) and then captured by streptavidin-coated magnetic beads prior to electrochemiluminescent detection.[48] Detection of DNA using this technique was more sensitive than visualization of the product by ethidium bromide staining. The PCR products can also be sized, detected, and purified using high-performance liquid chromatography (HPLC), and HPLC systems have been described for quantification of specific nucleic acid sequences.[49] Reversed-phase HPLC with ultraviolet detection at 254 nm was used for quantitative and qualitative analysis of RT-PCR products;[50] however, HPLC systems are not routinely used because the equipment is costly and not practical for the analysis of large numbers of samples.

Through the addition of the DNA intercalating dye, ethidium bromide, or other dyes to the PCR, simultaneous amplification and detection of DNA sequences can be achieved.[51] Drawbacks of this technique include the fact that both specific and non-specific products generate a signal and the amount of signal may be dependent on the mass of the DNA generated. Oligonucleotide primers labeled with fluorescent dyes may be employed in the PCR reaction, and after amplification, purification of products, and gel electophoresis it is possible to detect product using a fluorescent imaging system. Holland et al.[52] developed a detection system that makes use of the 5′ to 3′ exonuclease activity of *Taq* DNA polymerase to generate a sequence-specific probe signal. During amplification, the exonuclease activity of the enzyme degrades a probe specifically hybridized to the target sequence and releases smaller fragments that can be differentiated from uncleaved probe. Cleavage of the probe was measured using thin-layer chromatography. Commercially available, fluorescence-based detection systems utilizing this technology are described below.

VIII. COMMERCIALLY AVAILABLE PCR ASSAY SYSTEMS

The BAX® pathogen detection systems, manufactured by Qualicon (Wilmington, DE), are simple PCR-based assays currently available for detection of *E. coli* O157:H7, *Salmonella,* and *Listeria.* After enrichment and a short pretreatment of the sample, a simplified PCR assay is performed using pre-packaged reagent tablets. The PCR is followed by gel electrophoresis and analysis. Several reports have indicated that the BAX® systems are considerably faster, simpler, and more sensitive than cultural methods.[53–55]

A fluorogenic 5′ nuclease assay (Taqman™) for detection of *Salmonella* is available from PE Applied Biosystems (Foster City, CA), and Taqman™ kits for detection of other pathogens are in development. Fluorogenic probes are used to quantitatively detect specific nucleic acid sequences, and data analysis is automated. The samples are subjected to overnight enrichment culturing, then to a simple DNA extraction step. The PCR mixture, which includes a *Salmonella*-specific probe with a 5′ fluorescent reporter dye and a 3′ quencher dye, is added to template DNA. During the PCR, the labeled probe anneals specifically to the target region between the two primer binding sites. As *Taq* DNA polymerase, which has 5′ to 3′ activity, extends the primers, it cleaves the probe only if it has hybridized to the target region; the enzyme does not digest free probe. As the probe is cleaved, the reporter dye and quencher dyes are separated, resulting in a measurable increase in fluorescence intensity. Post-PCR processing is not necessary using fluorogenic 5′ nuclease assays, thus eliminating this step as a potential source of error. Because the reaction wells are kept closed the entire time, there is also a reduced risk of carryover contamination. The fluorescent signal can be detected in one of several instruments; however, the GeneAmp® 5700 and ABI PRISM® 7700 Sequence Detection Systems (PE Applied Biosystems) are designed to detect fluorescence during PCR cycling, allowing for real-time detection of PCR product accumulation. An attractive feature of these systems is that they permit real-time quantitation of DNA and RNA. A number of reports have appeared describing the use of fluorogenic PCR-based assays for detection of foodborne pathogens.[56–58] Another PCR-based procedure used for rapid fluorometric real-time detection of product and reaction kinetics is performed using the LightCycler™ (Idaho Technology, Inc.; Idaho Falls, ID). A DNA intercalating dye is added to the PCR reaction and increase in fluorescence is detected by the instrument.

The PROBELIA™ PCR system for detection of *E. coli* O157:H7, *Salmonella,* and *L. monocytogenes* was developed at the Institut Pasteur in Paris and is available through BioControl Systems, Inc. (Bellevue, WA). The test components include amplification and detection kits. A colorimetric microplate assay is performed on the PCR products, and results are read using a microplate reader.

IX. ADDITIONAL PCR TECHNIQUES, REAGENTS, AND KITS

Various techniques have been described and many companies market molecular biology reagents, kits, and equipment that can enhance PCR performance and PCR-based detection assays. For low copy number detection, to reduce the formation of non-specific products, and to increase sensitivity, a "hot start" PCR technique was developed in which at least one essential PCR component is added after the initial warm-up temperature or after the first denaturation step. Alternatively, a wax layer is used to physically separate critical PCR components. During the heating step, the wax melts and the reaction components are then allowed to mix together. Use of modified DNA polymerases such as Amplitaq Gold® (PE Applied Bioystems), HotStarTaq™ (Qiagen; Valencia, CA), or PLATINUM® *Taq* DNA polymerase (GIBCO BRL/Life Technologies; Rockville, MD), which are inactive until heated during the PCR before thermal cycling, eliminates the extra steps of addition of wax or a "manual" hot start. Use of these modified enzymes prevents extension of non-specifically annealed primers and formation of primer dimers which can occur at low temperatures during the initial PCR cycle. The HotStarTaq™ system comes with a unique buffer with a balanced combination of KCl and ammonium sulfate which should improve PCR performance by promoting specific primer-template annealing.

Also available are PCR optimization kits that simplify determination of optimal PCR conditions for individual assays; ready-to-use PCR mixes to which only template and primers need to be added by the user; enzyme sample kits containing a selection of thermostable DNA polymerases, buffers, and PCR enhancers; and PCR systems for amplification of very large (e.g., 30 to 40 kilobases) templates. These systems provide unique enzyme combinations: a non-proofreading polymerase with 5′ to 3′ polymerase activity and an enzyme with 3′ to 5′ proofreading ability to provide high fidelity and the capability to amplify very long sequences.

X. QUANTITATION IN THE PCR

Because the PCR involves the exponential amplification of target nucleic acid sequences, it is commonly assumed that quantitation of the number of input target molecules cannot easily be achieved. Although reports on the use of PCR-based techniques for quantitating pathogens in foods are lacking, several approaches for determining the level of input copy number have been described, and several reports have appeared recently.[28,59,60] Competitive PCR involves the addition of varying amounts of a DNA sequence bearing the same primer annealing sites as the template. The degree to which the added DNA competes will depend on its initial concentration relative to that of the test sequence. The amount of initial target sequence can be obtained from the point on the curve where target and standard values are equal.

The endogenous standard assay utilizes an endogenous sequence that is expressed at a relatively constant level in all of the samples. The level of amplification of the target sequence is then compared to that of the standard. Drawbacks with this method are that it is generally only a comparative technique and the actual copy number cannot easily be determined and that the endogenous standard may be expressed at levels far higher than the target sequence, making dilutions of the sample necessary.

The synthetic internal standard method involves the addition of synthetic standards designed to carry very slight sequence variations that make them readily distinguishable from the sequences of interest. Usually, a unique restriction site is created in the synthetic standard, such that after amplification and restriction enzyme digestion only the standard will yield two small digestion products.

A system was developed allowing for "real-time" measurement of the accumulation of PCR products at each cycle by various techniques.[51,61] Because quantitation can be performed during the early PCR cycles when product is first detected (i.e., kinetic PCR), rather than determining the amount of product accumulated after a fixed number of cycles, the level of precision of this approach is increased. Quantitation of the amount of target nucleic acid in the unknown samples is performed by comparison to a standard curve. The ABI PRISM® 7700 and GeneAmp® 5700 sequence detection systems (PE Applied Biosystems) or LightCycler™ (Idaho Technology, Inc.) are instruments designed for "real-time" detection of PCR product accumulation, and these systems eliminate the need for post-PCR processing of amplification products. Several ELISA systems employing biotinylated primers with the PCR or involving hybridization of product to capture oligonucleotides have also been described for quantifying PCR products.[62]

XI. REVERSE TRANSCRIPTION PCR

Reverse transcription PCR is a technique used for analysis and detection of RNA molecules. Because RNA cannot serve as a template for the PCR, reverse transcription is combined with the PCR to convert RNA into a complementary DNA (cDNA) molecule suitable for PCR amplification. The combination of reverse transcription and PCR is referred to as RT-PCR. As discussed above, food or other types of samples containing high numbers of non-viable cells may give positive results with conventional PCR. RT-PCR is an alternative technique that permits differentiation between viable and non-viable cells if mRNA serves as the target of amplification. Bacterial mRNA has a very short half-life and is rapidly degraded with processes that render the cells non-viable. Klein and Juneja[63] used RT-PCR for detection of viable *L. monocytogenes*. The RT-PCR assay, based on amplification of *L. monocytogenes* mRNA (the *iap* gene), was used to detect only viable cells. An amplification product was detected in cooked meat samples initially inoculated with 3 CFU/g. A 2-hour enrichment was necessary to increase the sensitivity of the RT-PCR, and the results showed that non-viable cells would have been detected following short enrichment periods if PCR had been used. Sheridan et al.[64] studied the relationship between viability and detection of specific *E. coli* mRNAs. They found that the type of cell inactivation treatment and subsequent holding conditions influenced the ability to detect mRNA targets; however, the presence of 16S rRNA was detected in samples containing dead cells, and it persisted for long periods. A number of RT-PCR assays for detection of RNA viruses, such as hepatitis A virus, have also been reported.[35,37,65]

XII. OTHER TYPES OF GENE AMPLIFICATION PROTOCOLS

A number of other *in vitro* gene amplification techniques have been developed; however, these methods have not yet been as widely applied for microbial analysis of foods as PCR. A procedure called the ligase chain reaction (LCR) involves the formation of new target DNA molecules through the joining, by a thermostable DNA ligase, of probe molecules which anneal to the target DNA such that they are immediately adjacent to each other. If there is a mismatch at the ligation site, the primers are not ligated by the enzyme. The LCR is cyclic, with ligated product serving as the template for the next reaction; product is formed without DNA replication. LCR is useful for discriminating between DNA sequences differing in a single base pair. Wiedmann et al.[66] used a PCR-coupled LCR assay to distinguish *L. monocytogenes* from other *Listeria* species by targeting a single base pair difference in the *Listeria* 16S rRNA gene. A similar approach was used to distinguish between the food spoilage yeasts *Zygosaccharomyces bisporus* and *Z. bailii* and to differentiate these species from other related species.[67]

The Qb replicase amplification method was named after the enzyme that replicates the RNA genome of bacteriophage Qb and is used for amplification of RNA probe molecules. A section of the Qb genome is replaced with an RNA probe specific for the target sequence such that the unique folded structure of the genome is not altered. Probe molecules that specifically anneal to the target sequence are enzymatically replicated then detected.[68] The level of amplification can approach 10^9-fold in a 30-min incubation; however, non-specific amplification is a limitation of Qb replicase assays.

Another target amplification method is strand displacement amplification (SDA). This method is based on the ability of DNA polymerases to initiate DNA synthesis at a single-strand nick in a DNA molecule and displace the existing strand. The displaced DNA molecules serve as substrates for additional nicking by a restriction enzyme, polymerization, and displacement, resulting in exponential amplification. SDA allows simultaneous, isothermal amplification of DNA. Amplification takes place at 37°C, and temperature cycling is not required.

Self-sustaining sequence replication (3SR) or nucleic acid sequence-based amplification (NASBA) involves isothermal (37 to 42°C) amplification, usually of an RNA template. The reaction consists of continuous cycles of reverse transcriptase-mediated synthesis of cDNA from an RNA target sequence, followed by *in vitro* transcription by RNA polymerase from the double-stranded cDNA template. The NASBA method requires two primers, one that bears a bacteriophage T7 promoter sequence at the 3′ end and binds to the 3′ end of the target sequence and another that is derived from the 5′ end of the target sequence. The NASBA method also requires three enzymes: reverse transcriptase to make cDNA, RNase H to digest the RNA strand of RNA:DNA heteroduplexes, and T7 RNA polymerase to synthesize as many as 100 copies of RNA used as substrate in the next cycle. Using a NASBA-based method, Uyttendaele and co-workers[69] were able to detect *C. jejuni* in foods inoculated with as few as 3 CFU/10 g after 18 hours of selective enrichment. NASBA targeting mRNA of *L. monocytogenes* inoculated into dairy and egg products successfully detected the organism at an initial inoculum level of less than 10 CFU/g.[70] Although LCR, Qb replicase, SDA, and NASBA assay kits are available commercially, they have found application mainly in clinical diagnostics. With the availability of an increased number of kit formats and the appearance of an increased number of reports demonstrating the reliability of non-PCR methods, such methods may find wider application in testing of foods for the presence of undesirable microorganisms.

XIII. NUCLEIC ACID-BASED TYPING OF MICROORGANISMS

Many outbreaks of foodborne disease are due to exposure to a common source of the etiologic agent. Isolates are usually clonal, meaning they are derived from a common ancestor, and are found to be indistinguishable from one another by a variety of subtyping methods. Molecular typing methods targeting bacterial lipopolysaccharides, fatty acids, proteins, or nucleic acids are gradually replacing phenotype-based subtyping methods such as serotyping, phage typing, biotyping, or bacteriocin typing. The relatively high discriminatory ability, ease of performance, and reproducibility of many nucleic acid-based typing techniques that have been developed in recent years have led to their application to study the relatedness of bacterial isolates and have allowed epidemiological investigations to be conducted more rapidly and thoroughly.[71] Plasmid typing involves analysis of the sizes of plasmids (i.e., extrachromosomal, circular pieces of supercoiled double-stranded DNA) possessed by microorganisms.

Pulsed field gel electrophoresis (PFGE) is a commonly used genotypic typing method that involves the generation of large fragments of chromosomal DNA through the use of restriction enzymes that cut genomic DNA infrequently. The fragments are then separated by electrophoretic procedures that allow resolution of DNA molecules ranging in size from 100 to 200,000 base pairs. The technique of PFGE is very discriminating and reproducible and has been found to be very useful for epidemiological investigations.[72]

Ribotyping involves restriction digestion of chromosomal DNA and agarose gel electrophoresis to separate the fragments, followed by transfer of the fragments onto a nylon or nitrocellulose membrane. The DNA is then probed with labeled rRNA sequences, cDNA made from the rRNA, a recombinant plasmid in which the *rrn* operon of *E. coli* is inserted, or synthetic oligonucleotides made from 16S or 16S plus 23S gene sequences.[71] A "fingerprint" pattern of up to 15 bands is generated from binding of the probes to DNA fragments containing ribosomal genes. An automated ribotyping system, known as the RiboPrinter® Microbial Characterization System (Qualicon), is commercially available.[73]

Several PCR-based typing techniques have been reported, and the advantages of these compared to other typing methods include increased speed, simplicity, and the need for smaller amounts of DNA. One PCR-based typing technique known as random amplified polymorphic DNA (RAPD) or also as arbitrarily primed PCR (AP-PCR) involves the use of a single primer that has no known homology to the target DNA. The reaction occurs with low annealing temperatures, allowing the primer to bind to DNA at sites where the match is imperfect, thus at multiple locations on the two DNA strands. The sequences between the primer annealing sites are amplified, and fragments, which form a fingerprint of the genomic DNA, are visualized on ethidium bromide-stained agarose gels. RAPD assays have been successfully applied for molecular typing of a number of organisms, including *L. monocytogenes* and *Bacillus cereus*.[74,75]

Polymerase chain reaction amplification of the intergenic spacer regions between genes coding for 16S and 23S rRNA is known as PCR-ribotyping. In addition to variations in the sequences of the spacer regions, many bacteria possess multiple rRNA alleles, each differing in the length of the spacer regions. A variable number of amplified fragments can be generated, allowing differentiation of bacterial isolates based on the pattern of the fragments.[76]

Polymerase chain reaction–restriction fragment length polymorphism (PCR-RFLP) analysis involves PCR amplification of a known DNA sequence such as a virulence region, followed by digestion of the amplification product with a restriction enzyme. Analysis of the DNA fingerprint is performed by agarose gel electrophoresis. Fields et al.[77] used PCR-RFLP analysis of the *E. coli fliC* gene to differentiate flagellar antigen groups and found that, in conjunction with serotyping, PCR-RFLP allowed identification of *E. coli* O157:H7 and related strains.

A typing technique referred to as rep-PCR entails the use of primers complementary to highly conserved, repetitive sequences present in multiple copies on the genome. Because the distance between the repetitive elements varies among strains, PCR amplification of the DNA sequences found between them results in generation of a distinct fingerprint. As is true for other genomic typing methods, the resulting "fingerprint" is analogous to UPC codes used in supermarkets. Families of repetitive sequences that have been identified in bacteria include repetitive extragenic palindromic (REP) sequences, enterobacterial repetitive intergenic consensus (ERIC) sequences, and BOX elements. The rep-PCR typing protocols employing these sequences are referred to as REP-PCR, ERIC-PCR, and BOX-PCR, respectively. The rep-PCR was used to type *L. monocytogenes* strains isolated form humans, animals, and foods[78] and to determine the genetic relatedness of *Bacillus sphaericus* strains.[79]

The amplified restriction fragment length polymorphism (AFLP) typing technique is based on detection of DNA restriction fragments by PCR amplification. The DNA is subjected to restriction enzyme digestion followed by amplification of the fragments by ligation of adapter sequences to the ends of the restriction fragments, which serve as sites for primer annealing. To reduce the number of fragments generated, the primers can have a number of selective bases at their 3′ ends extending into the restriction fragments, allowing amplification of only those fragments in which the nucleotides flanking the restriction sites match the 3′ primer extensions.[80]

The genotyping method selected for use will depend on the application, and the advantages of the method must be weighed against the disadvantages. The PFGE technique is a commonly used typing method that has excellent discriminatory power, is highly reproducible, and generates results (i.e., fingerprints) that are easy to interpret; however, some protocols that have been described are time consuming and require the use of relatively costly equipment and materials. The ribotyping method is amenable to automation; however, the automated system is costly, and the method is somewhat less discriminatory than PFGE. Advantages of PCR-based subtyping methods over PFGE and other nucleic acid typing techniques include rapidity, simplicity, and the need for fewer cells of the organism being analyzed. With more extensive comparisons to established typing methods such as PFGE, to determine reproducibility and discriminatory power, PCR-based methods, especially those amenable to automation, may replace more conventional (phenotypic and genotypic) but costly and time-consuming typing procedures.

NOTE

Mention of brand or firm names does not constitute an endorsement by U.S. Department of Agriculture over others of a similar nature not mentioned.

REFERENCES

1. Council for Agricultural Science and Technology Task Force Report, *Foodborne Pathogens: Review of Recommendations,* No. 122, Council for Agricultural Science and Technology, 1998.
2. Dooley, J.S.G., Nucleic acid probes for the food industry, *Biotech. Adv.,* 12, 669, 1994.
3. Olsen, J.E., Aabo, S., Hill, W., Notermans, S., Wernar, K., Granum, P.E., Popovic, T., Rasmussen, H.N., and Olsvik, Ø., Probes and polymerase chain reaction for detection of foodborne bacterial pathogens, *Int. J. Food Microbiol.,* 28, 1, 1995.
4. Koch, W.H., Payne, W.L., Wentz, B.A., and Cebula, T.A., Rapid polymerase chain reaction method for detection of *Vibrio cholerae* in foods, *Appl. Environ. Microbiol.*, 59, 556, 1993.
5. Relman, D.A., Schmidt, T.M., MacDermott, R.P., and Falkow, S., Identification of the uncultured bacillus of Whipple's disease, *N. Engl. J. Med.*, 327, 293, 1992.
6. Deng, M.Y., Cliver, D.O., Day, S.P., and Fratamico, P.M., Enterotoxigenic *Escherichia coli* detected in foods by PCR and an enzyme-linked oligonucleotide probe, *Int. J. Food Microbiol.*, 30, 217, 1996.
7. Tsen, H.Y., Jian, L.Z., and Chi, W.R., Use of a multiplex PCR system for the simultaneous detection of heat labile toxin I and heat stable toxin II genes of enterotoxigenic *Escherichia coli* in skim milk and porcine stool, *J. Food Prot.,* 61, 141, 1998.
8. Rossen, L., Nørskov, P., Holmstrøm, K., and Rasmussen, O.F., Inhibition of PCR by components of food samples, microbial diagnostic assays and DNA-extraction solutions, *Int. J. Food Microbiol.,* 17, 37, 1992.
9. Monteiro, L., Bonnemaison, D., Vekris, A., Petry, K.G., Bonnet, J., Vidal, R., Cabrita, J., and Mégraud, F., Complex polysaccharides as PCR inhibitors in feces: *Helicobacter pylori* model, *J. Clin. Microbiol.*, 35, 995, 1997.
10. Josephson, K.L., Gerba, C.P., and Pepper I.L., Polymerase chain reaction detection of nonviable bacterial pathogens, *Appl. Environ. Microbiol.*, 59, 3513, 1993.
11. Arnal, C., Crance, J.M., Gantzer, G., Schwartzbrod, L., Deloince, R., and Billaudel, S., Persistence of infectious hepatitis A virus and its genome in artificial seawater, *Zentralbl. Hyg. Umweltmed.,* 201, 279, 1998.
12. Longo, M.C., Berninger, M.S., and Hartley, J.L., Use of uracil DNA glycosylase to control carry-over contamination in polymerase chain reactions, *Gene,* 93, 2847, 1990.
13. Cimino, G.D., Metchette, K.C., Tessman, J.W., Hearst, J.E., and Isaacs, S.T., Post-PCR sterilization: a method to control carryover contamination for the polymerase chain reaction, *Nucl. Acids Res.*, 19, 99, 1991.
14. Persing, D.H. and Cimino, G.D., Amplification product inactivation methods, in *Diagnostic Molecular Microbiology, Principles and Applications*, Persing, D.H., Smith, T.F., Tenover, F.C., and White, T.J., Eds., American Society for Microbiology, Washington, D.C., 1993, p. 105.
15. Greenfield, L. and White, T.J., Sample preparation methods, in *Diagnostic Molecular Microbiology, Principles and Applications,* Persing, D.H., Smith, T.F., Tenover, F.C., and White, T.J., Eds., American Society for Microbiology, Washington, D.C., 1993, p. 122.

16. Lantz, P., Hahn-Hägerdal, B., and Rådström, P., Sample preparation methods and PCR-based detection of food pathogens, *Trends Food Sci. Technol.*, 5, 384, 1994.
17. Jinneman, K.C., Trost, P.A., Hill, W.E., Weagant, S.D., Bryant, J.L., Kaysner, C.A., and Wekell, M.M., Comparison of template preparation methods from foods for amplification of *Escherichia coli* O157 Shiga like toxins type I and II by multiplex polymerase chain reaction, *J. Food Prot.*, 58, 722, 1995.
18. Starbuck, M.A.B., Hill, P.J., and Stewart, G.S.A.B., Ultra sensitive detection of *Listeria monocytogenes* in milk by the polymerase chain reaction (PCR), *Lett. Appl. Microbiol.*, 15, 248, 1992.
19. Oyofo, R.A. and Rollins, D.M., Efficacy of filter types for detecting *Campylobacter jejuni* and *Campylobacter coli* in environmental water samples by polymerase chain reaction, *Appl. Environ. Microbiol.*, 59, 4090, 1993.
20. Tsai, Y-L. and Olson, B.H., Rapid method for separation of bacterial DNA from humic substances in sediments for polymerase chain reaction, *Appl. Environ. Microbiol.,* 58, 2292, 1992.
21. Anderson, M.R. and Omiecinski, C.J., Direct extraction of bacterial plasmids from food for polymerase chain reaction, *Appl. Environ. Microbiol.*, 58, 4080, 1992.
22. Lindqvist, R., Preparation of PCR samples from food by a rapid and simple centrifugation technique evaluated by detection of *Escherichia coli* O157:H7, *Int. J. Food Microbiol.*, 37, 73, 1997.
23. Fratamico, P.M. and Strobaugh, T.P., Simultaneous detection of *Salmonella* spp. and *Escherichia coli* O157:H7 by multiplex PCR, *J. Indust. Microbiol. Biotechnol.*, 21, 92, 1998.
24. Weagant, S.D., Jagow, J.A., Jinneman, K.C., Omiecinski, C.J., Kaysner, C.A., and Hill, W.E., Development of digoxigenin-labeled PCR amplicon probes for use in the detection and identification of enteropathogenic *Yersinia* and Shiga toxin-producing *Escherichia coli* from foods, *J. Food Prot.*, 62, 438, 1999.
25. Waage, A.S., Vardund, T., Lund, V., and Kapperud, G., Detection of small numbers of *Campylobacter jejuni* and *Campylobacter coli* cells in environmental water, sewage, and food samples by a seminested PCR assay, *Appl. Environ. Microbiol.*, 65, 1636, 1999.
26. Rijpens, N., Herman, L., Vereecken, F., Jannes, G., DeSmedt, J., and DeZutter, L., Rapid detection of stressed *Salmonella* spp. in dairy and egg products using immunomagnetic separation and PCR, *Int. J. Food Microbiol.,* 46, 37, 1999.
27. Simon, M.C., Gray, D.I., and Cook, H., DNA extraction and PCR methods for the detection of *Listeria monocytogenes* in cold smoked salmon, *Appl. Environ. Microbiol.,* 62, 822, 1996.
28. Wang, C. and Hong, C., Quantitative PCR for *Listeria monocytogenes* with colorimetric detection, *J. Food Prot.*, 62, 35, 1999.
29. Fratamico, P.M., Sackitey, S.K., Wiedmann, M., and Deng, M.Y., Detection of *Escherichia coli* O157:H7 by multiplex PCR, *J. Clin. Microbiol.*, 33, 2188, 1995.
30. Bhaduri, S. and Cottrell, B., Direct detection and isolation of plasmid-bearing virulent serotypes of *Yersinia enterocolitica* from various foods, *Appl. Environ. Microbiol.*, 63, 4952, 1997.
31. Pearson, B. and McKee, R., Rapid identification of *Saccharomyces cerevisiae, Zygosaccharomyces bailii* and *Zygosaccharomyces rouxii*, *Int. J. Food Microbiol.*, 16, 63, 1992.
32. Ibeas, J.I., Lozano, I., Perdigones, F., and Jimenez, J., Detection of *Dekkera-Brettanomyces* strains in sherry by a nested PCR method, *Appl. Environ. Microbiol.,* 62, 998, 1996.

33. Vaitilingom, M., Gendre, F., and Brignon, P., Direct detection of viable bacteria, molds, and yeasts by reverse transcriptase PCR in contaminated milk samples after heat treatment, *Appl. Environ. Microbiol.*, 64, 1157, 1998.
34. Shapira, R., Paster, N., Eyal, O., Menasherov, M., Mett, A., and Salomon, R., Detection of aflatoxigenic molds in grains by PCR, *Appl. Environ. Microbiol.*, 62, 3270, 1996.
35. Lopez-Sabater, E.L., Deng, M.Y., and Cliver, D.O., Magnetic immunoseparation PCR assay (MIPA) for detection of hepatitis A virus (HAV) in American oyster (*Crassostrera virginica*), *Lett. Appl. Microbiol.*, 24, 101, 1997.
36. Reynolds, K.A., Gerba, C.P., and Pepper, I.L., Detection of infectious enteroviruses by an integrated cell culture-PCR procedure, *Appl. Environ. Microbiol.*, 62, 1424, 1996.
37. Deng, M.Y., Day, S.P., and Cliver, D.O., Detection of hepatitis A virus in environmental samples by antigen-capture PCR, *Appl. Environ. Microbiol.*, 60, 1927, 1994.
38. Hafliger, D., Gilgen, M., Luthy, J., and Hubner, P., Semi-nested RT-PCR systems for small round structured viruses and detection of enteric viruses in seafoods, *Int. J. Food Microbiol.*, 37, 27, 1997.
39. Johnson, D.W., Pieniazek, N.J., Griffin, D.W., Misener, L., and Rose, J.B., Development of a PCR protocol for sensitive detection of *Cryptosporidium* oocysts in water samples, *Appl. Environ. Microbiol.*, 61, 3849, 1995.
40. Laberge, I., Ibrahim, A., Barta, J.R., and Griffiths, M.W., Detection of *Cryptosporidium parvum* in raw milk by PCR and oligonucleotide probe hybridization, *Appl. Environ. Microbiol.*, 62, 3259, 1996.
41. Warnekulasuriya, M.R., Johnson, J.D., and Holliman, R.E., Detection of *Toxoplasma gondii* in cured meats, *Int. J. Food Microbiol.*, 45, 211, 1998.
42. Jinneman, K.C., Wetherington, J.H., Hill, W.E., Adams, A.M., Johnson, J.M., Tenge, B.J., Dang, N.L., Manger, R.L., and Wekell, M.M., Template preparation for PCR and RFLP of amplification products for the detection and identification of *Cyclospora* spp. and *Eimeria* spp. oocysts directly from raspberries, *J. Food Prot.*, 61, 1497, 1998.
43. Hill., W.E., The polymerase chain reaction: applications for the detection of foodborne pathogens, *CRC Crit. Rev. Food Sci. Nutr.*, 36(1,2), 123, 1996.
44. Deng, M.Y. and Fratamico, P.M., A multiplex PCR for rapid identification of Shiga-like toxin-producing *Escherichia coli* O157:H7 isolated from foods, *J. Food Prot.*, 59, 570, 1996.
45. Saiki, R.K., Walsh, P.S., Levenson, C.H., and Erlich, H.A., Genetic analysis of amplified DNA with immobilized sequence-specific oligonucleotide probes, *Proc. Natl. Acad. Sci. USA,* 86, 6230, 1989.
46. Gutierrez, R., Garcia, T., Gonzalez, I., Sanz, B., Hernandez, P.E., and Martin, R., Quantitative detection of meat spoilage bacteria by using the polymerase chain reaction (PCR) and an enzyme linked immunosorbent assay (ELISA), *Lett. Appl. Microbiol.*, 26, 372, 1998.
47. Blais, B.W., Transcriptional enhancement of the *Listeria monocytogenes* PCR and simple immunoenzymatic assay of the product using anti-RNA:DNA antibodies, *Appl. Environ. Microbiol.*, 60, 348, 1994.
48. Yu, H., Bruno, J.G., Cheng, T.C., Calomiris, J.J., Goode, M.T., and Gatto-Menking, D.L., A comparative study of PCR product detection and quantitation by electrochemiluminescence and fluorescence, *J. Biolumin. Chemilumin.*, 10, 239, 1995.
49. Hayward-Lester, A., Chilton, B.S., Underhill, P.A., Oefner, P.J., and Doris, P.A., Quantification of specific nucleic acids, regulated RNA processing, and genomic polymorphisms using reversed-phase HPLC, in *Gene Quantification*, Ferré, F., Ed. Birkhäuser, Boston, 1997, p. 45.

50. Bachmair, F., Huber, C., and Daxenbichler, G., Quantitation of gene expression by means of HPLC analysis of RT-PCR products, *Clin. Chem. Acta*, 279(1–2), 25, 1999.
51. Higuchi, R., Dollinger, G., Walsh, P.S., and Griffith, R., Simultaneous amplification and detection of specific DNA sequences, *Bio/Technology*, 10, 413, 1992.
52. Holland, P.M., Abramson, R.D., Watson, R., and Gelfand, D.H., Detection of specific polymerase chain reaction product utilizing the 5′–3′ exonuclease activity of *Thermus aquaticus* DNA polymerase, *Proc. Natl. Acad. Sci. USA*, 88, 7276, 1991.
53. Johnson, J.L., Brooke, C.L., and Fritschel, S.J., Comparison of the BAX method for screening *E. coli* O157:H7 with conventional methods for detection of extremely low levels of *Escherichia coli* O157:H7 in ground beef, *Appl. Environ. Microbiol.*, 64, 4390, 1998.
54. Bailey, J.S., Detection of *Salmonella* cells within 24 to 26 hours in poultry samples with the polymerase chain reaction BAX system, *J. Food Prot.*, 61, 792, 1998.
55. Bennett, A.R., Greenwood, D., Tennant, C., Banks, J.G., and Betts, R.P., Rapid and definitive detection of *Salmonella* in foods by PCR, *Lett. Appl. Microbiol.*, 26, 437, 1998.
56. Bassler, H.A., Flood, S.J.A., Livak, K.J., Marmaro, J., Knorr, R., and Batt, C.A., Use of a fluorogenic probe in a PCR-based assay for the detection of *Listeria monocytogenes*, *Appl. Environ. Microbiol.*, 61, 3724, 1995.
57. Chen, S., Yee, A., Griffiths, M., Larkin, C., Yamashiro, C.T., Behari, R., Paszko-Kolva, C., Rahn, K., and Grandis, S.A., The evaluation of a fluorogenic polymerase chain reaction assay for the detection of *Salmonella* species in food commodities, *Int. J. Food Microbiol.*, 35, 239, 1997.
58. Oberst, R.D., Hays, M.P., Bohra, L.K., Phebus, R.K., Yamashiro, C.T., Paszko-Kolva, C., Flood, S.J.A., Sargeant, J.M., and Gillespie, J.R., PCR-based DNA amplification and presumptive detection of *Escherichia coli* O157:H7 with an internal fluorogenic probe and the 5′ nuclease (TaqMan) assay, *Appl. Environ. Microbiol.*, 64, 3389, 1998.
59. Atmar, R.L., Neill, F.H., Romalde, J.L., LeGuyader, F., Woodley, C.M., Metcalf, T.G., and Estes, M.K., Detection of Norwalk virus and hepatitis A virus in shellfish tissues with the PCR, *Appl. Environ. Microbiol.*, 61, 3014, 1995.
60. Arnal, C., Ferre-Aubineau, V., Migotte, B., Imbert-Marcille, B.M., and Billaudel, S., Quantification of hepatitis A virus in shellfish by competitive reverse transcription-PCR with coextraction of standard RNA, *Appl. Environ. Microbiol.*, 65, 322, 1999.
61. Higuchi, R., Fockler, C., Dollinger, G., and Watson, R., Kinetic PCR analysis: real-time monitoring of DNA amplification reactions, *Bio/Technology*, 11, 1026, 1993.
62. Lantz, O., Bonney, E., Umlauf, S., and Taoufik, Y., Kinetic ELISA-PCR: a versatile quantitative PCR method, in *Gene Quantification*, Ferré, F., Ed., Birkhäuser, Boston, 1997, p. 145.
63. Klein, P.G. and Juneja, V.K., Sensitive detection of viable *Listeria monocytogenes* by reverse transcription-PCR, *Appl. Environ. Microbiol.*, 63, 4441, 1997.
64. Sheridan, G.E.C., Masters, C.I., Shallcross, J.A., and Mackey, B.M., Detection of mRNA by reverse transcription-PCR as an indicator of viability in *Escherichia coli* cells, *Appl. Environ. Microbiol.*, 64, 1313, 1998.
65. Goswami, B.B., Koch, W.H., and Cebula, T.A., Detection of hepatitis A virus in *Mercenaria mercenaria* by coupled reverse transcription and polymerase chain reaction, *Appl. Environ. Microbiol.*, 59, 2765, 1993.
66. Wiedmann, M., Czajka, J., Barany, F., and Batt, C. A., Discrimination of *Listeria monocytogenes* from other *Listeria* species by ligase chain reaction, *Appl. Environ. Microbiol.*, 58, 3443, 1992.

67. Stubbs, S., Hutson, R., James, S., and Collins, M.D., Differentiation of the spoilage yeast *Zygosaccharomyces bailii* from other *Zygosaccharomyces* species using 18S rDNA as target for a non-radioactive ligase detection reaction, *Lett. Appl. Microbiol.*, 19, 268, 1994.
68. Wolcott, M.J., Advances in nucleic acid-based detection methods, *Clin. Microbiol. Rev.*, 5, 370, 1992.
69. Uyttendaele, M., Schukkink, R., VanGemen, B., and Debevere, J., Detection of *Campylobacter jejuni* added to foods by using a combined selective enrichment and nucleic acid sequence-based amplification (NASBA), *Appl. Environ. Microbiol.*, 61, 1341, 1995.
70. Blais, B.W., Turner, G., Sooknanan, R., and Malek, L.T., A nucleic acid sequence-based amplification system for detection of *Listeria monocytogenes hlyA* sequences, *Appl. Environ. Microbiol.*, 63, 310, 1997.
71. Farber, J.M., An introduction to the hows and whys of molecular typing, *J. Food Prot.*, 59, 1091, 1996.
72. Barrett, T.J., Lior, H., Green, J.H., Khakhria, R., Wells, J.G., Bell, B.P., Greene, K.D, Lewis, J., and Griffin, P.M., Laboratory investigation of a multistate foodborne outbreak of *Escherichia coli* O157:H7 by using pulsed field-gel electrophoresis and phage typing, *J. Clin. Microbiol.*, 32, 3013, 1994.
73. Oscar, T.P., Identification and characterization of *Salmonella* isolates by automated ribotyping, *J. Food Prot.*, 61, 519, 1998.
74. Czajka, J. and Batt, C.A., Verification of causal relationships between *Listeria monocytogenes* isolates implicated in foodborne outbreaks of listeriosis by randomly amplified polymorphic DNA patterns, *J. Clin. Microbiol.*, 32, 1280, 1994.
75. Nilsson, J., Svensson, B., Ekelund, K., and Christiansson, A., A RAPD-PCR method for large-scale typing of *Bacillus cereus, Lett. Appl. Microbiol.*, 27, 168, 1998.
76. Cartwright, C.P., Stock, F., Beekmann, S.E., Williams, E.C., and Gill, V.J., PCR amplification of rRNA intergenic spacer regions as a method for epidemiologic typing of *Clostridium difficile, J. Clin. Microbiol.*, 33, 184, 1995.
77. Fields, P.I., Blom, K., Hughes, H.J., Helsel, L.O., Feng, P., and Swaminathan, B., Molecular characterization of the gene encoding H antigen in *Escherichia coli* and development of a PCR-restriction fragment length polymorphism test for identification of *E. coli* O157:H7 and O157:NM, *J. Clin. Microbiol.*, 35, 1066, 1997.
78. Jersek, B., Gilot, P., Gubina, M., Klun, N., Mehle, J., Tcherneva, E., Rijpens, N., and Herman, L., Typing of *Listeria monocytogenes* strains by repetitive element sequence-based PCR, *J. Clin. Microbiol.*, 37, 103, 1999.
79. Miteva, V., Selenska-Pobell, S., and Mitev, V., Random and repetitive primer amplified polymorphic DNA analysis of *Bacillus sphaericus, J. Appl. Microbiol.*, 86, 928, 1999.
80. Restrepo, R., Duque, M., Tohme, J., and Verdier, V., AFLP fingerprinting: an efficient technique for detecting genetic variation of *Xanthomonas axonopodis* pv. *manihotis, Microbiology*, 145, 107, 1999.

9 Application of Automated Ribotyping To Improve Food Safety and Quality

S.J. Fritschel

CONTENTS

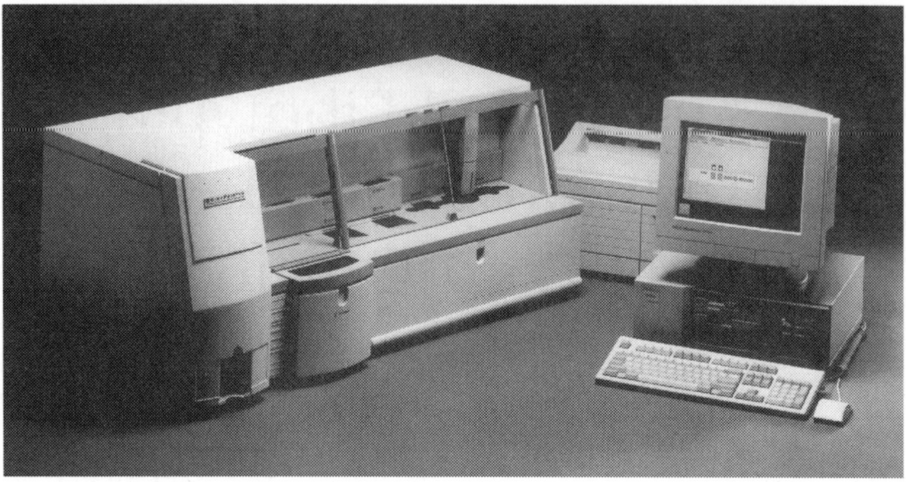

FIGURE 9.1 RiboPrinter® Microbial Characterization System.

I. INTRODUCTION: HOW AUTOMATED RIBOTYPING WORKS

Traditionally, there has been a lack of appropriate tools to help food microbiologists identify and characterize pathogenic and food spoilage organisms. Standard tests are subjective, time-consuming, and operator-dependent processes. The various existing tests lack a shared "language" for accurate, efficient electronic communication, and there is no shared standard for molecular biology.

Genetics-based ribotyping tests provide a universal (organism-independent) method that differentiates beyond the species level, provides exceptional accuracy, and allows for efficient data communication. Ribotyping is based on Southern blot analysis of restriction digests of bacterial DNA using probes to find fragments with information from the ribosomal RNA genes. The benefits of this technology include:

- Highly conserved regions
- Excellent stability
- Excellent discriminatory power

The RiboPrinter® Microbial Characterization System (Figure 9.1) automates well-established laboratory practices for ribotyping. The steps (Figure 9.2) consist of:

1. Cell lysis and restriction enzyme digestion to produce specific DNA fragments
2. Electrophoretic separation by molecular weight
3. Transfer to a membrane
4. Hybridization, probing, and binding of a chemiluminescent substrate to the specific fragments
5. Detection and capture of the fragment images with a charge-coupled device (CCD) camera

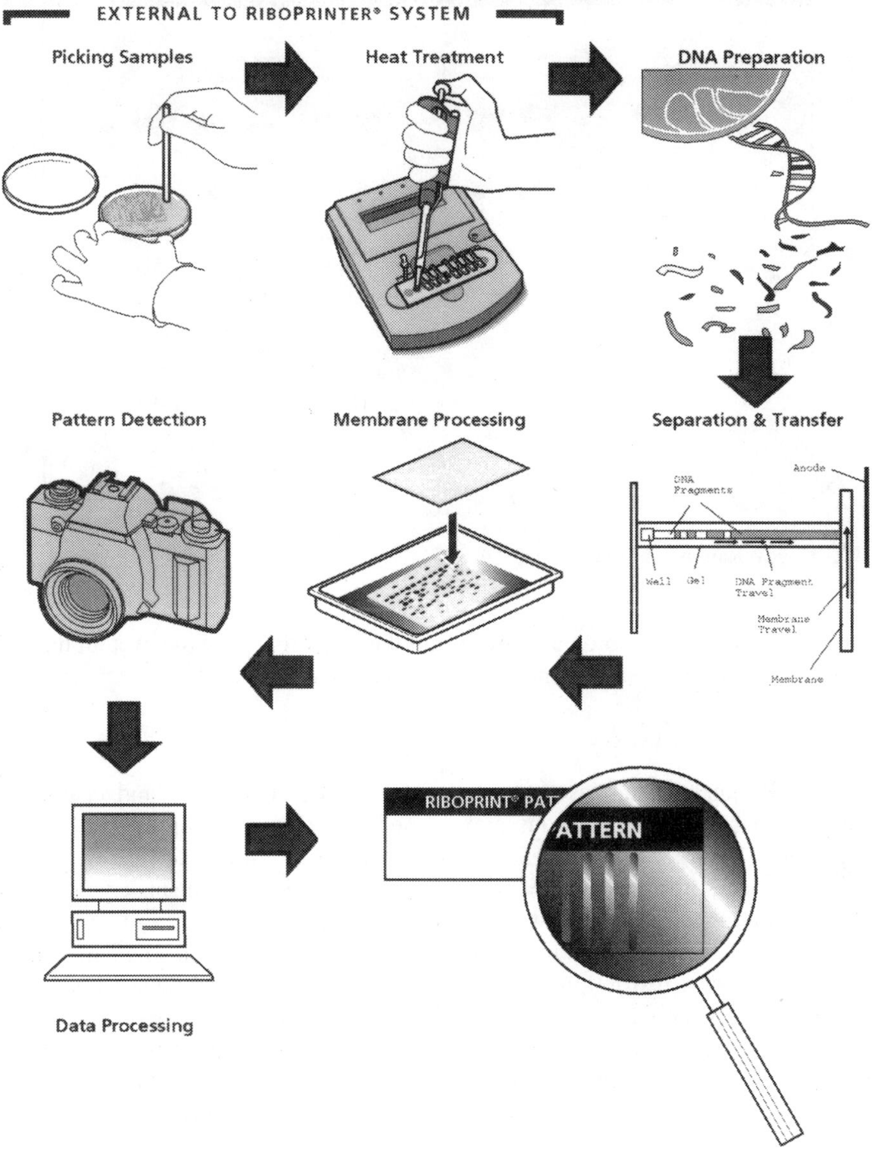

FIGURE 9.2 RiboPrinter® system process.

The system, which can be applied to virtually all bacteria, processes a batch of eight isolates within 8 hours, completely standardizing the technical and interpretive aspects of bacterial analysis. The RiboPrinter® system uses sophisticated computer algorithms to group similar isolates (characterization) and compare them against a database of known isolates (identification), eliminating the need for subjective interpretation of results. This discussion will explain how the system

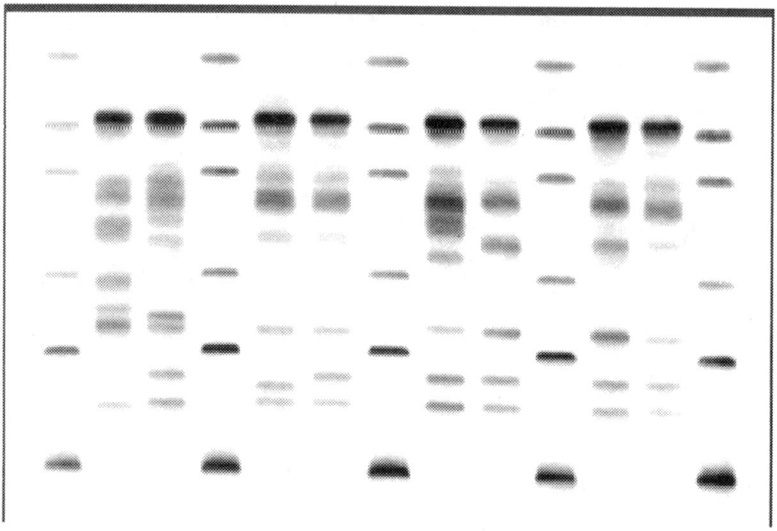

FIGURE 9.3 Raw batch image data.

automates ribotyping and data analysis and then describe how the system has been used to improve food safety and quality throughout the world.

A. Process Automation

The RiboPrinter® system uses genetic "fingerprints" to identify and characterize microorganisms. A simple colony pick is used to collect pure colony samples from agar plates (Figure 9.2). Each sample is suspended in buffer and heat treated in a sample carrier. After the carrier is transferred to the machine, all subsequent actions are fully automated. In the first step within the characterization unit, samples are treated with a lysing agent to release the DNA. The DNA is then digested to completion with a restriction enzyme (*Eco*RI, *Pst*I, *Pvu*II, etc.) The RiboPrinter® system transfers the resulting DNA restriction fragments to an agarose gel cassette containing 13 wells. Samples are placed in eight of the wells; marker DNA of known molecular weights occupies the remaining five wells. Using a process known as direct blot electrophoresis, the DNA fragments are size-separated and transferred to a moving nylon membrane.

After denaturation, each membrane is hybridized with a chemically labeled rRNA operon from *Escherichia coli*. The membrane is washed and treated with blocking buffer and an antisulfonated DNA antibody/alkaline phosphatase conjugate. Unbound conjugate is removed and then a chemiluminescent substrate applied.[1] This step makes each electrophoresis band containing genetic information from the rRNA genes visible to a custom CCD camera in the RiboPrinter® system. The camera detects the light intensity of the targeted bands and converts the patterns from luminescent DNA fragments to digital information. The image data (Figure 9.3) are stored on the system computer's hard drive.

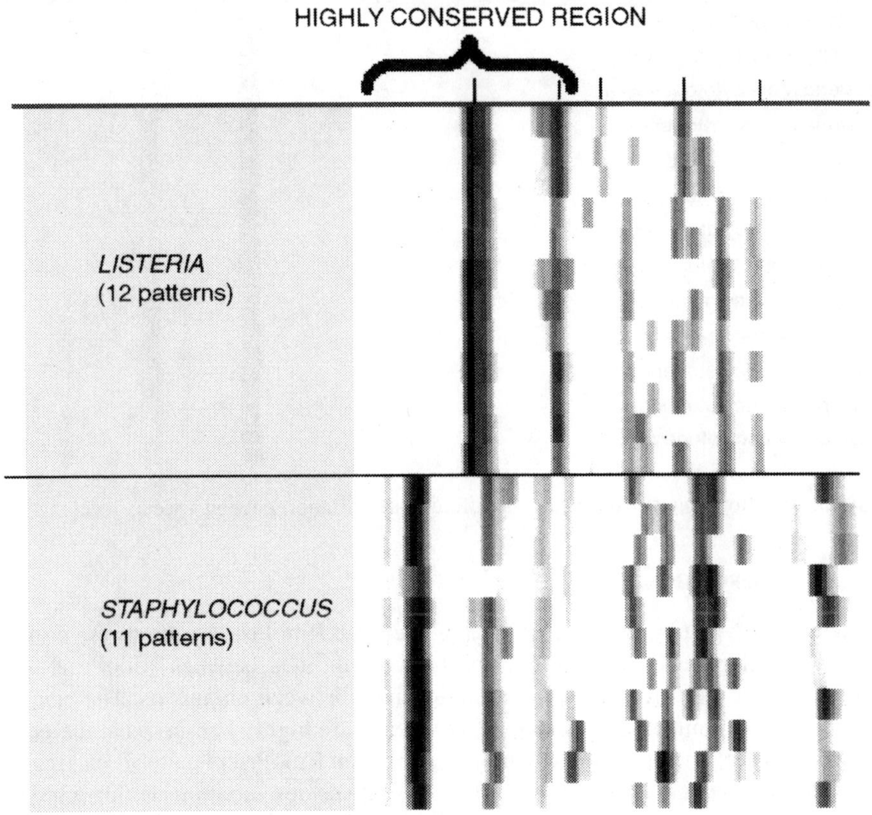

FIGURE 9.4 RiboPrint® patterns for different genera show pronounced differences.

B. Analysis Automation

The RiboPrinter® Microbial Characterization System analysis software uses the position and intensity of the five well-characterized marker fragments, run alongside the samples, to normalize the resulting output data. The positions of these standard marker bands are used to correct for lane-to-lane and membrane-to-membrane variations in band position. Then, the system generates a RiboPrint® pattern (Figure 9.4) for each sample and marker lane using proprietary algorithms. The RiboPrint® pattern for each lane consists of a series of light and dark bands. The system statistically compares the output pattern to patterns obtained previously— either known organisms or simply samples previously processed by the RiboPrinter® system. From a statistical analysis of similarity, the software concludes whether the unknown sample can be grouped with existing sample patterns or whether it must start a new pattern set (characterization). By comparison to a database of known standards, the software also decides whether or not the system can provide a taxonomic name (identification). The system then automatically produces a report that presents characterization and identification information for each sample.

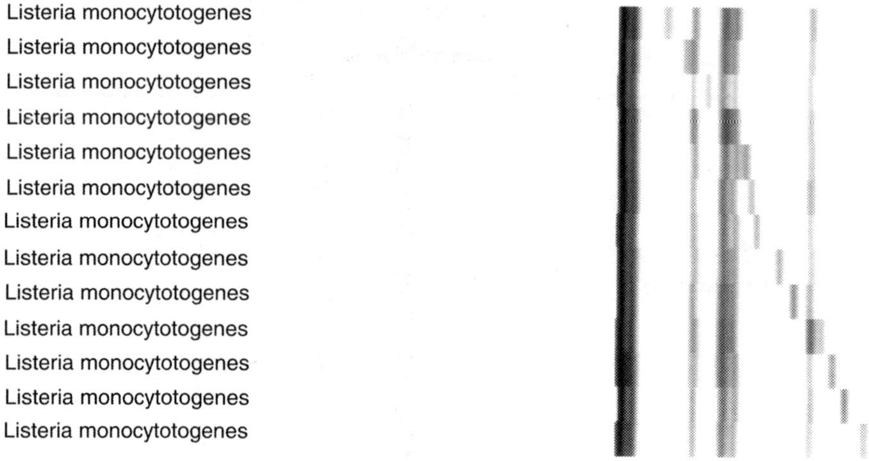

FIGURE 9.5 Non-conserved characters differentiate variation beyond species level.

C. Characterization

The RiboPrint® pattern is extremely stable. Because RiboPrint® patterns are derived from essential genetic information, observed similarities between RiboPrint® patterns are direct indicators of genetic relationships between organisms. The banding patterns of fragments from ribosomal RNA genes are highly conserved at the genus and species level.[2,3] For example, variations between RiboPrint® patterns can readily define different genera (Figure 9.4) or show the subspecies variations within a species (Figure 9.5). Conserved rRNA fragments and their associated polymorphic fragments can be used to characterize samples for which the genus and species are not reliably known. For example, RiboPrint® patterns obtained from unidentified isolates can be analyzed by the system. If the pattern does not match that of an existing group in the database, it is stored as a new RiboGroup. A RiboGroup is a collection of closely related sample patterns that are indistinguishable based on the statistical index of similarity used by the system. Each time a new sample is added to an existing RiboGroup, the system software incorporates the sample's pattern data into a new composite pattern representative of the entire group. Subsequent samples are then compared to the new RiboGroup. If the software finds the new sample pattern indistinguishable from an existing pattern in the database, it assigns the new sample to that group. In effect, this action declares the sample to be equivalent to the species or strain previously characterized, based on the similarity index used to distinguish between RiboGroups. Table 9.1 lists a few examples of the many organisms that have generated reproducible RiboPrint® characterization patterns.

D. Identification

The RiboPrint® Pattern Identification Database, included with the software for each RiboPrinter® system, contains the digitized patterns for a number of species and strains, including both pathogenic and spoilage organisms from 79 genera as well

TABLE 9.1
Some Representative Genera that Produce RiboPrint® Patterns for Characterization Patterns

Brochothrix	*Enterobacter*	*Morganella*	*Shigella*
Carnobacterium	*Erysipelothrix*	*Proteus*	*Stomatococcus*
Citrobacter	*Hafnia*	*Providencia*	*Streptococcus*
Clostridium	*Klebsiella*	*Rhodococcus*	*Yersinia*
Corynebacterium	*Kurthia*	*Serratia*	Others

TABLE 9.2
Representative Genera that Produce RiboPrint® Patterns for Identification Database

Genus	Number of Patterns Available in Database
Bacillus	209
Enterococcus	48
Escherichia	76
Lactobacillus	34
Lactococcus	40
Leuconostoc	10
Listeria	81
Pediococcus	8
Pseudomonas	103
Salmonella	97
Staphylococcus	260
Vibrio	59

as some Gram-positive and Gram-negative surrounds. These identifications are based on the use of the restriction enzyme *Eco*RI. Surrounds are species frequently found in the same environments as the identified genera or are very similar biochemically to those genera. Table 9.2 shows the number of RiboPrint® patterns available in the current identification database. Using proprietary algorithms, the RiboPrinter® Microbial Characterization System compares the distribution and intensity of fragment bands from the sample with all those in the Identification Database. Based on the similarity index between the isolate sample and species in the existing database, the system automatically reports if the sample is a known genus, species, and strain.

E. Bringing Microbiology into the Information Age

RiboPrinter® system technology allows communication of data within an organization, sharing data with others, and application of relevant publications to a business.

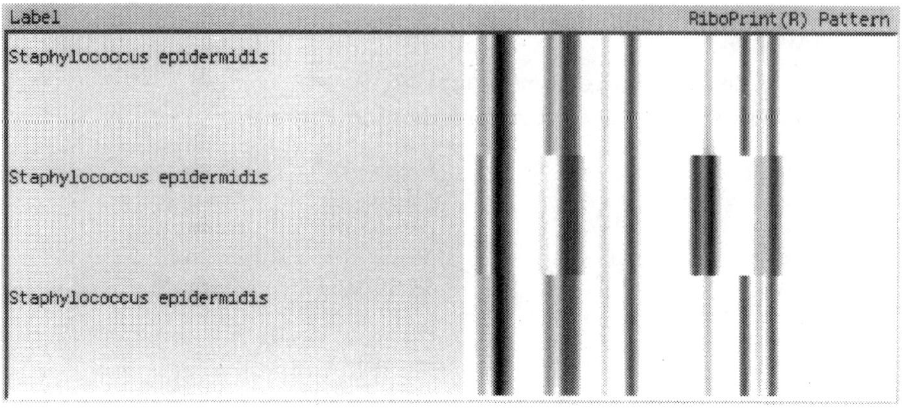

FIGURE 9.6 *Staphylococcus epidermidis* RiboPrint® patterns.

The RiboPrinter® system's Data Sharing system allows linking of a RiboPrinter® system or systems through modems to a central guest workstation for automated data pooling and analysis or to let many guest workstations share data from a single RiboPrinter® system.

II. CASE HISTORIES: USING THE RIBOPRINTER® SYSTEM TO IMPROVE FOOD SAFETY AND QUALITY

A. Tracking Contamination in Food Production

Routine testing of a finished product showed *Staphylococcus* spp. contamination. An investigative study was started using RiboPrint® pattern analysis (Figure 9.6) on isolates from finished product, environment samples, raw materials, personnel, and equipment. As a result of these tests, *Staphylococcus epidermidis* was isolated from the finished product, raw material, and hands of personnel. The RiboPrint® patterns showed that *S. epidermidis* from the raw material and the finished product did not have matching patterns. The RiboPrint® pattern of the finished product isolate matched that of an isolate from the hands of one of the production workers. Based on these findings, appropriate corrective actions were taken:

- Mandatory use of gloves in high-risk, post-processing areas
- Mandatory handwashing when entering a high-risk area from other areas within the plant

B. Using Other Restriction Enzymes with the RiboPrinter® System

The options now available for using optimized enzymes with automated ribotyping significantly enhance the utility of the system to address the major requirements of any typing strategy: accurately grouping discrete species into manageable related units and accurately splitting species into finer categories. The reagents used for

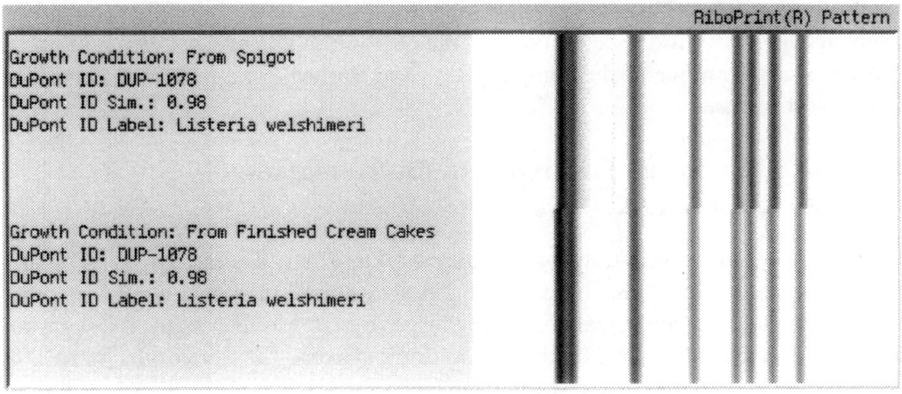

FIGURE 9.7 RiboPrint® patterns of non-pathogenic *Listeria welshimeri.*

sample DNA preparation and digestion by the RiboPrinter® system are potentially compatible with a variety of discrete restriction endonucleases.

Of those enzymes tested, *Pvu*II has been particularly useful for additional characterization of *Salmonella* and *E. coli* O157:H7 isolates. Discrete, individual *Pvu*II ribotypes have been determined for various pathogenic *Salmonella* serovars (e.g., *S.* ser. *Typhimurium* and *S.* ser. *Heidelberg*). *Pvu*II ribotyping with the RiboPrinter® system has also revealed stable *Pvu*II fragment length polymorphisms in other pathogenic *Salmonella* serovars (e.g., *S.* ser. *Enteritidis* and *S.* ser. *Choleraesuis)* and in *E. coli* O157:H7. *Pvu*II has shown excellent utility and cost-effectiveness with *E. coli* O157:H7 and *Salmonella* spp. Other enzymes (such as *Eco*RV, *Pst*I, *Sal*I, and *Nco*I) extend the functionality of automated ribotyping.

The RiboPrinter® Identification Database includes 66 discrete *Eco*RI patterns for *E. coli*, each representative of one or more *E. coli* serovars. *E. coli* strains of known serotype and pathogenic characteristics were ribotyped on the RiboPrinter® system using *Eco*RI, *Pvu*II, *Mlu*I, and *Cla*I. Automated processes with extended restriction digestion times were developed to facilitate cost-effective use of restriction enzymes. While all isolates were identified as *E. coli* on the basis of their *Eco*RI RiboPrint® patterns, RiboGroups (ribotypes) generated with *Mlu*I revealed greater structural diversity of the ribosomal operons. Further characterization with *Cla*I yielded additional discrimination. The results show that automated ribotyping of *E. coli* with *Eco*RI, *Mlu*I, and, as required, *Cla*I has significant value for identifying and discriminating pathogenic isolates of *E. coli.*

C. Controlling *Listeria* Contamination in Ready-To-Eat Foods

Testing indicated that a number of fancy, ready-to-eat cream cakes contained cream that was contaminated with the non-pathogenic organism, *Listeria welshimeri.* The manufacturer feared a potential for *Listeria monocytogenes* growth. Automated ribotyping analysis traced the RiboPrint® pattern (Figure 9.7) to a vat of cream above the production floor. This vat was kept very cold, and condensation that formed on a pipe leading from it dripped onto the cakes below. RiboPrint® patterns confirmed that this was the source of the contamination and allowed for appropriate corrective actions to be taken.

D. What's in a Name? Standardizing Characterization of Waterborne Contaminants

In 1996, a round-robin study was begun by Dr. Clive Thompson of Yorkshire Environmental in the U.K. It involved 27 laboratories in the U.K., 12 in other European countries, and Qualicon. The objective of the trial was to assess reproducibility between labs in identifying coliform bacteria. Study results were also intended to check for misidentifications of aeromonads as coliforms. A report midway through the study clearly showed the failings of methods available up to now for identifying bacteria. In contrast, these same mid-study results also clearly demonstrated the definitiveness of the RiboPrinter® system.

Of the 23 isolates analyzed in the first 6 months of the trial, only three were consistently identified by all 40 participants. Up to nine different genera were reported for a single sample. Also among the 23 isolates, duplicate samples were circulated among 36 laboratories, 58% of which could not tell that the strains were the same. Ten laboratories determined that the samples were of different genera; 10 others obtained the same genus but a different species. With a second pair of duplicate isolates, 42% of the participants incorrectly determined the samples were from different genera, and another 8% mistakenly showed the samples were of the same genus but different species. The RiboPrinter® system correctly showed that the samples had a common pattern (Figure 9.8) for both sets of duplicates.

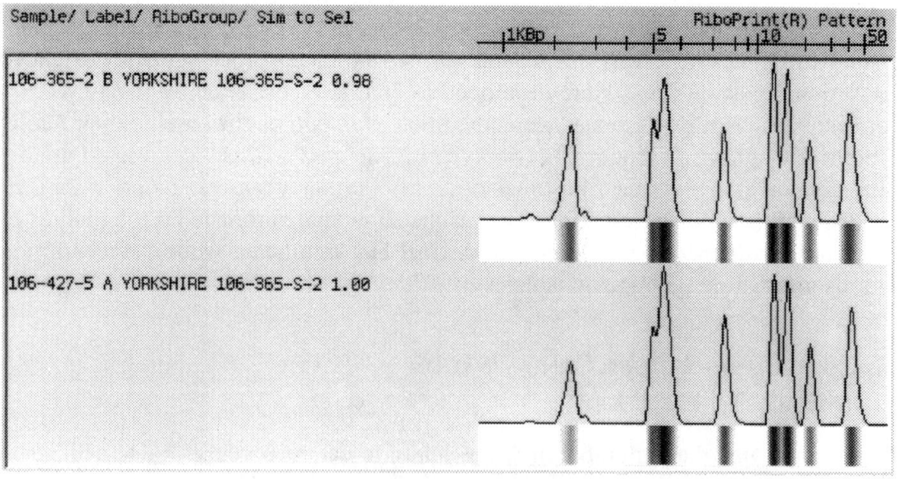

FIGURE 9.8 RiboPrinter® system reveals common pattern of contaminants.

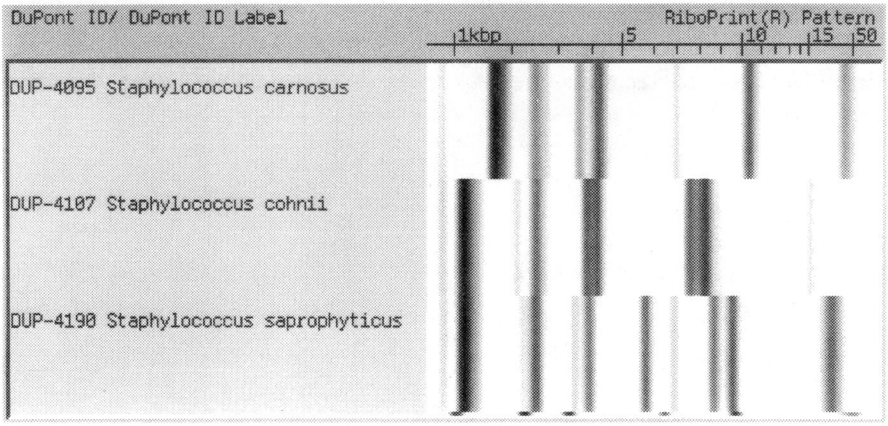

FIGURE 9.9 Detection of *Staphylococcus carnosus* using RiboPrint® patterns.

E. Tracking Waterborne *Pseudomonas* Contamination

The teaching hospital of a large eastern university became alarmed by an increasing number of patients infected by an antibiotic-resistant strain of *Pseudomonas aeruginosa.* Of these infected patients, 85% were found to be infected with identical organisms (Ribotype 1) and had been involved in an identical procedure. Further tests isolated the organism to a sink in one unit. Replacing the faucet eliminated the organism temporarily, but *P. aeruginosa* soon turned up again. Cultures of the interior of a pipe near the faucet were also positive for *Pseudomonas aeruginosa* Ribotype 1. A chlorine wash of the pipe eradicated the organism, and subsequent culturing showed that it did not return.

F. Solving Problems with Sausage Starter Culture — *Staphylococcus carnosus*

The cost benefits of buying starter cultures appealed to a meat products manufacturer. Their desired mix in purchased cultures was *Micrococcus* species, *Lactobacillus,* and *Staphylococcus carnosus*, each in quantities of greater than 10^9 colony-forming units (CFU)/gram. When the sensory quality of the company's raw, fermented goods began declining significantly, the producer made a detailed analysis of each item in the product ingredients. Assessing the culture and its quality required an in-depth profile of each item. Authenticating the culture was difficult using traditional biochemical methods that gave only subjective analysis and could not differentiate strains definitively. Samples of the *Staphylococcus* cultures were turned over to the RiboPrinter® Microbial Characterization System for scrutiny.

The RiboPrint® pattern for *Staphylococcus carnosus* (Figure 9.9) could be linked to samples of *Staphylococcus,* but other laboratory tests showed that the organism was present at only 0.01% of the needed concentration. In addition, other species of *Staphylococci* that had not been requested were present in much higher numbers.

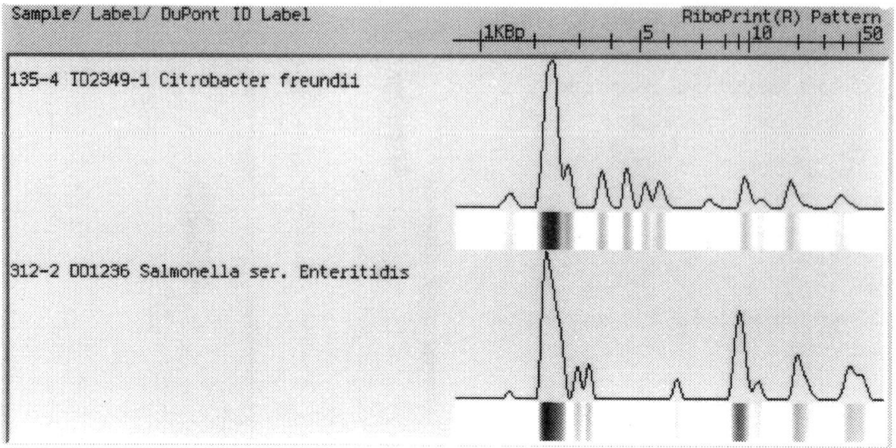

FIGURE 9.10 RiboPrinter® system detects false positives.

Samples of these unwanted organisms were matched to RiboPrint® patterns for *Staphylococcus cohnii* and *Staphylococcus saprophyticus.*

With this information in hand, the meat products manufacturer challenged the culture supplier on the formulation. The supplier admitted running into problems obtaining high yields of the needed *Staphylococcus carnosus*. The low concentrations of *S. carnosus* found in the starter culture confirmed the viability problems of the organism. Additionally, the confirmed presence of *S. cohnii* suggested that the culture was being contaminated by the supplier's personnel. The RiboPrinter® system characterization gave both the manufacturer and the culture supplier the definitive microbial profile they needed to control the quality of their products and the processes by which they were made.

G. Four Popular Tests Show *Salmonella* Contamination; BAX® and RiboPrinter® Systems Prove Them Wrong

The makers of high-value, refrigerated, ready-to-use foods learned at 2:00 p.m. on a Friday that they could have a problem so serious that it would be necessary to recall one of their products from four European countries. ELISA, latex agglutination, biochemical screening (with a 97% certainty), and a serology assay all indicated *Salmonella* contamination. But, use of the BAX® system — a highly sensitive polymerase chain reaction (PCR)-based assay from Qualicon, Inc. — provided an overnight answer of no *Salmonella*. Within 8 hours, RiboPrint® patterns (Figure 9.10) had identified the organism as *Citrobacter freundii*, a close cousin of *Salmonella* that often confounds biochemical assays.

H. RiboPrinter® System Networking Project

In this study, 15 strains of extended-spectrum-beta-lactamase (ESBL)-producing *Klebsiella pneumoniae* and one strain of *E. coli* were ribotyped on an electronically

TABLE 9.3
Sample Distribution of *Klebsiella pneumoniae* Isolates by RiboGroup

RiboGroup	Contains	Mean	Range of Similarity[a]	
NETWORK 4-1184-S-3	KP3/KP14	34 of 34 (100%)	0.95	0.88–0.98
NETWORK 4-1184-S-4	KP4/KP8	34 of 34 (100%)	0.96	0.91–0.99
NETWORK 4-1184-S-5	KP5	18 of 18 (100%)	0.97	0.95–0.99
NETWORK 4-1184-S-6	KP6	18 of 18 (100%)	0.95	0.87–0.98
NETWORK 4-1184-S-7	KP7/KP9	36 of 36 (100%)	0.96	0.92–0.99
NETWORK 4-1185-S-2	KP10	17 of 17 (100%)	0.97	0.93–0.98
NETWORK 4-1185-S-3	KP11	14 of 14 (100%)	0.97	0.94–0.99
NETWORK 4-1185-S-4	KP12/KP15	33 of 33 (100%)	0.96	0.93–0.99
NETWORK 4-1185-S-5	KP13	18 of 18 (100%)	0.94	0.85–0.97
NETWORK 4-1185-S-6	KP16	18 of 18 (100%)	0.97	0.95–0.99

[a] Within group.

linked network of eight RiboPrinter® systems. The isolates formed ten RiboGroup pattern sets that were identical at all sites (Table 9.3). Two laboratories performed pulsed field gel electrophoresis (PFGE) on the entire set of 15 strains. This resulted in 13 to 14 PFGE pattern types, depending on the laboratory. Because PFGE analysis took more than one week, a subset of eight strains was blind-coded and sent to four sites (one in Delaware, one in Iowa, and two in Maryland) to expedite PFGE analysis and throughput. The study concluded that:

- RiboPrint® patterns had sufficient discriminatory power to be useful in tracking this organism.
- Automated ribotyping had a more rapid turnaround and results were more reproducible than with PFGE.
- Pattern analysis was automatic and could be shared electronically by distant laboratories.

I. Epidemiological Tracking of *Campylobacter* in Poultry (Using *Pst*I)

Detailed insight into the sources and routes of *Campylobacter* contamination is important in the campaign to eliminate this human pathogen from poultry. TNO Laboratories in The Netherlands ribotyped large numbers of *Campylobacter* isolates as part of a comprehensive project monitoring broilers by using the RiboPrinter® system (Figure 9.11). Automated ribotyping was performed with *Pst*I as the restriction enzyme, because *Eco*RI does not cut *Campylobacter* DNA adequately. More than 60 *Pst*I RiboGroups were identified in the strains. These RiboGroups were composed of:

- 57 reference strains isolated from geographically "dispersed" origin of poultry flocks
- 800 field isolates generated during the monitoring project

On the basis of the more than 550 RiboPrint® patterns within the *Pst*I RiboGroups, static RiboGroups were created and used to automatically identify and count the various *Campylobacter* isolates (Figure 9.11).

III. CONCLUSION

The RiboPrinter® system offers a fully automatic, technique-independent, standardized method for characterizing and identifying bacteria. The RiboPrinter® system provides:

- Fully automated operation
- 32 samples/day throughput
- Fast, accurate, and reproducible characterization and identification of bacteria
- In-depth microbiological information at the genus, species, and subspecies level
- Standardized molecular typing information

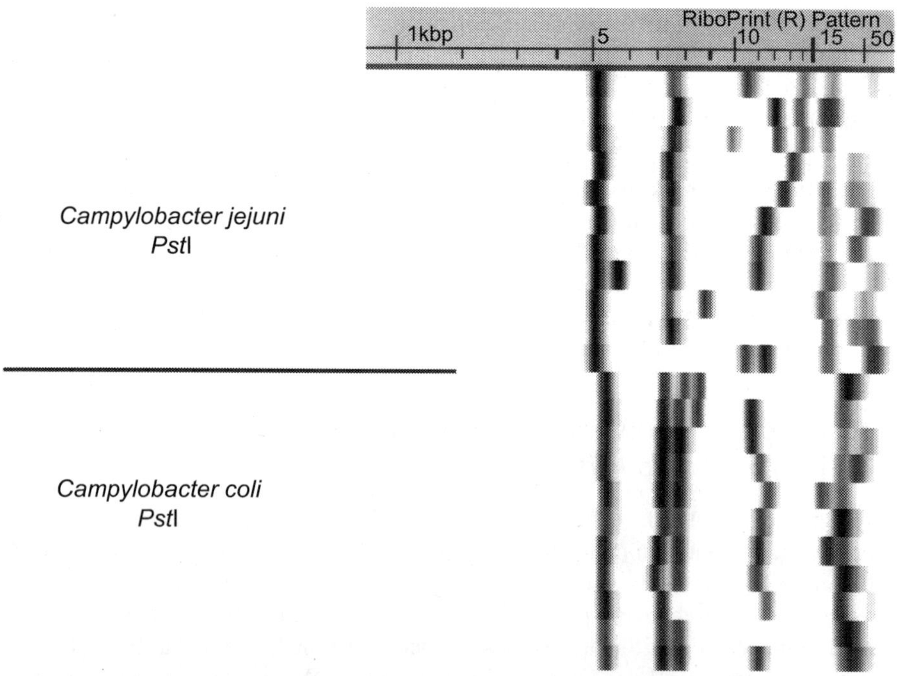

FIGURE 9.11 Tracking *Campylobacteria* in poultry using RiboPrint® patterns.

- Integration of genetic, geographical, and historical data
- Custom database capability
- Flexible restriction enzyme options

The system can address areas of concern such as:

- Tracking sources of contamination
- Examining shelf-life issues
- Proactively understanding the link between specific spoilage organisms and the products/environment
- Monitoring and tracking beneficial organisms
- Linking sites electronically to establish a network of ribotype databases that contain historical and geographical as well as genetic data
- Epidemiology

ACKNOWLEDGMENTS

- TNO
- CCFRA
- Kraft General Foods
- University of Iowa College of Medicine
- Johns Hopkins University
- University of Maryland Medical School
- Baltimore VA MHCS
- Christiana Care Health Services

NOTE

BAX®, RiboPrinter®, and RiboPrint® are registered trademarks of Qualicon, Inc., a subsidiary of E. I. du Pont de Nemours and Company.

REFERENCES

1. Kobos, R.K., Blue, B.A., Robertson, C.W., and Kielhorn, L.A., Enhancement of enzyme-activated 1,2-dioxetane chemiluminescence in membrane-based assays, *Analyt. Biochem.*, 224, 128, 1995.
2. Bruce, J.L., Hubner, R.J., Cole, E.M., McDowell, C.I., and Webster, J.A., Sets of *Eco*RI fragments containing ribosomal RNA sequences are conserved among different strains of *Listeria monocytogenes, Proc. Nat. Acad. Sci.,* 92 (May), 5229, 1995.
3. Hubner, R.J., Cole, E.M., Bruce, J.L., McDowell, C.I., and Webster, J.A., Types of *Listeria monocytogenes* predicted by the positions of *Eco*RI cleavage sites relative to rRNA sequences, *Proc. Nat. Acad. Sci.,* 92 (May), 5234, 1995.

Control of Microbial Food Contaminants

10 A Brief Introduction to the Kosher Laws and Possible Implications for Food Safety

Joe M. Regenstein and Carrie E. Regenstein

CONTENTS

I. INTRODUCTION

The kosher dietary laws determine which foods are "fit or proper" for consumption by Jewish consumers who observe these laws. The laws are Biblical in origin, coming mainly from the original five books of the Holy Scriptures. Over the years, the details have been interpreted and extended by the rabbis to protect the Jewish people from violating any of the fundamental laws and to address new issues and technologies. For example, the rabbis are currently dealing with issues related to biotechnology (see below).

Why do Jews follow the kosher dietary laws? Many explanations have been given. The following by Rabbi Grunfeld is possibly the best written explanation and probably summarizes the most widely held ideas about the subject (Grunfeld, 1972):

> "'And ye shall be men of a holy calling unto Me, and ye shall not eat any meat that is torn in the field' (Exodus XXII:30). Holiness or self-sanctification is a moral term; it is identical with ... moral freedom or moral autonomy. Its aim is the complete self-mastery of man.
>
> "To the superficial observer it seems that men who do not obey the law are freer than law-abiding men, because they can follow their own inclinations. In reality, however, such men are subject to the most cruel bondage; they are slaves of their own instincts, impulses, and desires. The first step towards emancipation from the tyranny of animal inclinations in man is, therefore, a voluntary submission to the moral law. The constraint of law is the beginning of human freedom. ...Thus, the fundamental idea of Jewish ethics, holiness, is inseparably connected with the idea of Law; and the dietary laws occupy a central position in that system of moral discipline which is the basis of all Jewish laws.
>
> "The three strongest natural instincts in man are the impulses of food, sex, and acquisition. Judaism does not aim at the destruction of these impulses, but at their control and indeed their sanctification. It is the law which spiritualizes these instincts and transfigures them into legitimate joys of life."

These laws do affect dietary consumption patterns, but they are not health laws. For a more complete discussion, see Regenstein (1994). The most cited example of the health benefits of kosher dietary laws is probably the issue of trichinosis in the banned meat, pork. Current archeological data suggest that trichina cysts were not observed in "mummified" pork from the period when the Jews were building pyramids in Egypt. Also, the incubation time for trichinosis is over 10 days, so it can be difficult to make the link. (Ask any scientist currently trying to do 3-day dietary recalls!)

II. THE KOSHER MARKET

The kosher market covers almost 40,000 products in the U.S. In dollar value, about $40 billion worth of products have a kosher marking on them. The actual consumers of kosher food (those who specifically look for the kosher mark) are estimated to be about 8 to 10 million Americans, who purchase almost $3 billion worth of kosher product (*Kosher Today,* 1998). Only about one third of the kosher consumers are Jewish; other consumers include Muslims, Seventh Day Adventists, vegetarians, people with various types of allergies (particularly dairy, grain, and legume allergies), and general consumers who value the quality of kosher products: "We report to a higher authority" is the highly successful slogan for Hebrew National products. *AdWeek Magazine* called kosher the "Good Housekeeping Seal of Approval for the '90s." By undertaking kosher certification, food companies can incrementally expand their market by opening up new markets. A recent presentation at the Institute of Food Technologists' annual meeting specifically dealt with improving the interface between kosher supervision and the needs of other groups using kosher foods (J. Regenstein, personal communication).

Although limited market data are available, the most dramatic data about the impact of kosher have been provided by Coors beer when the company made their products kosher. According to their market analysis, their share of the market in Philadelphia went up 18% upon going kosher; somewhat less dramatic increases were observed in other cities in the Northeast. Many other big-name American food manufacturers have gone kosher in recent years, including Nabisco, Entenmanns, Keebler, and Budweiser.

III. THE KOSHER DIETARY LAWS

The kosher dietary laws predominantly deal with three issues, all in the animal kingdom:

1. Allowed animals
2. Prohibition of blood
3. Prohibition of mixing of milk and meat

For the week of Passover (in late March or April), however, restrictions on "chometz," the prohibited grains, and the rabbinical extensions of this prohibition lead to a whole new set of regulations, focused in this case on the plant kingdom. In addition there is a separate set of laws dealing with grape juice, wine, and alcohol derived from grape products. Basically, these must be handled by Sabbath-observing Jews; however, if the juice is pasteurized (heated or "mevushal," in Hebrew), then this juice can be handled as an ordinary kosher ingredient.

A. Allowed Animals and Prohibition of Blood

Ruminants with split hoofs, most domestic birds, and fish with fins and removable scales are generally permitted. Pigs, wild birds, sharks, dogfish, catfish, monkfish, and similar species, along with all crustacean and molluscan shellfish, are prohibited. Insects are also prohibited, so carmine and cochineal (natural red pigments) are not used in kosher products. A few rabbis do permit cochineal/carmine but most mainstream products do not have this color, thus this potential allergen, as we have only recently learned, is not found in mainstream kosher products. By "mainstream" we are referring to the normative standard as reflected by the bulk of the kosher products produced in the U.S. at this time.

Furthermore, ruminants and fowl must be slaughtered according to Jewish law by a specially trained religious slaughterman. These animals are also subsequently inspected by the rabbis for various defects. In the U.S., the desire for stringent meat inspection requirements has led to the development of a more strict inspection requirement for kosher meat, referred to as "glatt kosher." The U.S. Department of Agriculture (USDA) will reject animals that are acceptable for kosher, while the rabbis will reject animals accepted by the USDA. Generally, more animals are rejected by the rabbis.

The meat and poultry must be further prepared by properly removing certain veins, arteries, prohibited fats, blood, and the sciatic nerve. In practical terms, this means that only the front quarter cuts of meat are generally used. Thus, kosher meat is subject to more extensive handling than regular meat, so the possibility of pathogen transmission from humans is increased. On the other hand, the process removes the visible blood, which spoils easily. Furthermore, to remove additional blood, the meat is soaked and salted within a specified time period. Also, any materials that might be derived from animal sources are generally prohibited because of the difficulty of obtaining them from kosher animals. For this reason, many products that might be used in the dairy industry, such as emulsifiers, stabilizers, and surfactants (particularly those that are fat derived), require careful rabbinical supervision to assure that no animal-derived ingredients are used. Almost all such materials are also available in a kosher form derived from plant oils.

Between the time of slaughter and the soaking and slaughtering, the rabbis are concerned about how the meat is treated. During this period, the meat is subject to additional handling to remove the prohibited veins and arteries. In addition, during this period any of the current steam/hot water and/or acid treatments are not permitted, as these would religiously "cook" or "coagulate" the blood, thus making it impossible to perform the soaking/salting process properly. In Jewish law, acids are in a special category and are of more concern than other chemical treatments. Interestingly, one of the other food safety treatments being proposed is the use of trisodium phosphate, a compound that is an anticoagulant. At this time, the rabbis have not yet dealt extensively with whether this might be used or not, although their instinctive reaction is to reject it.

The salting and soaking process has recently been shown to have an impact on meat pathogens. Obviously, one hour of contact with coarse salt will have an impact. The USDA had earlier recognized that poultry might experience a favorable reduction in *Campylobacter.* Current work at Kansas State University (Hajmeer, 1998) suggests that the process may be beneficial in beef, also. Thus, some folks are beginning to include soaking and salting with NaCl as another potential food safety treatment.

Another approach to dealing with the specific problems of kosher meat has to do with on-farm pathogen reduction. Such programs are being developed at various research centers, including Cornell, where the needs of the kosher consumer are being included as justifications for the work. Improvements in early animal health may also benefit the "glatt" kosher meat in particular. A reduction in the number of lung adhesions would permit more meat to fall into this higher quality value-added category.

B. Prohibition of Mixing of Milk and Meat

"Thou shalt not seeth the kid in its mother's milk." This passage appears three times in the Torah (the first five books of the Holy Scriptures) and is thus taken religiously as a very serious admonition. The meat side of the equation has been extended to include poultry. The dairy side includes all milk derivatives. To keep meat and milk separate requires that the processing and handling of all kosher products fall into one of three categories:

1. Meat products
2. Dairy products
3. Pareve (parve) or neutral products

The last category includes all products that are not classified as meat or dairy. All plant products, along with eggs, fish, honey, and lac resin (shellac), are pareve. These pareve foods can be used with either meat products or dairy products, except that fish cannot be mixed directly with meat. Once a pareve product is mixed with either meat or dairy products, they take on the status of meat or dairy, respectively. Some kosher-observant Jews are concerned with the possible adulteration of kosher milk with the milk of other animals (e.g., mare's milk) and as such require that the milk be watched from the time of milking. This "Cholev Yisroel" milk and products derived from the milk are required for all dairy ingredients by some of the more strict kosher supervision agencies.

In order to assure the complete separation of milk and meat, all equipment, utensils, etc. must be of the proper category. Thus, if plant materials (e.g., a fruit juice) are run through a dairy plant, it would become a dairy product religiously. Some kosher supervision agencies do permit such a product to be listed as "dairy equipment" (DE) rather than "dairy." The DE tells the consumer that the product does not contain dairy but was made on dairy equipment (see allergy discussion below). With the DE listing, the consumer can use the product immediately after a meat meal, while a significant wait would be required to use a true dairy product. In either case, the dishes would be switched from meat dishes to dairy dishes before consumption.

Kosher-observant Jews must wait a fixed time between meat and dairy consumption. Customs vary, but generally the wait before consuming dairy after meat is much longer (3 to 6 hours) than the wait from dairy to meat (0 to 1 hour). However, when a hard cheese (defined as a cheese that has been aged for over 6 months) is eaten, the wait is the same as that for meat. Thus, most companies producing cheese for the kosher market age their cheese for less than 6 months. In order for cheese to meet mainstream kosher standards, the coagulating agent (usually the biotechnology-derived chymosin) must be added by a Sabbath-observing Jew. However, in many cases, if the coagulating agent and all other aspects are kosher, then the whey would still be kosher even if the cheese does not meet mainstream standards. Thus, there is much more kosher whey than kosher cheese.

If one wants to make the product truly pareve, the dairy plant can usually be made pareve by the process of equipment kosherization (see below); however, being religiously pareve does not mean the product is free of all dairy allergens. Trace amounts, that the most sensitive people would react to, are not of concern with respect to kosher. Consumers must be aware of this limitation of pareve, but then again most people with dairy allergies will generally be able to tolerate these trace amounts of dairy.

C. Passover

During this holiday, which occurs in the spring, no products made from the five prohibited grains — wheat, rye, oats, barley, and spelt ("chometz") — can be used, except for the specially supervised production of unleavened bread ("matzos") prepared especially for the holiday. Special care is taken to assure that the matzos do not have any time to rise. In addition, products derived from corn, rice, legumes, mustard seed, buckwheat, and some other plants ("kitnyos") are prohibited, thus items such as corn syrup, corn starch, etc. would generally be prohibited. Many rabbis, however, permit the oil from kitnyos materials, and some permit liquid kitnyos products such as corn syrup. The major sources of sweeteners and starches used for Passover production of sweet items are real sugar or potato-derived products (some potato syrup is used). Even though Passover is a time for large family gatherings, some kosher consumers may not use any dairy products at this time because of the need for separate Passover dairy dishes. Overall, 40% of kosher sales for the traditional kosher companies occur during the week of Passover. People with allergies are able to purchase appropriate kosher products during or just before Passover (including some that specifically advertise no grain content), but, again, the extremely sensitive allergenic person must still maintain some caution in purchasing these products.

IV. EQUIPMENT KOSHERING

There are three ways to make equipment kosher and/or to change its status, depending on the prior production history of the equipment. (Note that after a plant, or a line, has been used to produce kosher pareve products, it can be switched to either kosher dairy or kosher meat without a special equipment kosherization step.) The simplest equipment kosherization occurs with equipment that has only been handled cold, as such equipment requires simply a good caustic/soap cleaning. Materials such as ceramics, rubber, earthenware, and porcelain cannot be koshered, so if these materials are found in a processing plant new materials may be required for production, and switching between different status conditions will be difficult.

Most food-processing equipment undergoes a heat treatment, generally above 105 to 110°F, which is defined rabbinically as "cooking;" however, the exact temperature for this cooking depends on the rabbi. To kosher these items, the equipment must be thoroughly cleaned with caustic/soap. The equipment must be left idle for 24 hours, and then the equipment must be flooded with boiling water (defined as between 190°F and 212°F) in the presence of a kosher supervisor. In the case of ovens or other equipment using "fire," kosherization involves heating the metal until it glows. Again, the rabbi will generally be present while this process is taking place. (We have published a fire safety and sanitation guide for the rabbis; see Scott and Regenstein, 1989.) The procedures that must be followed for equipment kosherization can be quite extensive, so the fewer status conversions, the better. Careful formulation of products and good production planning can minimize the inconvenience.

V. JEWISH COOKING

Depending on what is being cooked, it may be necessary for the rabbi to do the cooking. In practical terms, this is often accomplished by having a rabbi light the pilot light, which is then left on continuously. In the case of cheese-making, a similar concept usually requires the rabbi to add the coagulating agent into the vat. One rabbi is currently exploring whether or not this could be done remotely using modern technology. However, if the ingredients used during cheese-making are all kosher, even though a rabbi has not added the coagulant, the whey derived from such cheese (as long as the curds and whey have not been heated above about 110°F before the whey is drained off) would be considered kosher. Thus, there is much more kosher whey available than kosher cheese.

VI. DEALING WITH KOSHER SUPERVISION AGENCIES

Kosher supervision is taken on by a company in order to expand its market opportunities. It is a business investment, which, like any other investment, should be examined critically. In an era of Total Quality Management, Just-in-Time production, Strategic Suppliers, etc., it is appropriate for companies to look carefully at how they handle kosher supervision. Price alone may not be the best criterion for selecting a kosher supervision agency. When looking for a kosher supervision agency, a company must determine its priorities and attempt to find a kosher supervision agency that is compatible with these requirements. Like any purchasing decision, time spent in qualifying the vendor before purchasing is usually rewarded. The agency's name recognition may also not be the most important company consideration. Other important considerations should include: (1) how responsive the agency is to the company, in terms of both paperwork handling and providing rabbis at the plants as needed; (2) how willing the agency is to work with the company to solve problems; (3) how willing the agency is to explain its kosher standards and fee structure; (4) whether the "personal" chemistry is right (would the company feel comfortable working with the agency); and, finally, (5) what the religious standards of the agency are (whether or not they meet the company's needs in the marketplace).

One of the most difficult issues for the food industry to deal with in day-to-day kosher activities is the existence of so many different kosher supervision agencies, most of which have a trademark symbol that will appear on food products under their supervision. How does this impact the food companies? How do kosher consumers perceive these different groups? Because there has not been a central Jewish religious authority for many years, different rabbis follow different traditions with respect to their kosher standards. Some rabbis tend to follow the more lenient standards, while others follow more stringent standards. Given the availability of choices, the trend in the mainstream kosher community today is toward a more stringent standard.

One can generally divide the kosher supervision agencies into three broad categories. First, there are the large organizations that dominate the supervision of larger food companies, such as the OU, the OK, the Star-K, and the Kof-K. All four of them are nationwide and mainstream. Two of these, the OU and the Star-K, are communal organizations; that is, they are part of a larger community religious organization. This provides them with a wide base of support but also means the organizations are potentially subject to other priorities and needs of the organization. On the other hand, the Kof-K and the OK are private companies. Their only function is to provide kosher supervision.

In addition to these national companies, there are smaller private organizations and many local community organizations that provide equivalent religious standards of supervision. As such, products accepted by any of these mainstream organizations will be accepted by all other similar organizations. The local organizations may have a bigger stake in the local community. They may be more accessible and easier to work with. Although often having less technical expertise, they may be backed up by one of the national organizations. For a company marketing nationally, a limitation to using a local agency could be whether or not consumers nationwide would know and recognize that kosher symbol. With the advent of *KASHRUS* magazine, and its yearly review of symbols, this has become somewhat less of a problem. (*KASHRUS* magazine does not try to evaluate the standards of the various kosher supervision agencies, but simply reports their existence.) It is the responsibility of the local congregational rabbi to inform his congregation of his standards; if he does not know enough about the "far-away" organization, he may be uncomfortable recommending it.

The second category of kosher supervision includes individual rabbis, generally associated with the "Hassidic" communities. These are often affiliated with the ultra-orthodox communities of Williamsburg and Borough Park in Brooklyn and Monsey, NY, and Lakewood, NJ. There are special food brands that cater to their needs. Many more products used in these communities require continuous rabbinical supervision rather than the occasional supervision used by the mainstream organizations. The symbols of the kosher supervisory agencies representing these consumers are not as widely recognized as those of the major mainstream agencies in the kosher world beyond these communities. These rabbis will often perform special supervisions of products in a facility that is normally under mainstream kosher supervision, often without any changes other than continuous rather than occasional supervision, but sometimes with additional special needs reflecting their customs.

The third level are individual rabbis who are more "lenient" than the mainstream standard. Many of these rabbis are Orthodox; some may be Conservative. Their standards are based on their interpretation of the kosher laws. The more lenient such a rabbi, the more the food processor cuts out the mainstream and more strict consumer markets, but that is a retail marketing decision the company must make for itself.

Ingredient companies should try to use a mainstream kosher supervision agency. Selling ingredients to most kosher food-producing companies will require such supervision. The ability to sell to as many customers as possible requires a broadly acceptable standard. Unless an ingredient is acceptable to the mainstream, it is almost impossible to gain any benefit of having a kosher ingredient. In a few circumstances, if the company makes a product that would not be acceptable to the mainstream kosher supervision agencies, no matter what the company does, then the company might as well use one of the more lenient kosher supervision agencies willing to recognize that ingredient.

With respect to interchangeability between kosher supervision agencies, a system of certification letters is used to provide information to others from the certifying rabbi about the products he has approved. The supervising rabbi certifies that a particular plant produces kosher products or that only products with certain labels or certain codes are kosher under his supervision. Such letters should be renewed every year and should be dated with both a starting and ending date. These letters are the mainstay of how companies establish the kosher status of ingredients as ingredients move in commerce, and consumers may also ask to see such letters. Obviously, a kosher supervision agency will only accept letters from agencies they consider satisfactory, although they may also accept letters for specific products under the supervision of another rabbi with which they are comfortable.

In addition, the kosher symbol of the certifying agency or rabbi may appear on the packaging. (In some industrial situations, where kosher and non-kosher products are similar, some sort of color coding of products may also be used.) Most of these symbols are "trademarks" that are duly registered; however, in a few cases, the trademark is not registered and more than one rabbi has been known to use the same kosher symbol.

With respect to kosher markings on products, three issues should be highlighted:

1. It is the responsibility of the food company to show its labels to its kosher certifying agency prior to printing labels to ensure the label is marked correctly. This responsibility includes both the agency symbol and the documentation establishing its kosher status (e.g., dairy or pareve for most food plant items). Many agencies currently do not require that "pareve" be marked on products; others do not use the "dairy" marking. The kosher supervision agencies, the food companies, and the consumer would be better served if all kosher products had the status marked. In addition to providing the proper information, it would challenge everyone to pay more attention to marking products properly, thus avoiding the many recalls/announcements of mismarked products. This would also make the markings more trustworthy for other users, particularly those buying kosher for allergenic reasons.
2. Labels for private-label products with specific agency symbols on them cannot be moved easily between plants. This is why some companies, both private-label and others, use the generic "K." Thus, if the kosher supervision agency changes, the label can still be used. The sophisticated kosher consumer, however, is more and more uncomfortable with this symbol and questions will be asked. By paying for a respected symbol and then only using the "K," a company dilutes the value of its investment in kosher certification. In particular, if a company uses the "K," the customer service and sales departments and those people representing the company at trade shows need to know the identity of the certifying rabbi. In New Jersey, there is an active program challenging the supermarkets

each year to justify the kosher symbols, particularly the "K," on their store brands.

3. For many years, the only major company that did not mark their kosher products as kosher was the Hershey Corporation. In the last few years, Hershey's has been adding the kosher marks to its products — a significant improvement for its kosher consumers, including children.

VII. GELATIN

An important component of many food products, gelatin is probably the most controversial of all modern kosher ingredients. Gelatin can be derived from pork skin, beef bones, or beef skin. In recent years, some fish gelatins have also appeared. We are currently involved in research in this area. As a food ingredient, fish gelatin has many similarities to beef and pork gelatin, such as having a similar range of bloom strengths and viscosities; however, depending on the species from which the fish skins were obtained, the melting point can vary over a much wider range of melting points than does beef or pork gelatin. This attribute may offer some unique opportunities to the food industry, especially for margarine, water gels, icings, candy, ice cream, and yogurt. These gelatins would be fully kosher and acceptable to almost all of the mainstream kosher supervision organizations.

The only other currently available mainstream kosher gelatin is a very expensive beef-hide gelatin, used mainly in niche markets but also in Girl Scout cookies so the cookies retain their mainstream kosher status. Other currently available gelatins — even if called "kosher" — are not acceptable to the mainstream kosher supervision organizations.

Among the lenient kosher supervision agencies, one finds a wide range of attitudes toward gelatin. The most liberal view holds that gelatin, being made from bones and skin, is not being made from a food (flesh). Further, during the process used to make the product, it goes through a stage where it is so "unfit" that it is not edible by man or dog and as such becomes a new entity. Rabbis holding this view even accept pork gelatin. Most gelatin desserts with a generic "K" follow this ruling, which makes the product unacceptable to other groups, particularly Muslims. Other rabbis only permit gelatin from beef bones and hides and not pork. Other rabbis will only accept "India dry bones" as a source of beef gelatin. These bones, found naturally in India (because of the Hindu custom of not using cattle) are aged for over a year and are "dry as wood;" additional religious laws exist for permitting these materials. However, to repeat, none of these products is accepted by the "mainstream" kosher supervisions, thus products with these gelatins are not accepted by a significant part of the kosher community.

VIII. BIOTECHNOLOGY

Rabbis currently accept products resulting from simple genetic engineering; for example, chymosin (rennin) was accepted by the rabbis about a half year before it was accepted by the Food and Drug Administration! The production conditions in the fermentors must still be kosher; that is, the fermentor and any subsequent processing must use kosher equipment and ingredients of the appropriate status. A product produced in a dairy medium would be dairy. We believe that the rabbis may soon approve porcine lipase produced through biotechnology, if all the other conditions are kosher. (Any product produced by cattle by excretion in the milk would be dairy.) The rabbis have not yet determined the status of more complex genetic manipulations.

IX. FEDERAL AND STATE REGULATIONS

Making a claim of kosher on a product is a legal claim; 21CFR101.29 includes a paragraph indicating that such a claim must be appropriate, and approximately 20 states, some counties, and some cities have laws specifically regulating the claim of being kosher. Many of these laws refer to "Orthodox Hebrew Practice" or some variant of this term, and their legality in the 1990s became subject to further court interpretation. (Note the recent court rulings in New Jersey and Baltimore, MD, where the courts have ruled against parts of the kosher enforcement law.)

New York State probably has the most extensive set of kosher laws, including a requirement to register kosher products with the Kosher Enforcement Bureau of the Department of Agriculture and Markets (55 Hanson Place, Brooklyn, NY 11217). The laws in New Jersey, however, having been written after the state's original laws were in part declared unconstitutional by the state supreme court, probably have the clearest focus and, it is hoped, no constitutional issues. They focus specifically on "consumer right to know issues" and "truth in labeling." They avoid having the state of New Jersey define kosher. Rather, the rabbis providing supervision declare the information that consumers need to know to make an informed decision. We hope that a similar approach will be adopted by other states, particularly New York State.

X. KOSHER AND ALLERGIES

Although it is helpful for many consumers to use the kosher markings as a guideline for determining whether products might meet their special needs, there are also limitations that the particularly sensitive consumer needs to be aware of. With respect to all kosher products, two important limitations should be recognized:

1. A process of equipment kosherization is used to convert equipment from one status to another. This is a well-defined religious procedure but may not lead to 100% removal of materials previously run on the equipment.
2. Religious law does permit certain *ex post facto* (after the fact) errors to be negated, thus trace amounts (less than 1/60 by volume under very specific conditions) can be nullified. Many kosher supervision agencies, in deference to companies' desires to minimize negative publicity, do not announce when they have used this procedure to make a product acceptable, although many of the agencies are more aware of the allergy implications of their decisions.

Products that one might expect to be made in a dairy plant (pareve substitutes for dairy products and some other liquids such as teas and fruit juices) may be produced in plants that have been kosherized but may not meet a very critical allergy standard. A product that can be problematic is chocolate; many plants make both milk chocolate and pareve chocolate. Getting every last trace of dairy out of the pareve chocolate can be difficult. These plants cannot use water for equipment kosherization, so special treatments involving oils that are liquid at room temperature have been developed.

When the use of dairy and meat equipment is indicated on a label, the product was produced on a dairy or meat line without any equipment kosherization; however, there are no intentionally added dairy or meat ingredients. The product is considered pareve with some use restrictions in a kosher home. In a few instances where pareve or dairy products contain small amounts of fish (e.g., anchovies in Worcestershire sauce), this ingredient *may* be marked as part of the kosher supervision symbol, but many certifications will not specifically mark this.

For Passover, there is some dispute about derivatives of both chometz and kitnyos materials, with a few rabbis permitting items such as corn syrup, soybean oil, peanut oil, and similarly derived materials. In general, the "proteinaeous" part of these materials is not used, thus people with allergies to these items could purchase these special Passover products from supervision agencies that do *not* permit "kitnyos" derivatives. With respect to equipment kosherization, supervising rabbis tend to be very strict about the clean-up of the prohibited grains (wheat, rye, oats, barley, and spelt) but may not be as critical with respect to the extended kitnyos prohibition. Consumers should not assume that kosher markings ensure the absence of trace amounts of the ingredient to which they are allergic.

XI. TESTING FOR DAIRY PRODUCTS

The presence of dairy ingredients in nominally pareve products has become a concern for people, often children, who are very sensitive to even trace amounts of the allergen. Pareve products are routinely showing up with just enough of the allergen (mainly dairy allergens) to cause a reaction in these people — sometimes a life-threatening reaction. If you suspect that someone has been affected by a mislabeled product, do not destroy the product but send a sample of it to the company and to the kashrus agency, requesting that they test the sample. If no product remains, at least save the packaging, which has the lot number. In many cases, a particular lot may be all that is affected.

The senior author has visited with Dr. Stephen Taylor of the University of Nebraska, a food scientist specializing in allergies whose research deals with the development of a series of very sensitive immunological assays for various dairy components. He is trying to determine whether there really is a need for them and, if so, how frequently they would be required for kosher purposes by the food industry, the rabbinical supervision agencies (possibly as a relatively routine supplemental monitoring system), and by the medical profession.

REFERENCES AND ADDITIONAL READINGS

Chaudry, M.M. and Regenstein, J.M., 1994. Implications of biotechnology and genetic engineering for kosher and halal foods, *Trends Food Sci. Technol.,* 5:165–168.

Grunfeld, I., 1972. *The Jewish Dietary Laws,* The Soncino Press, London, pp. 11–12.

Hajmeer, M.N., Marsden, J.L., Crozier-Dodson, B.A., and Higgins, J.J., 1998. Microbial reduction due to the koshering process, *IFT Annu. Meet. Tech. Program Abstr.,* 17–18:23.

Kosher Today, 1998. New study reveals: 10 million Americans buy kosher, Sept., p. 1.

Regenstein, J.M., 1994. Health aspects of kosher foods, *Activities Report and Minutes of Work Groups and Sub-Work Groups of the R & D Associates*, 46(1):77–83.

Regenstein, J.M. and Regenstein, C.E., 1979. An introduction to the kosher (dietary) laws for food scientists and food processors, *Food Technol.,* 33(1):89–99.

Regenstein, J.M. and Regenstein, C.E., 1988. The kosher dietary laws and their implementation in the food industry, *Food Technol.,* 42(6):86, 88–94.

Regenstein, J.M. and Scott, D.L., 1989. *Sanitation and Fire Safety for Mashgiachs,* Food Science Report, Cornell University, Ithaca, NY, 2 pp.

11 Control of Foodborne Pathogens and Spoilage Microorganisms by Naturally Occurring Antimicrobials

Larry R. Beuchat

CONTENTS

I. ABSTRACT

Naturally occurring chemicals in foods of animal and plant origin often play an important role in controlling the growth of spoilage and pathogenic microorganisms. Numerous antimicrobials act as defense mechanisms against microbial invasion of animal and plant tissues or foods processed from them. Antimicrobials in animal tissues include cationic proteins, lytic enzymes such as lysozyme, and hydrolases such as lipases and proteases. Plant barks, stems, leaves, flowers, and fruits contain a wide range of phenolic compounds with various levels of antimicrobial activities. Among the most potent are eugenol from clove, cinnamic aldehyde (cinnamon), allicin (garlic, onion), and allyl isothiocyanate (mustard). Naturally occurring organic acids (e.g., citric and malic acids) also contribute substantially to protecting plant tissues and foods of plant origin against growth of pathogens. Phytoalexins are low molecular weight compounds produced in plant tissue in response to microbial infection or naturally occurring elicitors and have broad-spectrum antimicrobial activity. Their role in controlling human pathogenic bacteria is not fully understood. Clearly, naturally occurring antimicrobials are abundant in foods. Consumer demand for minimally processed, safe foods not preserved by synthetic chemicals is evident on a global scale. Opportunities for increased commercial application of natural antimicrobials exist but must be preceded by further research to assure the efficacy of single and multifactor preservative systems that include natural antimicrobials, without compromising sensory quality of foods.

II. INTRODUCTION

Prevention of human illnesses caused by foodborne pathogens is a major goal of public health agencies, food manufacturers, and food service facilities. The importance of food safety in many countries is evidenced by recent federal and local regulations based on the principles of hazard analysis and critical control point (HACCP) systems. Although industrially synthesized food antimicrobials are perceived by some consumers to have toxicological problems, their safety is ensured by regulatory authorities. Nevertheless, synthetic antimicrobial compounds may be considered less desirable by a segment of the consuming public than are naturally occurring antimicrobials. This has resulted in widespread interest in the food industry to exploit natural antimicrobials for the purpose of controlling the growth of spoilage microorganisms as well as microorganisms known to produce toxins or cause human infections. There are numerous natural compounds in plants and animals that are known to prevent the growth of these microorganisms. For ease of presentation here, these antimicrobials have been placed in two groups, those of plant origin and those of animal origin. For more extensive coverage of naturally occurring antimicrobials, the reader is referred to reviews by Davidson and Branen (1993), Dillon and Board (1994), Walker (1994), and Sofos et al. (1998).

III. ANTIMICROBIALS OF PLANT ORIGIN

Naturally occurring antimicrobial compounds are present in plant leaves, stems, barks, roots, flowers, and fruits. Information on the antimicrobial activity of plant substances, particularly spices and herbs, has been available for centuries (Beuchat, 1994). In many instances, however, concentrations of compounds in spices and herbs necessary for inhibiting microorganisms exceed those resulting from normal usage in foods (Shelef, 1984). Nevertheless, naturally occurring substances in plants do indeed play a role in controlling the growth of foodborne spoilage and pathogenic microorganisms in foods. The following is a brief review of naturally occurring plant substances that have antimicrobial activity. For a more extensive review, the reader is referred to a recent publication authored by Sofos et al. (1998).

A. *Allium* Species

Garlic (*Allium sativum*), onion (*A. cepa*), and leek (*A. porrum*) are probably the most widely consumed foods that have substantial antimicrobial activity. Garlic has been used for medicinal purposes for centuries, but it was not until the 1940s that scientific evidence confirmed that garlic does indeed possess antimicrobial and medicinal properties. Cavallito and Bailey (1944) and Cavallito et al. (1945) were the first to isolate the major antimicrobial component from garlic bulbs. They identified allicin, a diallyl thiosulfinate (2-propenyl-2propenethiol sulfinate), as an antimicrobial and described it as an extremely pungent colorless oil responsible for the principal odor and taste of garlic and onion. Allicin was reported to be bactericidal, at concentrations of 1:85,000 in laboratory broth, to a wide range of Gram-negative and Gram-positive organisms.

Tissues of garlic and other *Allium* species do not contain allicin while in an intact state but do contain the precursor, alliin (S-allyl-L-cysteine-S-oxide). Alliin undergoes hydrolysis to yield allicin, pyruvate, and ammonia by the action of the phosphopyridoxal enzyme, allinase, when tissue of the bulb is disrupted (Stoll and Seebeck, 1949). Garlic extracts inhibit alkaline phosphatase, invertase, urease, and papain activities (Wills, 1956). Allicin inhibits sulfhydral enzymes but very few non-sulfhydral enzymes. Sulfhydral enzymes can be protected by cysteine or glutathione against inhibition by allicin, but only partial recovery of enzyme activity is obtainable after the allicin and enzyme have been in contact. Inhibition of sulfhydryl enzymes is thought to be associated with the presence of the –SO–S– grouping and not –SO–, SS, or –S– groups; Wills (1956) demonstrated that most of the sulfhydryl enzymes were inhibited by 0.0005-*M* allicin. These included succinic dehydrogenase, urease, papain, xanthine oxidase, choline esterase, hexokinase, choline oxidase, glyoxylase, triose phosphate dehydrogenase, and alcohol dehydrogenase. Other sulfhydryl enzymes (carboxylase, adenosine triphosphatase, and ß-amylase) were unaffected by 0.0005-*M* allicin. Among the non-sulfhydryl enzymes inhibited were dehydrogenase, alkaline phosphatase, and tyrosinase.

Work on the antimicrobial activity of garlic and other *Allium* species has been done using foodborne pathogenic bacteria, mycotoxigenic molds, and spoilage microorganisms in general. Because sulfhydryl enzymes are common to all of these microorganisms, the spectrum of activity of *Allium* extracts is broad. Early studies revealed that allicin, at a dilution of 1:125,000, prevented growth of *Staphylococcus aureus* and *Bacillus* species (Cavallito and Bailey, 1944). Mantis et al. (1978) studied the effect of garlic extract on *S. aureus* in culture media. They reported that a 5% garlic extract solution had a lethal effect, whereas concentrations of garlic extract equal to or greater than 2% were inhibitory and concentrations less than 1% were not inhibitory. Garlic extract was generally more inhibitory at 37°C and pH 7.4 than at 28°C and pH 6.0. These researchers concluded that sausage containing greater than 1% garlic was not a favorable environment for growth of *S. aureus*, especially after the pH is reduced as a result of fermentation. Tynecka and Gos (1973) observed that an aqueous dilution (1:256) of garlic juice was inhibitory to *S. aureus*, while a lethal effect was exhibited by a 1:32 dilution. Garlic extract prevented growth of *Streptococcus* at a much lower concentration (1:125,000 dilution) in a laboratory medium (Cavallito and Bailey, 1944).

At concentrations of 3, 5, and 10%, aqueous extracts form fresh garlic bulbs inhibit growth of *Bacillus cereus* on nutrient agar plates by 31.3, 58.2, and 100%, respectively (Saleem and Al-Delaimy, 1982). Extracts from garlic bulbs stored at –18°C were slightly more inhibitory than extracts from bulbs stored at 15 to 35°C for 6 months. Gamma irradiation of bulbs at a dose of 0.47 kGy, with subsequent freezing before extraction, decreased activity up to 50%. Exposing extracts to heat treatments at 80 to 90°C for 5 minutes completely destroyed activity.

DeWitt et al. (1979) investigated the effect of garlic and onion oils on toxin production by *Clostridium botulinum* type A in meat slurry. A concentration of 1500 μg/g inhibited toxin production at 20°C. Inhibition was incomplete, and toxin production by *C. botulinum* type B and type E was not inhibited. It was concluded that garlic and onion oils should not be applied in the meat industry for the purpose of inhibiting toxin production by clostridia because of their apparent ineffectiveness against non-type A *C. botulinum.*

Growth of *Lactobacillus plantarum* in laboratory broth containing higher than 1% extract was reported by Karaioannoglou et al. (1977) to be inhibited. Growth at 30°C and pH 6.6 was less inhibitory than at 37°C and pH 7.4. It was concluded that garlic extract concentrations greater than 1% are inhibitory to *L. plantarum* while concentrations less than 1%, under favorable conditions of pH and temperature, are less inhibitory and may permit growth if large populations (>10^6 cells/mL) are initially present.

A large number of Gram-negative bacteria are also inhibited by extracts from *Allium* species. Sirvastava et al. (1982) compared fresh garlic extract to ampicillin for activity against 21 strains of Gram-negative bacteria, including *Citrobacter*, *Enterobacter*, *Escherichia*, *Klebsiella*, *Proteus*, *Pseudomonas*, *Salmonella*, *Serratia*, and *Shigella*. Extract was more effective than 30 μg/mL ampicillin for inhibiting all test bacteria except *Salmonella arizonae* and *Shigella cloacae*. The inhibitory effect of boiled extract was similar to that of the unheated extract; however, the antibacterial properties of garlic extract diminished during storage.

The bactericidal activity of freshly reconstituted, dehydrated onion and garlic at concentrations of 1 and 5% (w/v), respectively, against *Salmonella typhimurium* and *Escherichia coli*, was reported by Johnson and Vaughn (1969). Maximal death rates occurred with concentrations of 5 and 10%. At the higher concentrations, decimal reduction times at 37°C were 1.1 and 1.2 hr, respectively, for resting *S. typhimurium* cells and 1.8 and 2.1 hr, respectively, for growing cells. Of the major volatile aliphatic disulfide compounds in onions, n-propyl allyl and di-n-propyl, at a concentration of 0.1%, showed a comparable activity against resting cells. Growing cells of *E. coli* were more susceptible than those of *S. typhimurium* to garlic, but apparently more resistant to onion. Tynecka and Gos (1973) reported that *E. coli* was inhibited by a 1:28 dilution and killed by a 1:64 dilution of garlic juice.

Juices and solvent extracts of garlic, onion, and *Allium kurrat* were reported by Abdou et al. (1992) to prevent growth of *E. coli*, *Pseudomonas pyocyaneus*, and *S.*

typhimurium. Inhibition by crude juices appeared to be greater than inhibition by ether, chloroform, or ethanol extracts. It was concluded that both garlic and onion are strong antiseptics and could be used for medicinal applications.

Yeasts and molds are also susceptible to the inhibitory action of *Allium* extracts. Tynecka and Gos (1973) reported that garlic juice was very potent against *Candida albicans* and speculated that fungal skin diseases could be effectively treated with garlic extract. The yeast was inhibited by a 1:512 dilution of juice and killed by juice diluted 1:256. Barone and Tansey (1977) attributed inhibition of 39 of 41 strains of *C. albicans* by aqueous extracts of garlic to destruction of thiols such as L-cysteine and glutathione. They hypothesized that additional lethal or inhibitory effects of allicin are due to its ability to inactivate proteins by oxidation of essential thiols to the disulfide and to competitively inhibit the activity of these sulfhydryl compounds by combining them. Extract heated at 121°C for 10 minutes retained no anti-candidal activity. Activity was unaffected at pH 2.0 to 6.0 but reduced at pH 9.0 and eliminated at pH 12.0.

Moore and Atkins (1977) reported that a variety of yeast-like fungi representing the genera *Cryptococcus*, *Rhodotorula*, *Torulopsis*, and *Trichosporon* were inhibited *in vitro* by aqueous extract of garlic at concentrations diluted as much as 1:1024. In addition, they observed that 22 actively pathogenic isolates of *C. albicans* were inhibited.

Conner and Beuchat (1984a) investigated the inhibitory effects of 32 essential oils from plants on 13 food spoilage yeasts. Garlic oil was a potent inhibitor of growth at concentrations as low as 25 μg/mL. Onion oil was also strongly inhibitory. Resuscitation of heat-stressed yeasts representing several food-spoilage yeasts (*Candida lipolytica*, *Debaryomyces hansenii*, *Hansenula anomala*, *Kloeckera apiculata*, *Lodderomyces elongisporus*, *Rhodotorula rubra*, *Saccharomyces cerevisiae*, and *Torulopsis glabrata*) was less inhibited by garlic oil than by onion oil (Conner and Beuchat, 1984b, 1985). Essential oil and oleoresin of garlic adversely affected ethanol production by *S. cerevisiae* (Conner et al., 1984), and garlic oil delayed sporulation of *H. anomala* and *L. elongisporus*.

Aflatoxin-producing molds, namely *Aspergillus flavus* and *Aspergillus parasiticus*, are effectively inhibited by various extracts of onion, including ether extracts, lachrymatory factor (thiopropanal-S-oxide), and steam-distilled onion oil (Sharma et al., 1979). Ethyl acetate was ineffective. Exposure of onions to gamma irradiation with a sprout-inhibiting dose (6 krad) did not alter the inhibitory potency of the onion extracts which, however, appeared to be heat labile. In another study, Sharma et al. (1981a) reported that germinated conidia were more susceptible to inhibition by onion extracts than were ungerminated conidia. The lethal effect of extract against conidia of *A. parasiticus* was lost by heating, freeze-drying, dehydration, and aeration. The effect of garlic on mycelial growth and aflatoxin production by *A. flavus* was studied by Mabrouk and El-Shayeb (1981). Minced garlic was more effective than garlic extract in inhibiting aflatoxin formation. Heat treatment at 121°C reduced the potency of minced garlic. Sharma et al. (1981b) demonstrated that sporicidal activity of onion extracts against *A. parasiticus* could be retained by freeze-drying.

Extracts of garlic and onion inhibit the growth of many other molds, some of which are known to spoil grains, legumes, and processed foods (Appleton and

Tansey, 1975; Coley-Smith and King, 1969; Tansey and Appleton, 1975). Even when growing in soil, garlic and onion can influence the microflora in the rhizosphere. Timonin and Thexton (1950), for example, observed that the rhizosphere soil in proximity to garlic and onion harbored 11 and 12 times, respectively, more bacteria and 6 and 13 times more actinomycetes than control soil. The microflora profile in soil adhering to garlic and onion bulbs and roots as they were taken from the field for processing may therefore be substantially different than profiles in soil adhering to the bulbs or roots of other plants.

Fliermans (1973) described the inhibition of *Histoplasma capsulatum* by extracts of the garlic plant. The mycelial phase of the organism was inhibited by both volatile and water-soluble components of garlic at concentrations of 254 ng/mL. The problem of *H. capsulatum* infectivity is most significant in soils that have been exposed to bird excreta, a situation common in soils adjacent to bird-roosting sites. Although they did not extrapolate laboratory results to a field situation, application of the active component of garlic to infected soils may cause inhibition of the organism. Inhibition of the growth of *H. capsulatum* has also been reported by Tansey and Appleton (1975).

B. Spices and Herbs

Many spices and herbs also exhibit antimicrobial activity (Table 11.1). Compounds present in spices and herbs that are largely responsible for antimicrobial activity include many simple and complex derivatives of phenol which are volatile at room temperature. Concentrations of these compounds necessary for inhibiting growth or various metabolic activities in microorganisms often exceed those normally used in foods. Nevertheless, the preservative effects of these seasoning agents should not be discounted.

Among the spices having highest antimicrobial activity are cinnamon, clove, and allspice. The antimicrobial principle is often present in the essential oil of these spices. Cinnamic aldehyde (3-phenyl-2-propenal) has been shown to be the major antimicrobial compound in cinnamon. In addition to exhibiting antibacterial activity (Deans and Richie, 1987), cinnamic aldehyde also inhibits mold growth and mycotoxin production. Hitokoto et al. (1987) reported that cinnamon bark had a strong inhibitory effect on fungi, including *A. parasiticus*. Bullerman (1974) reported that a 1 to 2% concentration of ground cinnamon in broth would allow some growth of *A. parasiticus* but reduced aflatoxin production by 99%. In a later study, Bullerman et al. (1977) demonstrated that the essential oil of cinnamon at a concentration of 200 μg/mL was inhibitory to growth and subsequent toxin production by *A. parasiticus* and that cinnamic aldehyde (150 μg/mL) was inhibitory. Other saturated aldehydes (citral and citronellol), an unsaturated alcohol (geraniol), and a terpene alcohol (menthol) exhibit various degrees of antimycotic activity (Moleyar and Narasimham, 1986). Geraniol and citronellol also inhibit growth of *Erwinia* (Scortichini and Rossi, 1991), a bacterium often responsible for spoilage of raw fruits and vegetables.

Eugenol (2-methoxy-4-[2-propenyl]phenol), a major constituent in clove oil and present in considerable amounts in the essential oil of allspice, possesses antimicrobial activity. Karapinar and Aktug (1987) reported that eugenol was more effective

TABLE 11.1
Plants Used as Spices and Herbs Containing Compounds Possessing Antimicrobial Activity

Achiote	Citronella	Licorice	Peppermint
Allspice (pimenta)	Clove	Lime	Pimento
Almond (bitter)	Cocoa	Mace	Rosemary
Angelica	Coffee	Mandarin	Sage
Anise	Coriander	Marjoram	Sassafras
Basil (sweet)	Dill	Musky bugle	Savory
Bay (laurel)	Elecampane	Mustard	Spearmint
Bergamot	Fennel	Nutmeg	Star anise
Calmus	Fenugreek	Onion	Tarragon (estragon)
Cananga	Garlic	Orange	Tea
Caraway	Ginger	Oregano	Thyme
Cardamom	Horseradish	Paprika	Turmeric
Celery	Leek	Parsley	Vanilla
Chenopodium	Lemon	Pennyroyal	Verbena
Cinnamon	Lemongrass	Pepper	Wintergreen

Source: Adapted from Boonchird and Flegel (1982), Conner and Beuchat (1984a), Deans and Richie (1987), Maruzzella and Liguori (1958), and Huhtanen (1980).

against *S. typhimurium*, *S. aureus*, and *Vibrio parahaeomolyticus* than was thymol, anethol, or menthol. Clove powder (1200 μg/mL) and eugenol (200 μg/mL) have been shown to adversely affect the rate of germination of *Bacillus subtilis* spores (Al-Khayat and Blank, 1985; Blank et al., 1987). Spores of the same bacterium had increased heat sensitivity when exposed to clove powder (64 to 98% relative humidity, 25 to 35°C) prior to treatment (Blank et al., 1988).

Clove oil at 250 μg/mL has been shown to inhibit growth and toxin production of *A. parasiticus* (Bullerman et al., 1977); eugenol was inhibitory at a concentration of 125 μg/mL. Hitokoto et al. (1980) observed ground cloves to completely inhibit the growth of toxigenic aspergilli. The essential oils and oleoresins of cinnamon, clove, and allspice are known to inhibit the growth of several food-spoilage and fermentation yeasts (Conner and Beuchat, 1984a). Resuscitation of heat-stressed yeast cells was impaired in media containing as little as 25 μg of essential oil of cinnamon per mL (Conner and Beuchat, 1984b, 1985).

Vanillin (4-hydroxy-3-methoxybenzaldehyde), a major constituent in vanilla beans, is structurally similar to eugenol and is also antimycotic. The compound has been shown to inhibit or retard the growth of yeasts (Boonchird and Flegel, 1982) and molds (Maruzzella and Liguori, 1958).

Thymol (5-methyl-2-[1-methylethyl] phenol), present in the essential oils of thyme, oregano, savory, sage, and several other herbs, is among the compounds having a wide spectrum of antimicrobial activity. Beuchat (1976) reported that essential oils of thyme and oregano are inhibitory to *V. parahaemolyticus*. Of 50 plant essential oils examined by Deans and Richie (1987), thyme oil was the most inhibitory against 25 genera of bacteria. Alcoholic extracts of these spices, as well as those of rosemary and turmeric, were shown to inhibit germination, growth, and toxin production by *C. botulinum* when used at a concentration of 500 μg/mL (Huhtanen, 1980). The presence of 500 μg/mL ethanolic extract of thyme inhibits the growth of *S. aureus* (Aktug and Karapinar, 1986); growth of *V. parahaemolyticus* was inhibited by 1000, 5000, and 6000 μg/mL of powdered thyme, bay leaves, and mint, respectively. Sage is inhibitory to *V. parahaeomolyticus* (Shelef et al., 1980), *B. cereus*, *S. aureus,* and *S. typhimurium* (Shelef et al., 1984).

Thymol has been shown to inhibit growth and toxin production by mycotoxigenic molds (Akgul and Kivanc, 1988; Benjilali et al., 1984; Ray and Bullerman, 1982). Thymol concentrations of ≥500 μg/mL were shown to completely inhibit growth of *A. parasiticus*, while lower concentrations caused partial or transitory growth and toxin production patterns (Buchanan and Shepherd, 1981). Hitokoto et al. (1980) reported that a 2% concentration of oregano in potato dextrose agar completely inhibited the growth of seven mycotoxigenic molds. Maruzzella and Liguori (1958) reported that the volatile oils (essential oils) from origanum (oregano), savory, and thyme possessed substantial antifungal activities against 18 pathogenic and nonpathogenic fungi when tested *in vitro* using a standard zone-of-inhibition test.

Thyme and oregano can have a stimulatory effect on lactic acid production by *L. plantarum* and *Pediococcus cerevisiae* (Zaika and Kissinger, 1981; Zaika et al., 1983). Manganese was identified as a factor in these and other spices responsible for the enhancement of acid production by meat starter bacteria (Zaika and Kissinger, 1984). Farbood et al. (1976) reported that rosemary spice extractive at 1000 μg/mL substantially inhibited the growth of *S. typhimurium* and *S. aureus*. A concentration of 3000 μg/mL of sage or rosemary in culture media inhibited the growth of 20 foodborne Gram-positive bacteria, whereas a concentration of 5000 μg/mL was considered bactericidal (Shelef et al., 1980).

Bhavanti Shankar and Screenivasa Murthy (1979) investigated the effect of turmeric on the growth of intestinal and pathogenic bacteria. They reported that the oil fraction of turmeric was inhibitory toward numerous bacteria, including *B. cereus*, *S. aureus*, *E. coli*, and *L. plantarum*. Changes in membrane permeability and interference with enzyme function may be involved in the lethal mode of action of phenolic compounds in spices and herbs. Energy depletion in yeasts caused by allyl hydroxycinnamates (Baranowski et al., 1980) and essential oils of allspice, clove, cinnamon, oregano, thyme, and savory (Conner et al., 1984) has been reported. Ethanol production, respiratory activity, and sporulation of yeasts can also be influenced by essential oils of spices and herbs.

Many other plant parts and extracts used as spices and herbs are known to possess antimicrobial activity. Foodborne pathogenic bacteria adversely affected by a wide range of compounds present in these seasoning agents include *C. botulinum*, *B. cereus*, *E. coli*, *L. monocytogenes*, *S. typhimurium*, *S. aureus*, and *V. parahaemolyticus*. Growth of the mycotoxigenic molds, such as *A. flavus*, *A. parasiticus*, *Aspergillus versicolor*, *Aspergillus ochraceus*, *Penicillium urticae,* and *Penicillium roquefortii* (Azzouz and Bullerman, 1982), as well as food-spoilage molds, yeasts, and bacteria, is also retarded or inhibited in the presence of many commonly used spices and herbs.

C. Plant Pigments

The antimicrobial properties of several compounds responsible for the color of plant tissues have been demonstrated. Anthocyanin pigments are present in almost all higher plants, being predominant in flowers and fruits. These pigments consist of an aglycone (anthocyanidin) portion esterified with one or more sugars. All anthocyanidins are derivatives of flavylium cation and have various degrees of hydroxylation and methyoxylation. Pelargonidin, cyanidin, delphinidin, plonidin, petunidin, and malvidin are among the most important anthocyanins in terms of contributing to the sensory quality of foods. Although anthocyanin pigments are better known for their food-coloring capabilities, they are also inhibitory to some bacteria. Hartman (1959) reported that pelargonidin 3-monoglucoside and its degradation products inhibit the growth of *E. coli* and *S. aureus*, and Powers et al. (1960) observed that certain anthocyanins were inhibitory to *E. coli*, *S. aureus*, and *Lactobacillus casei*. Similar observations were made by Zimmerman (1959) with *Lactobacillus acidophilus*. Pratt et al. (1960) showed that monoglucosides of cyanidin, perlargonidin, and delphinidin not only increased that lag phase of bacteria, but also decreased the maximum growth attained. Marwan and Nagel (1986) observed that flavanols and proanthocyanidins in cranberries are highly inhibitory to yeasts.

Researchers have studied the mechanism of the antimicrobial activity of anthocyanins. Carpenter et al. (1967) reported that anthocyanins had an inhibitory effect on certain bacterial enzymes. The chelating ability of anthocyanins described by Somaatmadja et al. (1964) may partially explain their inhibitory action on bacterial enzymes. The unavailability of these metals could render these enzymes inactive, thus resulting in growth inhibition, as the activity of many enzymes is dependent on metal ions. The role of anthocyanins as chelators was substantiated by Somaatmadja et al. (1964), who reported that the addition of magnesium and calcium ions reversed bacterial inhibition caused by malvidin 3-monoglucoside.

All higher plants contain chlorophylls *a* and *b*. Upon degradation, chorophyllides, phylophytins, pheaphorbides, and pyrrole compounds are formed. Beuchat et al. (1966) reported that chlorophyllide *a* inhibited growth of *E. coli*, *Pseudomonas fluorescens,* and *B. subtilis*.

D. Other Phenolic Compounds

There are at least six major phenolic components in ethyl acetate extracts of green olives (Fleming et al., 1969), among them a phenolic glycoside (oleuropein) and the aglycone of oleuropein. The hydrolysis products of oleuropein, which contains glucose, ß-3,4-dehydroxyphenylethyl alcohol, and an acid, are ß-3,4-dihydroxyphenylethyl elenolic acid and the oleuropein aglycons. Oleuropein may not be antimicrobial but the aglycone and elenolic acid inhibit the growth of lactic acid bacteria (Fleming et al., 1973). Inhibition of *Lactobacillus*, *Leuconostoc mesenteroides*, and fungi, including *Geotrichum candidum*, *Rhizopus* sp., and *Rhizoctonia solani*, has been reported (Juven and Henis, 1970). The presence of these compounds is thought to be responsible for the occasional abnormal fermentation patterns of green olives. Degradation products of oleuropein are surface active, causing leakage of cytoplasmic constituents such as glutamate, potassium, and inorganic phosphate from bacterial cells (Juven et al., 1972). Aflatoxin production by *A. parasiticus* and *A. flavus* is greatly reduced in the presence of oleuropein (Gourama and Bullerman, 1987) and ethanolic extracts of olive callus tissues (Paster et al., 1988). Reduction in the amount of aflatoxin produced is not correlated with reduction in mycelium formation.

Phenolic compounds such as caffeic, chlorogenic, p-coumaric, ferulic, and quinic acids are present in plant parts used as spices (Beuchat and Golden, 1989). Depending upon the botanical species, the antimicrobial activity of these and other hydroxycinnamic and cinnamic acids has been shown to retard microbial invasion and delay rotting of fruits and vegetables. Gram-positive and Gram-negative bacteria, molds, and yeasts commonly encountered as food-spoilage organisms are sensitive to hydroxycinnamic acid derivatives (Davidson and Branen, 1981). Caffeic, ferulic, and p-coumaric acids, for example, inhibit *E. coli*, *S. aureus*, and *B. cereus* (Herald and Davison, 1983). Baranowski et al. (1989) reported that ferulic acid and, to a lesser extent, p-coumaric acid inhibited growth of *S. cerevisiae*, suggesting that naturally occurring hydroxycinnamic acids may interfere with the fermentation of fruit juices by this yeast. Paster et al. (1988) observed that caffeic acid and o-coumaric acid inhibited aflatoxin production by *A. flavus* and coumaric inhibited growth.

Other phenolic compounds have been demonstrated to exhibit antimicrobial activity. Tannins and tannic acid are present in bark, rinds, and other structural tissues of plants and are known to possess antimicrobial activity. Beuchat and Heaton (1975) attributed the toxic effect of pecan packing tissue on *Salmonella senftenberg* to a high concentration of tannins. Tannic acid is inhibitory to *L. monocytogenes*, *E. coli*, *S. enteritidis*, *S. aureus*, *A. hydrophila*, and *S. faecalis* (Chung and Murdock, 1991). Singleton and Esau (1969) predicted that the antimicrobial effect of red and white wines should be proportional to flavonoid tannin content.

Extracts of blueberries, crabapples, strawberries, red wines, grape juice, apple juice, and tea were studied for their antiviral activity by Konowalchuk and Speirs (1976a,b; 1978a,b). Some of these extracts inactivated poliovirus, coxsackievirus, ECHO virus, reovirus, and herpes simplex virus. The primary inhibitors were thought to be tannins. Tannic acid was antiviral against ECHO virus, poliovirus, and herpes simplex virus. Cliver and Kostenbader (1979) reported that the antiviral activity of grape juice was reversible.

E. Organic Acids

A major factor influencing the survival and growth of microorganisms is the acidity of the environment. Foodborne bacteria capable of causing illness will not grow at a pH less than 3.9 to 4.0, so any food containing sufficient acid to result in lower pH values is essentially protected against bacterial pathogens. Fruits, which contain citric, malic, tartaric, and other acids, generally fall in this range. Benzoic acid occurs in cranberries, prunes, plums, apples, and strawberries (Chipley, 1993) and is synthesized commercially for use as a preservative in foods (Sofos, 1994; Sofos and Busta, 1992). Its antimicrobial activity is greater against yeasts and molds than against bacteria; however, benzoic acid does inhibit growth of bacterial pathogens, including *V. parahaemolyticus*, *S. aureus*, *B. cereus*, and *L. monocytogenes* (Chipley, 1993). Sorbic acid also occurs naturally in fruits (e.g., rowanberries) and is used extensively to control microbial growth in foods (Sofos, 1989; Sofos and Busta, 1981). Acetic, lactic, and propionic acids are produced in various fermented foods, often playing a major role in preventing the growth of pathogenic bacteria.

Hop (*Hummulus lupulus*) vine flowers are used to impart desirable bitter flavor to beer. Resins — α-acids, represented by humulone and its congeners (cohumulone, adhumulone, prehumulone, and posthumulone), and ß-acids, represented by lupulone and its congeners (colupulone, adlupulone, prelupulone, and postlupulone) — are the major compounds responsible for this flavor. Most of these compounds are also known to inhibit microbial growth. Gram-positive bacteria and some fungi are most sensitive (Mizobuchi and Sato, 1985; Schmalreck et al., 1975). *Lactobacillus* species that may contaminate pitching yeasts are known to develop resistance to humulone. Richards and Macrae (1964) reported that several strains of lactobacilli acquired an 8- to 20-fold increase in resistance to 100 μg/mL humulone after two to four subcultures. Undissociated molecules are mainly responsible for inhibition of *Lactobacillus brevis* (Simpson and Smith, 1992). Hops may contribute very little to the microbial stability of beer, as resistant populations retain a degree of stability upon subculture in the absence of humulone.

Germination of conidia and the rate of colony development by *Aspergillus niger, Aspergillus glaucus,* and a *Penicillium* species are adversely affected by 1.5 to 1.5% hop extract in a laboratory medium (Engelson et al., 1980). The effects are more pronounced as the a_w (water activity) of the medium is reduced by adding glycerol. These researchers suggested that the combined effect of reduced a_w and hop extract may be used to impart biological stability to intermediate moisture foods.

F. Coffee, Tea, Kola, and Cocoa

Several commonly consumed beverages are based on plant products containing antimicrobial compounds. Caffeine (1,3,7-trimethylxanthine) is present in coffee and cocoa beans, tea, and kola nuts and is antimycotic as well as antibacterial. Inhibition of growth of several mycotoxigenic *Aspergillus* and *Penicillium* species has been documented (Buchanan et al., 1981, 1983a; Lenovich, 1981). Caffeine adversely affects the production of aflatoxin, ochratoxin A, sterigmatocystin, citrinin, and patulin. The mechanism by which caffeine inhibits polyketide mycotoxin synthesis and other metabolic activities of fungal cells has not been clearly defined. Tortora et al. (1982) reported that caffeine, but not theophylline or papaverine, uncoupled the regulation of glycolysis and glucogenesis in *S. cerevisiae*. Buchanan et al. (1983b) reported that inhibition of growth of *A. parasiticus* by caffeine is due in part to an alteration in purine metabolism, but inhibition of aflatoxin synthesis apparently does not involve an inhibition of cyclic AMP phosphodiesterase or a chelation of key metal ions. Evidence does suggest that caffeine may restrict glucose uptake by *A. parasiticus*, resulting in decreased aflatoxin production (Buchanan and Lewis, 1984). In a preliminary examination of the effect of caffeine on patulin production by *Penicillium urticae*, Buchanan et al. (1983a) suggested that activity does not involve a generalized inhibition of lipid synthesis.

The effect of caffeine on lactobacilli, the predominant group of microflora isolated from a commercial instant-coffee processing facility, was studied by Vanos and Bindschedler (1985). *L. plantarum* was the dominant species. Total inhibition of growth was obtained at 15 mg/mL caffeine and at 60% total non-decaffeinated coffee solids. *Staphylococcus aureus*, *Salmonella*, *E. coli*, *Streptococcus faecalis*, and *B. cereus* failed to grow in 2% (w/v) reconstituted decaffeinated and non-decaffeinated coffee. Pearson and Marth (1990c) investigated the effect of caffeine on *L. monocytogenes*. At a concentration of 0.5% in skim milk, growth at 30°C occurred, although the lag phase was extended to 6 to 9 hr compared to less than 3 hr in milk not containing caffeine.

Theophylline (1,3-dimethylxainthine) and theobromine (3,7-dimethylxanthine) are present in tea and cocoa. Theophylline is present in manufactured tea at a concentration range of 0.23 to 0.44 mg/100 g, whereas the theobromine content is about 50 mg/100 g in manufactured tea (Lunder, 1979). Theobromine is the principal alkaloid in cocoa beans (1.5 to 3%) and is also present in kola nuts. Pearson and Marth (1990b) observed that the addition of cocoa to milk enhanced the growth of *L. monocytogenes*. In another study (Pearson and Marth, 1990a), the pathogen was observed to have a longer lag phase in milk containing cocoa but reached a higher population than in milk without cocoa. The neutralization effect of casein on the anti-listerial activity of cocoa (Pearson and Marth, 1990a) is in agreement with observations reported by Zapatka et al. (1977) on *S. typhimurium*. They reported that the addition of 3 to 5% casein to 5% cocoa in distilled water resulted in 44 to 57% survival of the organism compared to no survival in the absence of casein. Buchanan and Fletcher (1978) reported that neither theophylline nor theobromine appear to have much effect on aflatoxin production.

Vanos et al. (1987) observed that the growth of *S. typhimurium*, *E. coli*, *S. aureus*, and *B. cereus* is adversely affected in a 2% (total solids) instant tea infusion. Inhibition of *L. plantarum* was observed in a 10% tea infusion. It was concluded that a flavanol (catechen) plus caffeine complex was the only natural inhibitor of lactobacilli in instant tea. The linden flower, used to prepare tea in the near East, has been reported to inhibit the growth of *S. aureus* and, to a lesser extent, *S. typhimurium* and *V. parahaemolyticus* (Gonul and Karapinar, 1987).

G. Phytoalexins

Phytoalexins are low molecular weight compounds produced by higher plants in response to microbial infection and naturally occurring elicitors (Dixon et al., 1983). Production of phytoalexins by plant cell cultures was reviewed by Whitehead and Threlfall (1992) and Walker (1994). The mechanism of action of phytoalexins is not fully understood, although evidence suggests that they alter properties of plasma membranes. Phytoalexins are known to occur in leaves, fruits, seeds, roots, and tubers of a wide range of plants. Glyceollin, coumestrol, and glycinol, all phytoalexins produced by soybeans, inhibit microbial membrane-associated processes. The major antibacterial phytoalexin produced in soybeans inoculated with *Erwinia carotovora* and *S. cerevisiae* is glycinol, which has been hypothesized to inhibit the growth of a wide range of other bacteria and fungi. Exposure of *E. coli* membrane vesicles to glycinol results in inhibition of respiration-linked transport (Weinstein and Albersheim, 1983). Numerous other phytoalexins are produced in soybeans and other legumes (Rizk and Wood, 1980). Low molecular weight antimicrobial compounds are also known to be produced by green beans, broad beans, garden peas, cowpeas, and alfalfa. A broad range of fungi and bacteria is sensitive to these stress metabolites. Capsidiol from bell peppers (Stoessel et al., 1972); rishitin, lubimin, and phytuberin from potatoes; and auberginone from eggplant (Dixon et al., 1983) are also known to exhibit antimicrobial activity.

The formation of dihydroisocoumarin, chromone, and scopoletin in carrot roots is induced by invasion with various spoilage (Coxin et al., 1973). Two other phytoalexins, ß-glucosides of 6-methoxymellein and 6-hydroxymellein, are also produced by carrot cells (Kurosaki et al., 1984). 6-Methoxymellein production is elicited by fungal invasion (Amin et al., 1986a; Kurosaki and Nishi, 1983) and also by partial hydrolysis of carrot cells (Kurusaki and Nishi, 1984). The compound is toxic to bacteria as well as fungi. Inhibition of Gram-positive (*S. aureus*, *Streptococcus pyogenes*, and *B. subtilis*) and Gram-negative bacteria (*S. typhimurium*, *Shigella sonnei*, *E. coli*, *Klebsiella pneumoniae*, *Serratia marcescens*, and *Proteus vulgaris*) has been reported when cells are exposed to 0.05 to 0.5 m*M* 6-methoxymellein (Kurusaki and Nishi, 1983). The mechanism of action of this phytoalexin apparently involves interference with membrane-associated functions. Alteration of membrane permeability (Amin et al., 1988; Weinstein and Albersheim, 1983), mitochondria (Boydston et al., 1983), cyclic nucleotide phosphodiesterase activity, and multilamella liposomes (Amin et al., 1986b) has been documented.

Beuchat and Brackett (1990) reported that viability of *L. monocytogenes* decreased upon contact with raw carrot but not with cooked carrot, indicating that the lethal compound(s) is heat labile. In a study using *L. monocytogenes*, *L. innocua*, *L. ivanovi*, *L. seeligeri*, and *L. welshimeri*, Nguyenthe and Lund (1991) confirmed that a lethal components existed in raw sliced and macerated carrots. It was subsequently shown that the anti-listerial activity of carrot slices was suppressed by anaerobioses, thiol compounds, and bovine serum albumin but not by sodium ascorbate, propyl gallate, catalase, superoxide dismutase, or chelating agents (Nguyenthe and Lund, 1992).

Differentiation and aflatoxin formation by *A. parasiticus* are inhibited by extracts of carrot roots (Batt et al., 1980). Geraniol, citrol, and terpineol in the volatile fraction of carrot seed oil prevents growth, while limonene and terpinene do not affect growth but do inhibit aflatoxin production (Batt et al., 1983). It is not known whether these compounds would also influence growth patterns of microflora naturally present on carrots and other fresh produce.

IV. ANTIMICROBIALS OF ANIMAL ORIGIN

Enzymes and other polypeptides with antimicrobial activity occur naturally in animals as well as plants. These polypeptides are often classified as functioning by oxygen-dependent or -independent mechanisms. The inhibition or killing of microorganism by oxygen-dependent means involves enzymes or metabolic processes generating toxic metabolites of oxygen, such as hydrogen peroxide, superoxide ions, or hydroxyl radicals, and decreasing production of halides (Sofos et al., 1998). Oxygen-independent mechanisms utilize peptides and proteins that react with the cell surface and disrupt structural surface layers or membranes. This group includes lytic enzymes, hydrolases, and cationic peptides and proteins.

A. Lytic Enzymes

Lysozyme, a mucopeptide glycohydrolase, and other lytic enzymes are found in many foods and in various secretions such as egg albumen, tears, and milk. Lysozymes act by cleaving the glycoside bond between C_1 of N-acetylmuramic acid and C_4 of acetylglucosine in the bacterial cell wall (Tranter, 1994). Some lysozymes hydrolyze chitin or possess esterase activity. Lysozyme in the albumen of chicken egg contributes to up to 3.5% of the protein content. Its lysing activity has been demonstrated against *L. monocytogenes* and *C. botulinum* (Hughey and Johnson, 1987). EDTA is known to act as a synergist with lysozyme to lyse pathogens and spoilage bacteria, including *Salmonella*, *E. coli*, *Yersinia*, and *Pseudomonas* (Repaske, 1958).

Enzymes other than lysozyme cleave carbohydrate or peptide linkages. These include glycosidases specific for the glycoside bond between alternating sugars in the polypeptide backbone of microbial cells and an amidase that releases the tetrapeptide from N-acetylmuramic acid, endopeptidases, and lipases. Nielsen (1991) reported that lytic enzymes lysed *Campylobacter*, *E. coli*, *Salmonella*, and *V. parahaemolyticus*, indicating their potential use as natural food antimicrobials to reduce the risk of foodborne illness.

B. Peroxidases and Oxidases

Peroxidases are widespread in nature and oxidize molecules at the expense of hydrogen peroxide. Lactoperoxidase is the most abundant enzyme in bovine milk and is also produced in salivary glands of mammals. Lactoperoxidase oxidizes thiocyanate or halogens, thereby producing toxic metabolites. In raw milk, the lactoperoxidase system inhibits lactic acid bacteria but is bactericidal to Gram-negative spoilage psychrotrophs and *Salmonella* (Leyer and Johnson, 1993; Reiter and Harnulv, 1984). The lactoperoxidase system is also active against Gram-positive pathogens such as *L. monocytogenes* (Siragusa and Johnson, 1989), *S. aureus* (Kamau et al., 1990b), and *B. cereus* (Zajac et al., 1981).

Glucose oxidase catalyzes a reaction between glucose and oxygen to yield gluconic acid or D-glucono-S-lactone and hydrogen peroxide. The enzyme is used commercially to remove glucose from eggs, prevent Maillard browning reactions, and remove oxygen from beverages (Frank, 1992). Inhibition of growth of *Salmonella infantis*, *S. aureus*, *Clostridium perfringens*, and *B. cereus* by glucose oxidase was reported by Tiina and Sandholm (1989).

C. Transferrins

Survival and growth of foodborne pathogens depend on the availability of iron ions, although the absolute requirement has yet to be determined (Sofos et al., 1998). Sequestration of iron through chelation by iron-binding polypeptides, especially the transferrins and related proteins, is the mechanism of control of iron-dependent microorganisms. Gvotransferrin (conalbumin) comprises about 13% of the protein in egg white and has been known to inhibit growth of *S. aureus*, *E. coli*, and *Shigella* since the early 1940s (Schade and Caroline, 1944). Lactoferrin, present in milk, is another transferrin known to inhibit or kill foodborne pathogenic and spoilage microorganisms (Arnold et al., 1980; Hoek et al., 1997; Tomita et al., 1992).

REFERENCES

Abdou, I.A., A.A. Abdou-Zeid, M.R. El-Sherbeeny, and Z.H. Abou-El-Gheat. 1972. Antimicrobial activity of *Allium sativum*, *Allium cepa*, *Barbanus sativus*, *Capsicum frutescens*, *Eruca sativa*, and *Allium kurrat* on bacteria. *Qual. Plant Mater. Veget.*, 22:29–35.

Akgul, A. and M. Kivanc. 1998. Inhibitory effects of Turkish spices and oregano components on some foodborne fungi. *Int. J. Food Microbiol.*, 6:263–268.

Aktug, S.E. and M. Karapinar. 1986. Sensitivity of some common food poisoning bacteria to thyme, mint and bay leaves. *Int. J. Food Microbiol.*, 3:349–354.

Al-Khayat, M.A. and G. Blank. 1985. Phenolic spice-components sporostatic to *Bacillus subtilis*. *J. Food Sci.*, 50:971–974, 980.

Amin, M., F. Kurosaki, and N. Nishi. 1986a. Extracellular pectinolytic enzymes of fungi elicit phytoalexin accumulation in carrot suspension culture. *J. Gen. Microbiol.*, 132:771–777.

Amin, M., F. Kurosaki, and N. Nishi. 1986b. Inhibition of cyclic nucleotide phosphodiesterase by carrot phytoalexin. *Phytochemistry*, 25:2305–2307.

Amin, M., F. Kurosaki, and N. Nishi. 1988. Carrot phytoalexin alters the membrane permeability of *Candida albicans* and multilamellar liposomes. *J. Gen. Microbiol.*, 134:241–246.

Appleton, J.A. and M.R. Tansey. 1975. Inhibition of growth of zoopathogenic fungi by garlic extract. *Mycologia*, 67:882–885.

Arnold, R.R., J.E. Russell, W.J. Champion, and J.J. Gauthier. 1980. Bactericidal activity of human lactoferrin: influence of physical conditions and metabolic state of the target microorganism. *Infect. Immun.*, 32:655–660.

Azzouz, M.A. and L.B. Bullerman. 1982. Comparative effects of selected herbs, spices, plant compounds, and commercial antifungal agents. *J. Food Prot.*, 45:1298–1301.

Baranowski, J.D., P.M. Davidson, C.W. Nagel, and A.L. Branen. 1980. Inhibition of *Saccharomyces cerevisiae* by naturally occurring hydroxycinnamates. *J. Food Sci.*, 45:592–594.

Barone, F.E. and M.R. Tansey. 1977. Isolation, purification, identification, synthesis and kinetics of activity of the anticandidal component of *Allium sativum* and a hypothesis for its mode of action. *Mycologia*, 69:793–825.

Batt, C., M. Solberg, and M. Ceponis. 1980. Inhibition of aflatoxin production by carrot root extract. *J. Food Sci.*, 45:1210–1213.

Batt, C., M. Solberg, and M. Ceponis. 1983. Effect of volatile components of carrot seed oil on growth and aflatoxin production by *Aspergillus parasiticus*. *J. Food Sci.*, 48:762–768.

Benjilali, B., A. Tantaoui-Elaraki, A. Ayadi, and M. Ihlal. 1984. Method to study antimicrobial effects of essential oils: application of six Moroccan essences. *J. Food Prot.*, 47:748–752.

Beuchat, L.R. 1976. Sensitivity of *Vibrio parahaemolyticus* to spices and organic acids. *J. Food Sci.*, 41:899–902.

Beuchat, L.R. 1994. Antimicrobial properties of spices and their essential oils. In R.G. Board and V. Dillon (Eds.), *Natural Antimicrobial Systems in Food Preservation*. CAB International, Wallingford, Oxon, U.K., pp. 167–179.

Beuchat, L.R. and R.E. Brackett. 1990. Inhibitory effects of raw carrots on *Listeria monocytogenes*. *Appl. Environ. Microbiol.,* 56:1734–1742.

Beuchat, L.R. and D.A. Golden. 1989. Antimicrobials occurring naturally in foods. *Food Technol.,* 43(1):134–142.

Beuchat, L.R. and E.K. Heaton. 1975. *Salmonella* survival on pecans as influenced by processing and storage conditions. *Appl. Microbiol.,* 29:795–801.

Beuchat, L.R., R.V. Lechowich, S.W. Schanderl, D.Y.C. Co, and R.F. McFeeters. 1966. Inhibition of bacterial growth by chlorophyllide a. *Quart. Bull. Mich. Agric. Exp. Sta.,* 48:411–416.

Bhavanti Shankar, T.N. and V. Screenivasa Murthy. 1979. Effect of turmeric (*Curcuma longa*) fractions on the growth of some intestinal and pathogenic bacteria *in vitro*. *J. Exp. Biol.,* 17:1363–1366.

Blank G., M. Al-Khayat, and M.A.H. Ismond. 1987. Germination and heat resistance of *Bacillus subtilis* produced on clove and eugenol based media. *Food Microbiol.,* 4:35–42.

Blank, G., G. Akomas, M. Henderson, and J. Zawistowski. 1988. Heat sensitization of *Bacillus subtilis* spores by selected spices. *J. Food Safety,* 9:83–96.

Boonchird, C. and T.W. Flegel. 1982. *In vitro* antifungal activity of eugenol and vanillin against *Candida albicans* and *Cryptococcus neoformans*. *Can. J. Microbiol.,* 28:1235–1241.

Boydston, R., J.D. Paxton, and D.E. Koeppe. 1983. Glyceollin, a site-specific inhibitor of electron transport in isolated soybean mitochondria. *Plant Physiol.*, 72:151–155.

Buchanan, R.L. and A.M. Fletcher. 1978. Methylxanthine inhibition of aflatoxin production. *J. Food Sci.,* 43:654–655.

Buchanan, R.L. and D.F. Lewis. 1984. Caffeine inhibition aflatoxin synthesis: probable site action. *Appl. Environ. Microbiol.,* 47:1216–1220.

Buchanan, R.L. and A.J. Shepherd. 1981. Inhibition of *Aspergillus parasiticus* by thymol. *J. Food Sci.,* 46:976–977.

Buchanan, R.L., G. Tice, and D. Marino. 1981. Caffeine inhibition of ochratoxin A production. *J. Food Sci.,* 47:319–321.

Buchanan, R.L., M.A. Harry, and M.A. Gealt. 1983a. Caffeine inhibition of sterigmatocystin, citrinin and patulin production. *J. Food Sci.,* 48:1226–1228.

Buchanan, R.L., D.J. Joover, and S.B. Jones. 1983b. Caffeine inhibition of aflatoxin production: mode of action. *Appl. Environ. Microbiol.,* 46:1193–1200.

Bullerman, L.B. 1974. Inhibition of aflatoxin production by cinnamon. *J. Food Sci.,* 39:1163–1165.

Bullerman, L.B., F.Y. Lieu, and S.A. Seier. 1977. Inhibition of growth and aflatoxin production by cinnamon and clove oils, cinnamic aldehyde and eugenol. *J. Food Sci.,* 42:1107–1109, 1116.

Carpenter, J.A., Y.P. Wang, and J.J. Powers. 1967. Effect of anthocyanin pigments on certain enzymes. *Proc. Soc. Exp. Biol. Med.,* 124:702–706.

Cavellito, C.J. and J.H. Bailey. 1944. Allicin, the antibacterial principle of *Allium sativum*. I. Isolation, physical properties and antibacterial action. *J. Am. Chem. Soc.,* 16:1950–1951.

Cavellito, C.J., J.H. Bailey, and J.S. Buck. 1945. The antibacterial principle of *Allium sativum*. III. Its precursor and "essential oil of garlic." *J. Am. Chem. Soc.,* 67:1032–1033.

Chipley, J.R. 1993. Sodium benzoate and benzoic acid. In P.M. Davidson and A.L. Branen (Eds.), *Antimicrobials in Foods*, 2nd ed. Marcel Dekker, New York, pp. 11–48.

Chung, K.T. and C.A. Murdock. 1991. Natural systems for preventing contamination and growth of microorganisms in foods. *Food Microstruct.*, 10:361.

Cliver, D.O. and K.D. Kostenbader. 1979. Antiviral effectiveness of grape juice. *J. Food Prot.,* 42:100–104.

Coley-Smith, J.R. and J.E. King. 1969. The production of species of *Allium* of alkyl sulphides and their effect on germination of sclerotia of *Sclerotium cepivorum* Berk. *Ann. Appl. Biol.,* 64:289–301.

Conner, D.E. and L.R. Beuchat. 1984a. Effects of essential oils from plants on growth of food spoilage yeasts. *J. Food Sci.,* 49:429–434.

Conner, D.E. and L.R. Beuchat. 1984b. Sensitivity of heat-stressed yeasts to essential oils of plants. *Appl. Environ. Microbiol.,* 47:229–233.

Conner, D.E. and L.R. Beuchat. 1985. Recovery of heat-stressed yeasts in media containing plant oleoresins. *J. Appl. Bacteriol.,* 59:49–55.

Conner, D.E., L.R. Beuchat, R.W. Worthington, and H.L. Hitchcock. 1985. Effects of essential oils and oleoresins on plants of ethanol production, respiration and sporulation of yeasts. *Int. J. Food Microbiol.,* 1:63–74.

Coxin, D.T., R.F. Curtis, K.R. Price, and G. Levett. 1973. Abnormal metabolites produced by *Daucas carota* roots stored under conditions of stress. *Phytochemistry,* 12:1881–1885.

Davidson, P.M. and A.L. Branen. 1981. Antimicrobial activity of non-halogenated phenolic compounds. *J. Food Prot.,* 44:623–632.

Davidson, P.M. and A.L. Branen, Eds. 1993. *Antimicrobials in Foods.* Marcel Dekker, New York.

Deans, S.G. and G. Ritchie. 1987. Antibacterial properties of plant essential oils. *Int. J. Food Microbiol.,* 5:165–180.

DeWitt, J.C., S. Notermans, N. Gorin, and E.H. Kempelmacher. 1979. Effect of garlic oil or onion oil on toxin production by *Clostridium botulinum* in meat slurry. *J. Food Prot.,* 42:222–224.

Dillon, V.M. and R.G. Board. 1994. *Natural Antimicrobial Systems and Food Preservation.* CAB International, Wallingford, Oxon, U.K., 328 pp.

Dixon, R.A., P.M. Dey, and C.J. Lamb. 1983. Phytoalexins: enzymology and molecular biology. *Adv. Enzymol.,* 55:1–136.

Engelson, M., M. Solbert, and E. Karmas. 1980. Antimycotic properties of hop extract in reduced water activity media. *J. Food Sci.,* 45:1175–1178.

Farbood, M.I., J.H. McNeil, and K. Ostovar. 1976. Effect of rosemary spice extractive on growth of microorganisms in meat. *J. Milk Food Technol.,* 39:675–679.

Fleming, H.P., W.M. Walter, Jr., and J.L. Etchells. 1969. Isolation of a bacterial inhibitor from green olives. *Appl. Microbiol.,* 18:856–860.

Fleming, J.P., W.M. Walter, Jr., and J.L. Etchells. 1973. Antimicrobial properties of oleuropein and products of its hydrolysis from green olives. *Appl. Microbiol.,* 26:777–782.

Fliermans, C.B. 1973. Inhibition of *Histoplasma capsulatum* by garlic. *Mycopath. Mycol. Applicata,* 50:227–231.

Frank, H.K. 1992. *Dictionary of Food Microbiology.* Technomic Publishing, Lancaster, Basel, Switzerland.

Gonul, S.E.A. and M. Karapinar. 1987. Inhibitory effect of linden flower (tilia flower) on the growth of foodborne pathogens. *Food Microbiol.,* 4:97–100.

Gourama, H. and L.B. Bullerman. 1987. Effects of oleuropein on growth and aflatoxin production *Aspergillus parasiticus. Lebensm. Wiss. Technol.,* 20:226–228.

Hartman, G.D. 1959. The Effect of Anthocyanin Pigment, Pelargonidin-3-Monoglucoside, and Its Heat-Degration Products on Growth of Selected Bacteria. Master's thesis, University of Georgia, Athens.

Herald, P.J. and P.M. Davidson. 1983. Antibacterial activity of selected hydroxycinnamic acids. *J. Food Sci.,* 48:1378–1379.

Hitokoto, H., S. Morozumi, T. Wauke, S. Sakai, and H. Kurata. 1980. Inhibitory effects of spices on growth and toxin production of toxigenic fungi. *Appl. Environ. Microbiol.,* 39:818–822.

Hitokoto, H., S. Morozumi, T. Wauke, S. Sakai, and I. Veno. 1978. Inhibitory effects of condiments and herbal drugs on the growth and toxin production of toxigenic fungi. *Mycopathologia,* 66:161–167.

Hoek, K.S., J.M. Milne, P.A. Grieve, D.A. Dionysius, and R. Smith. 1997. Antibacterial activity in bovine lactoferrin-derived peptides. *Antimicrob. Agents Chemother.,* 41(1):54–59.

Hughey, V.L. and E.A. Johnson. 1987. Antimicrobial activity of lysozyme against bacteria involved in food spoilage and foodborne disease. *Appl. Environ. Microbiol.,* 53:2165–2170.

Huhtanen, C.N. 1980. Inhibition of *Clostridium botulinum* by spice extracts and aliphatic alcohols. *J. Food Prot.,* 43:195–196.

Johnson, M.G. and R.H. Vaughn. 1969. Death of *Salmonella typhimurium* and *Escherichia coli* in the presence of freshly reconstituted dehydrated garlic and onion. *Appl. Microbiol.,* 17:903–905.

Juven, B. and Y. Henis. 1970. Studies on the antimicrobial activity of olive phenolic compounds. *J. Appl. Bacteriol.,* 33:721–732.

Juven, B., Y. Henis, and B. Jacoby. 1972. Studies on the mechanism of the antimicrobial action of oleuropein. *J. Appl. Bacteriol.,* 35:559–567.

Kamau, D.N., S. Doores, and K.M. Pruitt. 1990b. Enhanced thermal destruction of *Listeria monocytogenes* and *Staphylococcus aureus* by the lactoperoxidase system. *Appl. Environ. Microbiol.,* 56:2711–2716.

Karaioannoglou, P.G., A.J. Mantis, and A.G. Panetsos. 1977. The effect of garlic extract on lactic acid bacteria (*Lactobacillus plantarum*) in culture media. *Lebensm. Wiss. Technol.,* 10:148–150.

Karapinar, M. and S.E. Aktug. 1987. Inhibition of foodborne pathogens by thymol, eugenol, menthol and anethole. *Int. J. Food Microbiol.,* 4:161–166.

Konowalchuk, J. and J.I. Speirs. 1967a. Antiviral activity of fruit extract. *J. Food Sci.,* 14:1013–1017.

Konowalchuk, J. and J.I. Speirs. 1967b. Virus inactivation of grapes and wines. *Appl. Environ. Microbiol.,* 32:757–763.

Konowalchuk, J. and J.I. Speirs. 1978a. Antiviral effect of commercial juices and beverages. *Appl. Environ. Microbiol.,* 35:1219–1220.

Konowalchuk, J. and J.I. Speirs. 1978b. Antiviral effect of apple beverages. *Appl. Environ. Microbiol.,* 36:798–801.

Kurosaki, F. and A. Nishi. 1983. Isolation and antimicrobial activity of the phytoalexin 6-methoxymellein from cultured carrot cells. *Phytochemistry,* 22:669–672.

Kurosaki, F. and A. Nishi. 1984. Elicitation of phytoalexin production in cultured carrot cells. *Physiol. Plant Pathol.,* 24:169–176.

Kurosaki, F., K. Matsui, and A. Nishi. 1984. Production and metabolism of 6-methoxymellein in cultured carrot cells. *Physiol. Plant Pathol.,* 25:313–322.

Lenovich, L.M. 1981. Effect of caffeine on aflatoxin production of cocoa beans. *J. Food Sci.,* 46:655–656.

Leyer, G.L. and E.A. Johnson. 1993. Acid adaptation induces cross-protection against environmental stress in *Salmonella typhimurium. Appl. Environ. Microbiol.,* 59:1842–1847.

Lunder, T.L. 1979. *Tea: History, Botany, Chemistry, Manufacture, Nutritional Value.* Centre de Documentation Technique, Nestle, Vevey, Switzerland.

Mabrouk, S.S. and N.M.A. El-Shayeb. 1981. The effects of garlic on mycelial growth and aflatoxin formation by *Aspergillus flavus. Chem. Mikrobiol., Technol., Lebensm.* 7:37–41.

Mantis, A.J., P.G. Karaiannoglou, G.P. Spanos, and A.G. Panetsos. 1978. The effect of garlic extract on food poisoning bacteria in culture media. I. *Staphylococcus aureus. Lebensm. Wiss. Technol.,* 11:26–28.

Marwan, A.G. and C.W. Nagel. 1986. Microbial inhibitors in cranberries. *J. Food Sci.,* 51:1009–1013.

Maruzzella, J.C. and L. Liguori. 1958. The *in vitro* antifungal activity of essential oils. *J. Am. Pharm. Assoc.,* 47:250–254.

Mizobuchi, S. and Y. Sato. 1985. Antifungal activities of hop bitter resins and related compounds. *Agric. Biol., Chem.,* 49:399–403.

Moleyar, V. and P. Narasimham. 1986. Antifungal activity of some essential oil components. *Food Microbiol.,* 3:331–336.

Moore, G.S. and R.D. Atkins. 1977. The fungicidal and fungi-static effects of an aqueous garlic extract on medially important yeast-like fungi. *Mycologia,* 69:341–348.

Nguyenthe, C. and B.M. Lund. 1991. The lethal effect of carrot on *Listeria* species. *J. Appl. Bacteriol.,* 70:479–488.

Nguyenthe, C. and B.M. Lund. 1992. An investigation of the antibacterial effect of carrot on *Listeria monocytogenes. J. Appl. Bacteriol.,* 73:23–30.

Nielsen, H.K. 1991. Novel bacteriolytic enzymes and cyclodextrin glycosyl transferase for the food industry. *Food Technol.,* 45(1):102–104.

Paster, N., B.J. Juven, and H. Harshemesh. 1988. Antimicrobial activity and inhibition of aflatoxin B_1 formation by olive plant tissue constitutents. *J. Appl. Bacteriol.,* 64:293–297.

Pearson, L.J. and E.H. Marth. 1990a. Inhibition of *Listeria monocytogenes* by cocoa in a broth medium and neutralization of this effect of casein. *J. Food Prot.,* 53:38–46.

Pearson, L.J. and E.H. Marth. 1990b. Behavior of *Listeria monocytogenes* in the presence of cocoa, carrageenan and sugar in a milk medium incubated with and without agitation. *J. Food Prot.,* 53:30–32.

Pearson, L.J. and E.H. Marth. 1990c. Behavior of *Listeria monocytogenes* in the presence of methylxanthines: caffeine and theobromine. *J. Food Prot.,* 53:47–50.

Powers, J.J., D. Somaatmadja, D.E. Pratt, and M.K. Hamdy. 1960. Anthocyanins. II. Action of anthocyanin pigments and related compounds on the growth of certain microorganisms. *Food Technol.,* 14:626–632.

Pratt, D.E., J.J. Powers, and D. Somaatmadja. 1960. Anthocyanins. I. The influence of strawberry and grape anthocyanins on the growth of certain bacteria. *Food Res.,* 25:26–32.

Ray L.L. and L.B. Bullerman. 1982. Preventing growth of potentially toxic molds using antifungal agents. *J. Food Prot.,* 45:953–963.

Reiter, B. and B.G. Harnulv. 1984. Lactoperoxidase antibacterial system: natural occurrence, biological functions, and practical applications. *J. Food Prot.,* 47:724–732.

Repaske, R. 1958. Lysis of Gram-negative organisms and the role of versene. *Biochim. Biophys. Acta,* 30:225–232.

Richards, M. and R.M. Macrae. 1964. The significance of the use of hops in regard to the biological stability of beer. II. The development of resistance tohop resins by strains of lactobacilli. *J. Inst. Brew.,* 70:484–488.

Rizk, A.-F. and G.E. Wood. 1980. Phytoalexins of leguminous plants. *CRC Crit. Rev. Food Sci. Nutr.,* 13:245–295.

Saleem, Z.M. and K.S. Al-Delaimy. 1982. Inhibition of *Bacillus cereus* by garlic extracts. *J. Food Prot.,* 45:1007–1009.

Schade, A.L. and L. Caroline. 1944. Raw hen egg white and the role of iron in growth inhibition of *Shigella dysenteriae, Ataphyloccus aureus, Escherichia coli,* and *Saccharomyces cerevisia. Sciences (NY),* 100:14–15.

Schmalreck, A.F., M. Teuber, W. Reininger, and A. Hartl. 1975. Structural features determining the antibiotic potencies of nature and synthetic hop bitter resins, their precursors and derivatives. *Can. J. Microbiol.,* 21:205–212.

Scortichini, M. and M.O. Rossi. 1991. *In vitro* susceptibility of *Erwinia amylovora* (Burrill) Winslow et al. to geraniol and citronellol. *J. Appl. Bacteriol.,* 71:113–118.

Sharma, A., G.M. Tewari, A.J. Shrikhande, S.R. Padwal-Desai, and C. Bandyopadhyay. 1979. Inhibition of aflatoxin-producing fungi by onion extracts. *J. Food Sci.,* 44:1545–1547.

Sharma, A., S.R. Padwal-Desai, G.M. Tewari, and C. Bandyopadhyay. 1981a. Factors affecting antifungal activity of onion extractives against aflatoxin-producing fungi. *J. Food Sci.,* 46:741–744.

Sharma, A., G.M. Tewari, C. Bandyopadhyay, and S.R. Padwal-Desai. 1981b. Sporicidial activity of onion extracts against aflatoxin-producing fungi. *Lebensm. Wiss. Technol.,* 14:21–22.

Shelef, L.A., O.A. Naglik, and D.W. Bogen. 1980. Sensitivity of some common foodborne bacteria to the spices sage, rosemary and allspice. *J. Food Sci.,* 45:1042–1044.

Shelef, L.A., E.K. Jyothi, and M.A. Bulgarelli. 1984. Growth of enteropathogenic and spoilage bacteria in sage-containing broth and foods. *J. Food Sci.,* 49:737–740, 809.

Simpson, W.J. and A.R.W. Smith. 1992. Factors affecting antibacterial activity of hop compounds and their derivatives. *J. Appl. Bacteriol.,* 72:327–334.

Singleton, V.L. and P. Esau. 1969. *Phenolic Substances in Grapes and Wine and Their Significance.* Academic Press, New York.

Siragusa, G.R. and M.G. Johnson. 1989. Inhibition of *Listeria monocytogenes* growth by the lactoperoxidase-thiocyanate-H_2O_2 antimicrobial system. *Appl. Environ. Microbiol.,* 55:2802–2805.

Sofos, J.N. 1989. *Sorbate Food Preservatives.* CRC Press, Boca Raton, FL.

Sofos, J.N. 1994. Antimicrobial agents. In A.T. Tu and J.A. Maga (Eds.), *Handbook of Toxicology.* Vol. 1. *Food Additive Technology.* Marcel Dekker, New York, pp. 501–529.

Sofos, J.N. and F.F. Busta. 1981. Antimicrobial activity of sorbate. *J. Food Prot.,* 44:614–622.

Sofos, J.N. and F.F. Busta. 1992. Chemical food preservatives. In A.D. Russell, W.B. Hugo, and G.A.J. Ayliffe (Eds.), *Principles and Practice of Disinfection, Preservation and Sterilization,* 2nd ed. Blackwell Scientific, London, pp. 351–386.

Sofos, J.N., L.R. Beuchat, P.M. Davidson, and E.A. Johnson. 1998. *Naturally Occurring Antimicrobials in Food.* Task Force Report No. 132, Council for Agricultural Science and Technology, Ames, IA, 103 pp.

Somaatmadja, D., J.J. Powers, and M.K. Hamdy. 1964. Anthocyanins. VI. Chelation studies on anthocyanins and other related compounds. *J. Food Sci.,* 29:655–660.

Sirvastava, K.C., A.D. Perera, and H.O. Saridakis. 1982. Bacteriostatic effects of garlic sap on Gram negative pathogenia bacteria — an *in vitro* study. *Lebensm. Wiss. Technol.,* 15:74–76.

Stoessel, A., C.H. Unwin, and E.W.B. Ward. 1972. Post-infectional inhibitors from plants. I. Capsidol and antifungal compounds from *Capsicum frutescens. Phytopathol. Z.,* 74:141–152.

Stoll, A. and E. Seebeck. 1949. Uber aliin, die genuine muttersubstantz des knoblauchols. *Helv. Chim. Acta,* 31:189–210.

Tansey, M.R. and J.A. Appleton. 1975. Inhibition of fungal growth by garlic extract. *Mycologia,* 67:409–413.

Tiina, M. and M. Sandholm. 1989. Antibacterial effect of the glucose oxidase-glucose system on food poisoning organisms. *Int. J. Food Microbiol.,* 8:165–174.

Timonin, M.I. and R.H. Thexton. 1950. The rhizosphere effect of onion and garlic on soil microflora. *Soil Sci. Soc. Proc.,* 15:186–189.

Tomita, M., W. Bellamy, M. Takase, K. Yamaguchi, H. Wakabayashi, and K. Kawase. 1992. Potential antibacterial peptides generated from pepsin digestion of bovine lactoferrin. *J. Dairy Sci.,* 74:4137–4142.

Tortora, P., N. Burlini, G.M. Hanozet, and A. Guerritore. 1982. Effect of caffeine on glucose-induced inactivation of gluconeogenetic enzymes in *Saccharomyces cerevisiae*: a possible role of cyclic AMP. *Eur. J. Biochem.* 126:617–622.

Tranter, H.S. 1994. Lysozyme, ovotransferrin and avidin. In V.M. Dillon and R.G. Board (Eds.), *Natural Antimicrobial Systems and Food Preservation.* CAB International, Wallingford, Oxon, U.K., pp. 65–97.

Tynecka, Z. and Z. Gos. 1973. The inhibitory action of garlic (*Allium sativum* L.) on growth and respiration of some microorganisms. *Acta Microbiol. Pol. Ser. B,* 5:51–62.

Vanos, V. and O. Bindschedler. 1985. The microbiology of instant coffee. *Food Microbiol.,* 4:19–33.

Vanos, V., S. Hofstaetter, and L. Cox. 1987. The microbiology of instant tea. *Food Microbiol.,* 2:187–197.

Walker, J.R.L. 1994. Antimicrobial compounds in food plants. In V.M. Dillon and R.G. Board (Eds.), *Natural Antimicrobial Systems and Food Preservation.* CAB International, Wallingford, Oxon, U.K., pp. 181–204.

Weinstein, L.I. and P. Albersheim. 1983. Host-pathogen interactions. XXIII. The mechanisms of the antibacterial action of glycinol, a pterocarpan phytoalexin synthesized by soybeans. *Plant Physiol.,* 72:557–563.

Whitehead, I.M. and D.R. Threlfall. 1992. Production of phytoalexins by plant tissue cultures. *J. Biotechnol.,* 26:63–81.

Wills, E.D. 1956. Enzyme inhibition of allicin, the active principle of garlic. *Biochem. J.,* 63:514–520.

Zaika, L.L. and J.C. Kissinger. 1981. Inhibitory and stimulatory effects of oregano on *Lactobacillus plantarum* and *Pediococcus cerevisiae. J. Food Sci.,* 46:1205–1210.

Zaika, L.L. and J.C. Kissinger. 1984. Fermentation enhancement by spices: identification of active component. *J. Food Sci.,* 49:5–9.

Zaika, L.L., J.C. Kissinger, and A.E. Wasserman. 1983. Inhibition of lactic acid bacteria by herbs. *J. Food Sci.,* 48:1455–1459.

Zajak, M., L. Bjorck, and O. Claasson. 1981. Antibacterial effect of the lactoperoxidase system against *Bacillus cereus* (from milk). *Milchwissenschaft,* 38:417–418.

Zapatka, F.A., G.W. Varney, and J. Skinskey. 1977. Neutralization of the bactericidal effect of cocoa powder on salmonellae by casein. *J. Appl. Bacteriol.,* 42:21–25.

Zimmerman, V.C. 1959. The Influence of Anthocyanin Pigments on Certain Bacteria *In Vitro.* Master's thesis, University of Georgia, Athens.

12 Irradiation and Other Physically Based Control Strategies for Foodborne Pathogens

V.K. Juneja and D.W. Thayer

CONTENTS

I. INTRODUCTION

The ubiquity of potentially life-threatening pathogens in our environment and their contamination of our foods is an enormous problem. The ability of some of these pathogens to survive and/or proliferate under refrigeration and in reduced oxygen atmospheres and, for some of them, the low number necessary for food-poisoning outbreaks indicate a potential public health hazard. Moreover, microorganisms, previously unknown or not known to be causes of foodborne disease, have recently been linked with documented outbreaks of illness. The U.S. Public Health Service has estimated that foodborne diseases caused by pathogenic bacteria such as *Salmonella, Campylobacter, Escherichia coli*, and *Vibrio*, as well as *Toxoplasma gondii* and other parasites, may cause as many as 9000 deaths and 6.5 million to 81 million cases of diarrheal diseases in the U.S. annually.[1] The annual economic losses in relation to medical costs, loss of productivity, loss of business, and possible legal action associated with foodborne diseases may be as large as $5 billion to $6 billion.[2] These food-safety concerns are magnified because of consumer preferences for high quality and minimally processed convenient meals that require minimal preparation time. Accordingly, the need for better control of foodborne pathogens has become paramount in recent years.

Strategies for control of foodborne pathogens include established physical microbiocidal treatments such as ionizing radiation and heating. Microorganisms can also be destroyed by the emerging methods of new non-thermal treatments, such as the application of high hydrostatic pressure, high-intensity pulsed electric fields, oscillating magnetic fields, intense light pulses, or a combination of physical processes such as heat-irradiation, or heat-high hydrostatic pressure. Each of the non-thermal technologies has specific applications in terms of the types of food that can be processed. Mechanical removal of microorganisms from food can be accomplished by centrifugation, filtration, trimming, and washing. Cleaning and sanitation strategies can be used for minimizing the access of microorganisms to foods. This chapter deals with a variety of conventional and newly developed physical treatments for controlling foodborne pathogens and enhancing the safety and shelf-life of foods.

II. IRRADIATION OF FOOD

Food irradiation improves the safety of meat, poultry, and other foods by destroying indigenous microflora and thereby extending the shelf-life of these products during refrigerated storage. Sources of ionizing radiation include X-rays with a maximum energy of 5 million electron volts (MeV), electrons with a maximum energy of 10 MeV, and gamma rays emitted by the radioisotopes cobalt-60 or cesium-137. The food is exposed to doses of ionizing radiation sufficient to create positive and negative charges leading to the death of bacteria and other pathogens in foods. It is the rapidly growing cells of pathogenic and spoilage bacteria or parasites that are killed when food is irradiated. Ionizing radiation affects organisms by damaging the genetic material, such as DNA base damage, single-strand and double-strand DNA breaks, and cross-linking between bases. As a consequence of this damage,

TABLE 12.1
Ionizing Radiation Dose Requirements for Various Applications of Food Irradiation

Application	Dose Requirement (kGy)
Inhibition of sprouting potatoes and onions	0.03–0.12
Insect disinfestation of seed products, flours, fresh and dried fruits, etc.	0.2–0.8
Parasite disinfection of meats and other foods	0.1–3.0
Radurization of perishable food items (fruits, vegetables, meat, poultry, fish)	0.5–10
Radicidation of frozen meat, poultry, eggs, and other foods and feeds	3.0–10
Reduction or elimination of microbial populations in dry food ingredients (spices, starch, enzyme preparations, etc.)	3.0–2.0
Radappertization of meat, poultry, and fish products	25–60

Source: Adapted from Farkas.[41]

microorganisms are unable to replicate DNA and reproduce, leading to their death. In addition to DNA damage, ionizing radiation may damage bacterial membranes and cause other changes leading to sublethal injury.[3]

A. Absorbed Doses

There are several terms that must be known with regard to the application of radiation to foods. These terms include:

1. *Curie:* Quantity of radioactive substance in which 3.7×10^{10} radioactive disintegrations occur per second.
2. *Rad:* As used in the past is a unit equivalent to the absorption of 100 ergs energy/g of irradiated material.
3. *Gray (Gy):* Currently used unit of absorbed dose; one Gy is an energy absorption of one joule per kilogram (1 Gy = 100 rads = 1 joule/kg; 1 kGy = 10^5 rads; 1 joule = 10^7 ergs).

Depending upon the dose, a variety of desirable effects can be achieved (Table 12.1). Like all processing technologies, excessive doses can produce adverse effects. Sudarmadji and Urbain[4] estimated threshold doses of irradiation (5 to 10°C) for an organoleptically detectable "off-flavor" in foods of animal origin (Table 12.2). Higher doses can be used without adverse effects by exclusion of oxygen and or by irradiation with the product in the frozen state. Goresline et al.[5] devised the following terms to describe the applications of irradiation in food processing:

1. *Radacidation:* Considered as equivalent to pasteurization of milk and accordingly referred as "irradiation pasteurization." It is intended to

TABLE 12.2
Threshold Doses of Irradiation at 5 to 10°C for Foods of Animal Origin for an Organoleptically Detectable "Off-Flavor"

Food	Threshhold Dose (kGy)
Beef	2.5
Chicken	2.5
Turkey	1.5
Lamb	6.25
Pork	1.75
Shrimp	2.5

Source: Adapted from Sudarmadji and Urbain.[4]

reduce the number of specific, viable, non-spore-forming pathogens, including parasites other than viruses, to non-detectable levels as determined by any standard method (<10 kGy dose).

2. *Radurization:* May be considered as equivalent to pasteurization. It is intended to considerably reduce the population densities of specific, viable, spoilage microbes with an aim to extend the shelf-life of foods (<10 kGy dose).
3. *Radappertization:* Considered as equivalent to sterilization or rendering the food "commercially sterile," as it is known in the canning industry (>10 kGy dose). If the food has been enzyme inactivated and has been irradiated while hard-frozen *in vacuo,* it will be shelf-stable and of excellent quality.

B. Safety

Food exposed to ionizing radiation is never in contact with any radioactive material. None of the sources of radiation, such as gamma rays, X-rays, or electrons can render the food radioactive. There is little effect on the food itself, as the cells in the food are not multiplying. Extensive research using animal models has provided sufficient evidence that ingestion of irradiated foods is completely safe and that the nutritive value remains essentially unchanged.[2]

Some vitamins (e.g., Vitamins B_1 and C), however, are sensitive to radiation. Factors affecting the amount of vitamin loss include dose, temperature, presence of oxygen, and the type of food. Packaging of foods in the absence of oxygen and exposing them to radiations at low temperatures minimize any vitamin loss, and further loss can be prevented by storage at low temperatures in sealed packages.[6] It has been estimated that only 2.3% of vitamin B_1 in the American diet would be lost if all the pork in the U.S. were to be irradiated.[2] Also, irradiation causes a small amount of ascorbic acid in fruits to be converted (oxidized) to dehydro-ascorbic acid. This compound is as biologically active as its reduced form and is converted back to the reduced form during storage of the fruits or vegetables.

When molecules absorb ionizing energy, they become reactive and form ions or free radicals that react to form stable radiolytic products.[7] The Council for Agricultural Science and Technology (CAST)[8] estimated that a dose of 1 kGy will break fewer than 10 chemical bonds for every 10 million such bonds present (cooking produces similar changes in chemical bonds). Researchers have developed methodologies to detect irradiated foods and have identified alkylcyclobutones in some irradiated foods that were not detected in unirradiated foods; however, Crawford and Ruff[9] reported that no radiolytic product of toxicological significance have been found in irradiated foods. The committee on the wholesomeness of irradiated foods convened by the Food and Agriculture Organization of the United Nations, the World Health Organization, and the International Atomic Energy Agency concluded, based on decades of research, that irradiated foods are safe and wholesome at any radiation dose.[10]

C. Radiation Susceptibility of Microorganisms

Several factors influence the survival of microbial cells when exposed to ionizing radiation. First, the higher the dose of ionizing radiation, the greater is the destruction of microorganisms; however, the microbiocidal efficacy of irradiation is lower under anaerobic conditions than in the presence of oxygen. This effect is attributed to the slower rate of oxidizing reactions, such as the formation of radicals due to the interaction of ionizing energy with water molecules.

Second, food-processing treatments such as curing, high hydrostatic pressure, temperature, decreased pH, and added preservatives (sodium benzoate, potassium sorbate, sodium salt of methyl, propyl esters of parahydroxy benzoic acid, etc.) increase the efficiency of the ionizing radiation by decreasing the number of surviving organisms. However, the reduction of water activity or a decrease in the moisture content, a common preservation method, exhibits a protective effect against the lethal effect of ionizing radiation as a result of reduced free radical formation due to lower moisture content.[11] Similarly, freezing causes a substantial increase in the resistance of vegetative cells, due to reduced availability of reactive water molecules; the radical formation is practically inhibited. Microbial radiation resistance in frozen foods is about two- to threefold higher than at ambient temperature. The composition of the food, in addition to its thickness and particle size, also plays an important role in determining the survival of microorganisms and the extent of the dose required to achieve the desired microbiological lethal effect. Bacterial cells are more protected against the lethal effects of irradiation in solid foods than in phosphate buffer. This is because of greater competition of the medium components in food matrix for the free radicals formed from water and activated molecules, thereby protecting the microorganisms. Accordingly, it is not advisable to predict the dose required to kill the microorganism in one food based on the dose quantified in other foods.

TABLE 12.3
Some D_{10} Values of Foodborne Pathogens

Organism	Product	Irradiation Temperature (°C)	D_{10} (kGy)	Ref.
Aeromonas hydrophila	Ground beef	2	0.04–0.90	Palumbo et al.[42]
Campylobacter jejuni	Ground beef	18–20	0.14–0.16	Tarkowski et al.[43]
Escherichia coli O157:H7	Beef	2–4	0.24	Clavero et al.[14]
Listeria monocytogenes	Chicken	2–4	0.45	Hutanen et al.[44]
Salmonella spp.	Chicken	2	0.38–0.77	Thayer et al.[45]
Shigella dysenteriae	Oysters	5	0.40	Quinn et. al.[46]
Staphylococcus aureus	Chicken	0	0.40–0.46	Thayer et al.[47]
Vibrio parahemolyticus	Crab meat	24	0.053–0.357	Matches and Liston[48]
Yersinia enterocolitica	Ground beef	18–20	0.10–0.21	Tarkowski et al.[43]

Finally, microorganisms vary considerably in their sensitivity to ionizing radiation. In general, the simpler the organism, the more resistant it is to the effects of ionizing radiation. For example, viruses are more resistant than bacteria, which are more resistant than yeasts, which are more resistant than molds. Within bacteria, Gram-negative cells are more sensitive than Gram-positive bacteria, and rods are more sensitive than cocci. Spores are very resistant to irradiation because of their low water content.

D. Reduction/Elimination of Microorganisms

Foodborne pathogens and food spoilage microflora can be destroyed by irradiation. The D_{10} value is the radiation dose required to destroy 90% of a bacterial population. Of the Gram-negative bacterial pathogens of public health significance, such as *Escherichia coli*, *Yersinia enterocolitica*, *Aeromonas hydrophila,* and *Campylobacter* species, *Salmonella* is the most resistant to irradiation, with a D value of 0.6 kGy. Recommended doses for reduction of the most resistant serotype of *Salmonella* by about 3 log-cycles (99.9%) to 5 logs (99.999%) are 3 to 5 kGy for frozen poultry, and 1.5 to 2.5 kGy for chilled poultry.[12] Irradiation doses designed to eliminate *Salmonella* will also render the food product safe from other Gram-negative bacterial pathogens (Table 12.3). For *E. coli* O157:H7, elimination of 90% of the viable cells in mechanically deboned chicken meat was achieved using 0.27 kGy at 5°C.[13] Clavero et al.[14] reported D_{10} values of 0.241 to 0.307 kGy for *E. coli* O157:H7 in ground beef. Thus, irradiation at a dose levels of <2 kGy can effectively eliminate at least 6 logs of *E. coli* O157:H7 in ground beef.

Both Gram-positive and Gram-negative spoilage bacteria are easily destroyed by irradiation pasteurization doses. Irradiation doses to 2.5 kGy reduced the levels of aerobic and anaerobic bacteria by 4 and 5 $\log_{10}$, respectively, in chilled ground beef.[15] In an earlier study, Niemand et al.[16] reported that the shelf-life of vacuum-packaged beef was increased by 6 weeks after irradiation at 2 kGy (shelf-life of 4 weeks for non-irradiated product vs. 10 weeks for irradiated product). Lefebvre et al.[17] reported that ground beef irradiated at doses of 2.5 kGy exhibited a 3-$\log_{10}$ reduction in psychrotrophic aerobic bacteria, with a shelf-life extension of 9 days. Fish fillets treated with 1 kGy ionizing radiation had a refrigerated shelf-life 15 days longer than non-irradiated fillets.[18,19] Novak et al.[20] reported that oysters, when irradiated with 2 kGy, had a shelf-life of 23 days compared to non-irradiated oysters that began to spoil after 7 days.

E. Parasite Disinfestation

Doses of 0.15 to 0.30 kGy are required to eliminate the risk of contamination by *Trichinella spiralis*, a pork parasite. Ionizing radiations act by rendering the parasite sexually sterile and blocking the maturation of ingested larvae in the host's gut.[21] The U.S. Food and Drug Administration consequently approved the use of irradiation to control *T. spiralis* in pork at a minimum absorbed dose of 0.3 kGy, not to exceed 1.0 kGy.[22] Similarly, doses of 0.3 to 0.7 kGy are required to kill *Toxoplasma gondii* and render the pork safe for human consumption.[23] Gamma irradiation of *Cysticercus bovi*-infected beef, with a dose of 0.4 kGy, prevents development of this parasite in the human host.[24]

III. THERMAL INACTIVATION

The use of heat treatment to kill bacteria is the most common food preservation process in use today. Heat treatment designed to achieve a specific lethality for foodborne pathogens in the specific target food is fundamentally important to assure the shelf-life and microbiological safety of such thermally processed foods. The heat resistance of bacteria is described by two parameters, the D and z values. The D value is defined as the heating time required at a specific temperature to destroy 90% of the viable cells or spores of a specific organism. The z value is defined as the change in heating temperature needed to change the D value by 90% (1 log cycle). The z value provides information on the relative resistance of an organism to different destructive temperatures in a given substrate. D and z values are invaluable tools in the design of heat-processing requirements for desirable destruction of microorganisms in specific target food products.

During thermal processing, the rate of destruction of the microbial population was traditionally assumed to follow first-order kinetics; that is, at a given temperature, the reduction in the log number of survivors occurs in a linear manner with time.[25] However, the traditional log-linear thermal-death-time model is often a good representation of the actual inactivation data only in situations when inactivation is rapid. Significant deviations from the log-linear declines with time are frequently observed.[26,25] These deviations include survival curves exhibiting an initial lag period or shoulder before any death occurs — time period when the bacterial population remains at the inoculation level — followed by an exponential decline. In some instances, a tailing of a subpopulation of more resistant bacteria that decline at a slower rate than the majority of the cells is observed (Figure 12.1). Hansen

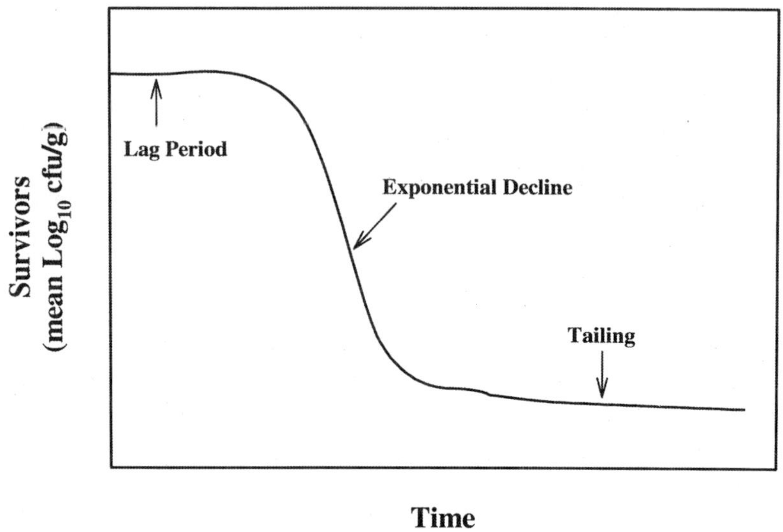

FIGURE 12.1 Heat inactivation of microorganisms showing a lag period, an exponential decline, and a tailing.

and Rieman[27] suggested that the deviations in linear survival curves may be due to variability of heat resistance within a population. Also, the "shoulder effect" observed may be attributed to poor heat transfer through the heating menstruum and may be due to an initial requirement for the bacterial cells to sustain sufficient injury before the bacterial destruction exhibits first-order inactivation kinetics. The "tailing effect" may be due to the clumping of a small number of cells in the heating menstruum, resulting in their protection and therefore an apparent increase in thermal resistance.[27]

A. Factors Affecting Heat Resistance

An appropriate heat treatment designed to achieve a specified lethality of microorganisms is influenced by many factors, some of which are due to the inherent resistance of microorganisms, while others are due to environmental influences. Examples of the inherent resistance include differences among species and the different strains of bacteria, as well as the differences between spores and vegetative cells. Environmental factors include those affecting the microorganisms during growth and formation of cells or spores (e.g., stage of growth, growth temperature, growth medium, previous exposure to stress, etc.) and those acting during exposure to heat, such as the composition of the heating menstruum (amount of carbohydrate, proteins, lipids, solutes, etc.), water activity, pH, added preservatives, method of heating, recovery procedures, etc. Similar to ionizing radiation doses required to inactivate a certain number of specific organisms in a specific substrate, thermal processes should be designed for the specific food and not adapted from information derived for other foods.

TABLE 12.4
Relative Heat Resistance of Some Foodborne Pathogens

Organism	Heating Medium	Heating Temperature (°C)	D Value (minutes)	Ref.
L. monocytogenes	Beef	62	2.9–4.2	Gaze et al.[49]
Y. enterocolitica	Physiological saline	60	0.40–0.51	Sörqvist[50]
A. hydrophila	Physiological saline	51	8.08	Palumbo[51]
Escherichia coli O157:H7	Beef	62.8	0.93	Juneja et al.[26]
Salmonella spp.	Beef	62.8	0.54	Goodfellow and Brown[52]
C. perfringens (spores)	Turkey	99	23.2	Juneja and Marmer[53]
C. botulinum (non-proteolytic, type B)	Turkey	75	32.5	Juneja et al.[54]

B. Heat Resistance of Pathogens

Table 12.4 depicts the heat resistance of some foodborne pathogens. Clearly, the heat resistance varies with the intrinsic properties of the heating menstruum. Among the non-spore-forming bacteria, *L. monocytogenes* is relatively more heat resistant. Among spores, proteolytic *Clostridium botulinum* type A and B are the most heat resistant. These spores are targeted for destruction to ensure the microbiological safety of low-acid foods. The canning industry adopted a D value at 121°C of 0.2 minutes and a 12-log reduction as the standards for designing a required thermal process for an adequate degree of protection against *C. botulinum*. The non-proteolytic *C. botulinum* strains produce less heat-resistant spores. Thus, it is even practically feasible to inactivate these spores by the type of mild heat treatment given to minimally processed foods, without negatively impacting the product quality.

It is worth mentioning that the heat resistance of pathogens is influenced by heat shock. In a study by Juneja et al.,[28] when beef gravy samples inoculated with *E. coli* O157:H7 were subjected to sublethal heating at 46°C for 15 to 30 minutes and then heated/cooked to a final internal temperature of 60°C, the organism survived longer than non-heat-shocked cells; the time to a 4-D (D being the time to inactivate 90% of the population) inactivation value at 60°C increased 1.56-fold. There is concern that a heat-shocking condition may be created in cook-chill processing, potentially facilitating an increase in the heat resistance of pathogens. Manufacturers must be aware of the heat-shock-induced thermotolerance of the pathogens and take into account this factor when designing the heating processes for their products. Likewise, hazard analysis and critical control point (HACCP) plans should include an adequate heat treatment designed to kill heat-sensitive microorganisms (e.g., spoilage bacteria, infectious pathogens, some spore-formers), cooling at a rapid rate, and subsequent chilled storage to control the growth of spores that have survived the heat treatment.

IV. HIGH-INTENSITY ELECTRIC FIELD PULSES

The lethal effect of a pulsed electric field against microorganisms can potentially be used for cold pasteurization and commercial sterilization of foods. Microbial cells are inactivated when a certain threshhold electric field intensity is exceeded. The antimicrobial effect is due to the rupture of the cell membrane. Exposure of the cell to high-voltage electric field pulses can produce a potential difference between the inside and outside of a cell membrane. When the transmembrane potential exceeds the critical value of approximately 1 V, the pore formation becomes irreversible, leading to the destruction of the membrane functions and subsequently the cell death. Pulses ranging for 2 to 20 μsec with an electric field strength of 15 to 25 kV/cm are necessary for the destruction of microorganisms. For bacterial and fungal spores, higher voltage and longer duration pulses are required. Electric field pulses can be combined with temperature and lysozyme for the inactivation of spores.

Factors affecting the extent of microbial inactivation include: (1) temperature, pH, and ionic strength of food; (2) electric field intensity and duration of exposure; and (3) the type of microorganisms and their growth stage.[29] Gram-positive bacteria and yeasts are more resistant to pulsed electric fields than Gram-negative bacteria.[30] While the inactivation increases with an increase in the electric field intensity, exposure time, and temperature of food, every effort should be made to maintain the temperature below 30 to 40°C.

V. INTENSE LIGHT PULSES

Microorganisms on food and packaging can be reduced by high-intensity, short-duration pulses (1 μsec to 0.1 sec) of white light. The intense light pulses can be generated using gas-filled flash lamps or spark gap discharge apparatus. Both full- and filtered-spectrum light are used.[31] The filtered spectrum is achieved with glass or liquid filters and is more effective for microbial inactivation than full-spectrum light. The light pulses have a wavelength spectrum between 170 and 2600 nm; thus, both ultraviolet and near infrared wavelengths are used to inactivate microorganisms, including bacterial and fungal spores. The light pulses for a fraction of a second can result in inactivation of a substantially high number of microorganisms. Among the organisms that have been shown to be inactivated by this method are *E. coli*, *Staphylococcus aureus*, *Bacillus subtilis,* and *Saccharomyces cerevisiae*. This method can reduce the need for chemical disinfectants and preservatives, and can clear packaging material of hydrogen peroxide residues that may result from some preservation methods. In addition to its use for the sterilization of packaging material used in aseptic processing, this method can also be used for surface sterilization of animal carcasses, fruits and vegetables, bakery goods, solid dairy products, bulk sterilization of transparent homogeneous liquids, etc.[32] Fruits and vegetables, such as tomatoes, potatoes, bananas, and apples, can have an extended shelf-life through light pulse treatment. Fresh cut potatoes and apples exposed to light pulses do not undergo enzymatic browning for an extended period of time.

VI. OSCILLATING MAGNETIC FIELD PULSES

Inactivation of microorganisms by magnetic fields requires a magnetic field intensity or magnetic flux of 5 to 50 telsa, frequency of 5 to 500 kHz, and an exposure time of from 25 μsec to a few milliseconds.[33] Some evidence exists to document that such treatments can reduce microbial population densities by 2 logs. In pasteurization processes using oscillating magnetic fields (OMFs), food is sealed in plastic bags and subjected to 1 to 100 pulses in an OMF with a specific frequency and temperature, depending upon the food type and the microbial lethality required. High electrical resistivity (greater than 10 to 25 ohms per cm) is essential for many foods that have electrical resistivity in this range.

The advantages of using OMFs include: (1) avoidance of post-process contamination, as foods are treated inside a flexible film package; (2) minimal thermal denaturation of nutritional and organoleptic attributes of foods; and (3) reduced energy requirements for adequate processing. Only flexible films or paper packaging materials can be used, and reflective metallic packaging materials cannot be used.[33] Foods tested for antimicrobial effectiveness and sensorial acceptability using oscillating magnetic fields include orange juice, milk, and yogurt.

VII. HIGH HYDROSTATIC PRESSURE

This process involves filling a sterilized container with the food product, sealing the container, and placing it in a pressure chamber. Thereafter, the water surrounding the food is compressed by pumping additional liquid into the closed chamber, thereby subjecting the food to very high pressure (4000 to 9000 atm[34]). Because of the uniform and instantaneous pressure throughout the food, no deformation on the package occurs, and the processing time is not a function of container size. Also, the temperature remains essentially unchanged. Microbial cell death has been attributed primarily to the damage and loss of activity of the cytoplasmic membrane.[35] Pressure-induced membrane function impairment causes inhibition of amino acid uptake, probably due to protein denaturation in the membrane.[36] Smelt et al.[37] reported that bacteria with a relatively high content of diphosphatidylglycerol are more susceptible to inactivation. Some studies have indicated that denaturation of enzymes, such as membrane-bound ATPases, plays an important role in pressure-induced injury and inactivation of microorganisms.[38,39] For spores, a combination of high pressure and high temperature is necessary for inactivation. Under high pressure, bacterial spores germinate to vegetative cells and are then inactivated due to the effect of temperature.

High pressure weakens or denatures protein molecules in the food components because the hydrophobic and ion-pair bonds are disrupted. Covalent bonds are not affected. However, changes in the tertiary structure from the breaking and reformation of chemical bonds can alter the coagulation or gelation characteristics of some foods, giving them a unique and novel texture. The flavor or nutrient content of a food is generally not altered. Some of the applications of the high-pressure technology include gelation of surimi; manufacture of food purees, jams, and jellies from strawberries and marmalade from oranges; and a shelf-life extension of juices and milk.

VIII. MICROBIAL CONTROL BY PHYSICAL REMOVAL

A variety of approaches can be used to physically remove microorganisms from solid or liquid foods or reduce population densities, thus increasing the efficiency of subsequent intervention steps. These methods include:

1. *Centrifugation:* This method can be used to remove undesirable particles such as dust, leucocytes, etc. For example, centrifugation can be used to remove thermoduric bacterial spores in milk that are not inactivated by the normal pasteurization time and temperatures.
2. *Filtration:* This method can be used to remove yeasts, molds, and most bacterial cells and spores from liquid foods. Also, filtration of air is performed for spray drying of milk. By using this method, the natural flavor and nutrient content of food are not altered.
3. *Trimming:* This method is used to physically remove the grossly visible damaged and spoiled portions of fruits and vegetables and meat. For example, trimming the outer leaves of cabbage or lettuce; visible mold growth from hard cheeses, fermented sausages, and bread; fecal stain marks and abscesses from animal carcasses, etc.
4. *Washing:* Fruits and vegetables, shell eggs, and animal carcasses including beef, pork, lamb, etc. are commonly washed during processing. Also, chicken and turkey carcasses during processing are exposed to water several times. The effectiveness of hot water, steam, ozonated water, and water containing chlorine, acetic and propionic acids, lactic acid, tripolyphosphates, or bacteriocins of lactic acid bacteria have been assessed in removing the bacterial contamination. While washing alone can reduce the bacterial numbers, the efficacy can be increased by a combination of two or more of the above-named agents.

IX. CLEANING AND SANITATION

Microorganisms can gain access to foods from a variety of sources; therefore, proper cleaning and sanitation of all food production and distribution facilities are important critical control points in the reduction of microbial levels and must be incorporated in HACCP plans. Before surfaces can be sanitized, they must be cleaned (dirt and soil removed). This is critical because bacteria can form biofilms on the surface of stainless steel or other food contact or equipment surfaces, floors, drains, and even food surfaces. To remove biofilms, adequate amounts of detergents and hot water must be applied, and mechanical action with a scrub brush or pressure sprayer must be used to loosen the surface biofilms, which can be 25 to 30 microorganisms "deep." After the cleaned surface is rinsed, a sanitizing agent can be applied. Sanitizing agents will be ineffective if the biofilms are not first removed from surfaces.

A. Cleaning Agents (Soaps and Detergents)

The minerals, commonly calcium and magnesium, present in hard water replace the sodium in regular soap to form an insoluble curd. As a result, the ability of soap to emulsify grease and free dirt and films from surfaces is diminished. Detergents that are surface-active agents are usually biodegradable alkyl sulfates, ethoxylates, and their sulfates or alkylbenzenesulfonates. The action of detergents lifts and suspends the oily or greasy portion of soil by reducing interfacial and surface tension.

B. Sanitizers

Sanitizers are chemical compounds used to reduce the bacterial count on or within surfaces to safe levels. Sanitizer activity or effectiveness is affected by exposure time, temperature, concentration, water hardness, and surface cleanliness.[40] Chemical sanitizers can be classified into two classes: (1) halogens, which include chlorine (as a hypochlorite) and iodine (iodophors) compounds; and (2) surfactants, which include quaternary ammonium compounds (quats) and acid ionic compounds. Chemical sanitizing is done in two ways: either by immersion in an appropriate concentration of sanitizers or by rinsing, swabbing, or spraying with double the immersion concentration (an exception would be that the quat concentration is the same for both methods). Water also acts as a sanitizer when hot (above 76°C) or in the form of steam.

X. CONCLUSIONS AND OUTLOOK TO THE FUTURE

Research continues to demonstrate that food irradiation is a suitable process to control and potentially eliminate foodborne pathogens in a number of raw and cooked foods. In view of the consumers' demand for high-quality, convenient meals that require minimal preparation time, irradiation as a non-thermal treatment holds promise in combination with other intervention techniques for ensuring the safety of these new generation foods. Heat treatment is the most common and effective method in use today for the inactivation of microorganisms and may be used in combination with irradiation. The non-thermal processes outlined here show promise as alternative methods for enhancing food safety. Both conventional and non-thermal processes can be used in combination and along with other preservative factors to control the pathogens and enhance the safety and shelf-life of foods.

Early findings suggest that non-thermal technologies will induce only minimal quality changes in food; however, the comparative efficacy of these non-thermal physical treatments or processes in inactivating a specific organism, as well as changes in organoleptic attributes and the quality and shelf-life of foods, need to be assessed to determine which method is superior for a specific food. Much research on these emerging technologies has focused on applications; however, additional mechanistic studies are still needed, as is research regarding the expansion of these technologies to an industrial scale.

NOTE

Mention of brand or firm name does not constitute an endorsement by the U.S. Department of Agriculture over others of a similar nature not mentioned.

REFERENCES

1. Lee, P.R., Irradiation to prevent foodborne illness (from the Assistant Secretary for Health, U.S. Public Health Service), *J. Am. Med. Assoc.,* 4, 261, 1994.
2. CAST, *Radiation Pasteurization of Food,* Issue Paper No. 7, Council for Agricultural Science and Technology, Ames, IA, 1996.
3. Grecz, N., Rowley, D.B., and Matsuyama, A., The action of radiation on bacteria and viruses, in *Preservation of Food by Ionizing Radiation*, Vol. 2., Josephson, E.S. and Peterson, M.S., Eds., CRC Press, Boca Raton, FL, 1983.
4. Sudarmadji, S. and Urbain, W.M., Flavor sensitivity of selected raw animal protein foods to gamma radiation, *J. Food Sci.,* 37, 671, 1972.
5. Goresline, H.E., Ingram, M., Macuch, P., Macquot, G., Mossel, D.H, Niven, C.F., and Thatcher, F.S., Tentative classification of food irradiation processes with microbiological objectives, *Nature,* 204, 237, 1964.
6. WHO, *Safety and Nutritional Adequacy of Irradiated Food,* World Health Organization, Geneva, 1994.
7. Woods, R.J. and Pikaev, A.K., Interaction of radiation with matter, in *Applied Radiation Chemistry: Radiation Processing*, John Wiley & Sons, New York, 1994.
8. CAST, *Ionizing Energy in Food Processing and Pest Control.* II. *Applications,* Task Force Report No. 115, Council for Agricultural Science and Technology, Ames, IA, 1989.
9. Crawford, L.M. and Ruff, E.H., A review of the safety of cold pasteurization through irradiation, *Food Control,* 7(2), 87, 1996.
10. WHO, *Food Irradiation* [press release], WHO/68, Sept. 19, World Health Organization, Geneva, 1997.
11. Thayer, D.W., Boyd, G., Fox, J.B, Jr., and Lakritz, L., Effects of NaCl, sucrose, and water content on the survival of *Salmonella typhimurium* on irradiated pork and chicken, *J. Food Prot.,* 58, 490, 1995.
12. Kampelmacher, E.H., Irradiation of food: a new technology for preserving and ensuring the hygiene of foods, *Fleischwirtschaft,* 64, 322, 1984.
13. Thayer, D.W. and Boyd, G., Elimination of *Escherichia coli* O157:H7 in meats by gamma irradiation, *J. Food. Sci.,* 57, 848, 1993.
14. Clavero, M.R.S, Monk, J.D., Beuchat, L.R., Doyle, M.P., and Brakett, R.E., Inactivation of *Escherichia coli* O157:H7, salmonellae, and *Campylobacter jujeni* in raw ground beef by gamma irradiation, *Appl. Environ. Microbiol.,* 60, 2069, 1994.
15. Niemand, J.G., van der Linde, H.J., and Holapfel, W.H., Shelf-life extension of minced beef through combined treatments involving radurization, *J. Food Prot.,* 46, 791, 1983.

16. Niemand, J.G., van der Linde, H.J., and Holapfel, W.H., Radurization of prime beef cuts, *J. Food Prot.,* 44, 677, 1981.
17. Lefebvre, N., Thibault, C., and Charbonneau, R., Improvement of shelf-life and wholesomeness of ground beef by irradiation. 1. Microbial aspects, *Meat Sci.,* 32, 203, 1992.
18. Nickerson, J.T.R., Licciardello, J.J., and Rosivalli, L.J., Radurization and radicidation: fish and shellfish, in *Preservation of Food by Ionizing Irradiation,* Vol. 3, Josephson, E.S. and Peterson, M.S., Eds., CRC Press, Boca Raton, FL, 1983, pp. 13–82.
19. Przybylski, L., Finerty, M., and Grodner, R., Extension of shelf-life of fresh channel catfish fillets under modified atmosphere packaging and low-dose irradiation, *J. Food Sci.,* 54, 269, 1989.
20. Novak, A.F., Liuzzo, J.A., Grodner, R.M., and Lovell, R.T., Radiation pasteurization of Gulf coast oysters, *Food Technol.,* 20, 201, 1966.
21. Sivinski, J.S., Control of trichinosis by low-dose irradiation of pork. *Food Irradiation Newsl.,* 2, 8, 1985.
22. FDA, Irradiation in the production, processing and handling of food. *Fed. Regist.*, 50(14), 21 CFR part 179, 29658, 1985.
23. Dubey, J.P. and Thayer, D.W., Killing different strains of *Toxoplasma gondii* tissue cysts by irradiation under defined conditions, *J. Parasitol.,* 80, 764, 1994.
24. King, B.L. and Josephson, E.S., Actions of radiations on protazoa and helminths, in *Preservation of Foods by Ionizing Irradiation,* Vol. 2, Josephson, E.S. and Peterson, M.S, Eds., CRC Press, Boca Raton, FL, 1983, pp. 245–267.
25. Tomlins, R.I. and Ordal, Z.J., Thermal injury and inactivation in vegetative bacteria, in *Inhibition and Inactivation of Vegetative Microbes,* Skinner, F.A. and Hugo, W.B., Eds., Academic Press, New York, 1976.
26. Juneja, V.K., Snyder, O.P., and Marmer, B.S., Thermal destruction of *Escherichia coli* O157:H7 in beef and chicken: determination of D and z values, *Int. J. Food Microbiol.,* 35, 231, 1997.
27. Hansen, N.H., and Riemann, H., Factors affecting the heat resistance of non-sporing organisms, *J. Appl. Bacteriol.,* 26, 314, 1963.
28. Juneja, V.K., Klein, P.G., and Marmer, B.S., Heat shock and thermotolerance of *Escherichia coli* O157:H7 in a model beef gravy system and ground beef, *J. Appl. Microbiol.,* 84, 677, 1997.
29. Qin, B., Vega-Mercado, H., Pothakamury, U.R., Barbosa-Cánovas, G.V., and Swanson, B.G., Application of pulsed electric fields for inactivation of bacteria and enzymes, *J. Franklin Inst.,* 332A, 209, 1995.
30. Hülsheger, H., Potel, J., and Niemann, E.G., Killing of bacteria with electric pulses at high field strength, *Radiat. Environ. Biophys.,* 20, 53, 1981.
31. Barbosa-Cánovas, G.V., Pothakamwry, U.R., Palou, E., and Swanson, B.G., Emerging technologies in food preservation, in *Nonthermal Preservation of Foods*, Marcel Dekker, New York, 1998.
32. Dunn, J.E., Clark, R.W., Asmus, J.F., Pearlman, J.S., Boyer, K., and Paurchaud, F., Methods and Apparatus for Preservation of Foodstuffs, International Patent Application No. WE 88/03369, 1988.
33. Hofmann, G.A., Deactivation of Microorganisms by an Oscillating Magnetic Field, U.S. Patent No. 4,524,079, 1985.
34. Pothakamury, U.R., Barbosa-Cánovas, G.V., Swanson, B.G., and Meyer R.S., The pressure builds for better food processing, *Chem. Eng. Prog.*, March, 45–55, 1995.
35. Morita, R.Y., Psychrophilic bacteria, *Bacteriol. Rev.,* 39, 144, 1975.

36. Paul, K.L. and Morita, R.Y., Effects of hydrostatic pressure and temperature on the uptake and respiration of amino acids by a facultative psychrophilic marine bacterium, *J. Bacteriol.*, 108, 835, 1971.
37. Smelt, J.P.P.M., Rijke A.G.F., and Hayhurst, A., Possible mechanism of high pressure inactivation of microorganisms, *High Press Res.,* 12, 199, 1994.
38. Mackey, B.M., Forestiere, K., and Isaacs N., Factors affecting the resistance of *Listeria* monocytogenes to high hydrostatic pressure, *Food Biotechnol.,* 9, 1, 1995.
39. Marquis, R.E., and Bender G.R., Barophysiology of prokaryotes and proton translocating ATPases, in *Current Perspectives in High Pressure Biology,* Jannisch, H.W., Marquis, R.E., and Zimmerman, A.M., Eds., Academic Press, London, pp. 65–73, 1987.
40. Bakka, R.L., Making the Right Choice: Chemistry of Cleaning, Nature of Soil, Water Quality, Detergent Terminology, seminar material, Klenzade, Division of Ecolab, Inc., St. Paul, MN, 1991.
41. Farkas, J., Principles of food irradiation, in *Handouts of IFFIT International Training Course on Food Irradiation,* International Facility for Food Irradiation Technology, Wageningen, The Netherlands, 1980, pp. 1–7.
42. Palumbo, S.A., Jenkins, R.K., Buchanan, R.L., and Thayer D.W., Determination of irradiation D values for *Aeromonas hydrophila, J. Food Prot.,* 49, 189, 1986.
43. Tarkowski, J.A., Stoffer, S.C.C., Beumer, R.R., and Kampelmacher, E.H., Low dose gamma irradiation of raw meat. I. Bacteriological and sensory quality effects in artificially contaminated samples, *Int. J. Food Microbiol.,* 1, 13, 1984.
44. Hutanen, C N., Jenkins R.K., and Thayer D.W., Gamma irradiation sensitivity of *Listeria monocytogenes, J. Food Prot.,* 9, 610, 1989.
45. Thayer, D.W., Boyd G., Mueller W.S., Lipson C.A., Hayne W.C., and Baer S.H., Radiation resistance of *Salmonella, J. Indust. Microbiol.,* 5, 383, 1990.
46. Quinn, D.J., Anderson A.W., and Dyer, J.F., The inactivation of infection and intoxication microorganisms by irradiation in seafood, in *Microbiological Problems in Food Preservation by Irradiation*, P1-199/9, IAEA, Vienna, 1967, pp. 1–13.
47. Thayer, D.W., Boyd, G., Fox, J.B, Jr., Lakritz, L., and Hampson, J.W., Variations in radiation sensitivity of foodborne pathogens associated with suspending meat, *J. Food Sci.,* 60, 63, 1995.
48. Matches, J.R. and Liston, J., Radiation destruction of *Vibrio parahaemolyticus, J. Food Sci.,* 36, 339, 1971.
49. Gaze, J.E., Brown, G.D., Gaskill D.E., and Banks J.G., Heat resistance of *L. monocytogenes* in homogenates of chicken, beef, steak and carrot, *Food Microbiol.*, 6, 251, 1989.
50. Sörqvist, S., Heat resistance of *Campylobacter* and *Yersinia* strains by three methods, *J. Appl. Bacteriol.*, 67, 543, 1989.
51. Palumbo, S.A., Williams, A.C., Buchanan, R.L., and Phillips, J.G., Thermal resistance of *Aeromonas hydrophila, J. Food. Prot.,* 50(9), 761, 1987.
52. Goodfellow, S.J. and Brown, W.L., Fate of *Salmonella* inoculated into beef for cooking, *J. Food Prot.,* 41, 598, 1978.
53. Juneja, V.K. and Marmer, B.S., Growth of *Clostridium perfringens* from spore inocula in *sous-vide* turkey products, *Int. J. Food Microbiol.,* 32, 115, 1996.
54. Juneja, V.K., Eblen, B.S., Marmer, B.S., Williams, A.C., Palumbo, S.A. and Miller, A.J., Thermal resistance of non-proteolytic type B and type E *Clostridium botulinum* spores in phosphate buffer and turkey slurry, *J. Food Prot.,* 58, 758, 1995.

13 Biologically Based Technology for the Control of Postharvest Diseases of Fruits and Vegetables

Samir Droby, Charles L. Wilson, Michael Wisniewski, and Ahmed El Ghaouth

CONTENTS

I. INTRODUCTION

Fresh fruits and vegetables are considered an important component of a healthy and balanced diet. They are an excellent source of dietary fiber, vitamins, carbohydrates, and antioxidants. Greater consumer awareness that diet and health are linked has resulted in a rapid increase in the demand for fresh fruits and vegetables. Consumption of whole and lightly processed fruits and vegetables is also increasing due to greater selection and year-round availability. Contamination of fruits and vegetables by pesticides, plant and human pathogens, and other pollutants has been and continues to be a major concern to both consumers and the fruit and vegetable industry.

Concerns over pesticide residues are at the top of the list of food safety issues for consumers. Due to questions about the potential carcinogenicity of fungicides, a number of key postharvest fungicides have been recently banned or are undergoing re-evaluation. In addition, some of the fungicides registered for postharvest use, particularly benzimidazole, are becoming ineffective due to the development of fungicide-resistant strains of postharvest pathogens

World trends are moving toward reduced pesticide use. In response, several physical and biological means have been evaluated as safer alternatives to the use of synthetic, chemical fungicides. The use of natural plant products (essential oils and plant extracts), biocontrol agents (yeast and bacterial antagonists), and nonselective biofungicides (sodium carbonate, sodium bicarbonate, active chlorine, and sorbic acid) are among the approaches currently being evaluated for the control of postharvest diseases.

In this chapter, special attention will be given to the use of naturally occurring microbial antagonists for the control of postharvest diseases of fruits and vegetables. Recent advancements in the search for biocontrol agents of postharvest diseases and the problems facing the widespread implementation of the technology will be discussed.

II. MICROBIAL CONTROL OF POSTHARVEST DECAY OF FRUITS AND VEGETABLES

The infection process responsible for the postharvest decay of fruits and vegetables can be initiated at anytime between flowering and fruit maturity or during harvesting and subsequent handling and storage. In the first case, the disease process remains quiescent until the physiological conditions of the produce are favorable for the proliferation of the pathogen. In the latter case, surface wounds inflicted during harvest and subsequent handling serve as the infection court for weak necrotrophic pathogens that only colonize damaged tissue.

Latent infections are difficult to control with antagonists, as infection may occur at different stages of fruit development. In order to control latent infections, microbial antagonists must be present on the fruit surface before infection occurs. Examples of latent infections on fruits include *Colletotrichum* infections of avocado, mangoes, bananas, and papayas (Eckert and Ogawa 1988); stem end rots of mangoes, avocado, and citrus; *Botrytis cinerea* infections of grapes and strawberries (Snowdon, 1990); and *Monilinia fructicola* on stone fruits (Kable, 1971). On the other hand, many pathogens responsible for postharvest decay are able to infect host tissue only through surface wounds resulting from either mechanical or physiological injury. Postharvest losses resulting from wound pathogens are very significant in most commodities (Snowdon, 1990).

Research on biocontrol of postharvest diseases has concentrated mainly on searching for microorganisms that are antagonistic to wound pathogens. Typically, infection of wounds by spores of the pathogen is very rapid (often within 24 hours). Thus, rapid colonization and growth of an antagonist in the wound site are key characteristics of a successful antagonist. Fokkema (1992) suggested that biological control of infection of fresh wounds has the advantage of the absence of resident microflora competing with the antagonists. In addition, favorable nutritional conditions, as well as a suitable microenvironment, also favor the use of an antagonist. In order to select a successful biocontrol agent, detailed knowledge of the epidemiology of the pathogen, the infection process, and disease development is critical to determine the type of antagonist needed to interrupt the establishment of the pathogen in the tissue.

The first step before initiating a screening program to identify potential antagonists should be a consideration of the features needed from the biocontrol agent. In general, they include: (1) good colonization of fruit surfaces and wounds, (2) rapid growth in surface wounds, (3) effective utilization of nutrients at low concentrations present in the wound, and (4) an ability to survive and develop better than the pathogen on the surface of the commodity and at infection sites under a wide range of temperature, pH, and osmotic conditions.

A method designed for optimizing the isolation and screening of yeast antagonists with potential for controlling postharvest diseases was developed by Wilson et al. (1993). The method is based on the application of fruit surface washings to fruit wounds, which are subsequently challenged with spores of a postharvest pathogen. After a few days of incubation, microorganisms are isolated from the non-infected wound sites. Interestingly, it was found that yeast colonies were usually predominant in the growth medium. Following random selection, according to color and morphological features, pure cultures are then tested for antagonism via *in vivo* tests, as described by Droby et al. (1989). These procedures favor the selection of yeasts that colonize wounds rapidly.

Several bacteria and yeasts have been shown to protect a variety of harvested commodities against a number of postharvest pathogens (Chalutz and Wilson, 1990; Chand-Goyal and Spotts, 1996; Droby et al., 1989, 1991; El Ghaouth et al., 1998; Gullino et al., 1991; Janisiewicz, 1988, 1994; Janisiewicz and Roitman, 1988; Leibinger et al., 1997; Lurie et al., 1995; Roberts, 1990a,b; Stretch, 1989; Teixidoi et al., 1998). In particular, yeasts naturally occurring on fruits and vegetables have been targeted by many workers as potential antagonists of postharvest diseases because they exhibit a number of traits that confer greater potential for colonizing wound sites. As indicated by Janisiewicz (1988), yeasts can colonize a plant surface for long periods under dry conditions and produce extracellular polysaccharides that enhance their survivability and restrict both colonization sites and the flow of germination cues to fungal propagules. Antagonistic yeasts are able to use available nutrients rapidly to proliferate and are affected minimally by postharvest fungicides. While the mechanism of action of yeast antagonists in inhibiting postharvest pathogens has not been fully understood, there are several lines of evidence to indicate that they do not rely on the production of antibiotic substances (Droby and Chalutz, 1994). This feature would be advantageous, as the use of antibiotic-producing antagonists on food could be problematic with regard to the development of antibiotic-resistant strains of animal and plant pathogens.

Microbial biocontrol agents possess a number of important advantages over traditional chemical pesticides which make their commercial outlook particularly promising, as, in general, they: (1) are considered nonhazardous to humans and animals; (2) are biodegradable and environmentally friendly; (3) attack specific target organisms, leaving other beneficial organisms unaffected; (4) are easy to genetically modify; and (5) can be commercially developed with relative ease. These advantages, however, are counterbalanced by a number of limitations which include: (1) the sensitivity of most of the currently marketed microbial control agents to adverse environmental conditions such as extreme dryness, heat, and cold; (2) limited shelf-life; (3) limited biocontrol efficacy in situations where several pathogens are involved in decay development; and (4) limited effectiveness under high disease pressure.

III. MICROBIAL BIOFUNGICIDES: FROM THE LABORATORY TO THE MARKETPLACE

The size of a market is a major deciding factor for a company investing its resources in the development and marketing of a biocontrol agent. For that reason, candidate organisms must exhibit activity against a range of pathogens on a number of commodities. Another consideration is the ease with which the antagonist can be mass produced in fermentors and packaged in a stable and dispensable form. These factors should be taken into consideration early at the initial stage of product development.

Antagonists should meet certain criteria to be developed successfully for commercial use on harvested crops (these traits have been described by Wilson and Wisniewski, 1989). Antagonists should: (1) be genetically stable; (2) be effective at low concentrations; (3) not be fastidious as far as nutrient requirements; (4) be able to survive adverse environmental conditions (including low temperatures and controlled atmosphere storage); (5) be effective against a wide range of pathogens on a variety of fruits and vegetables; (6) be amenable to production on an inexpensive growth medium; (7) be amenable to a formulation with a long shelf-life and that is easy to dispense; (8) not produce metabolites that are deleterious to human health; (9) be resistant to pesticides; (10) be compatible with commercial processing procedures; and (11) be nonpathogenic to the host commodity.

Registration is required by regulatory agencies (e.g., U.S. Environmental Protection Agency) before any biocontrol agent can be used commercially. Although the registration process is not as expensive or time consuming as it is for synthetic chemical fungicides, this requirement does affect the development process (Whitesides et al., 1994). As part of the development process, semi-commercial pilot tests need to be conducted using relatively large quantities of fruits treated in a manner that resembles commercial conditions as closely as possible. In the final stages of development, the product should be tested under packinghouse conditions. Registration, large-scale production, formulation of the biocontrol agent, and its semi-commercial and commercial testing are costly operations that can be conducted only by or in association with a private company. Indeed, in recent years, several private firms have been involved in the development of biocontrol products for the control of postharvest diseases of fruits and vegetables. Among these products is the Aspire™ biofungicide (Ecogen; Langhorne, PA), based on the yeast *Candida oleophila*, and the EcoScience (Orlando, FL) Biosave 111 and 110, both isolates of *Pseudomonas syringae* that are registered and recommended for the control of postharvest decay of citrus and pome fruit.

The development of Aspire™ involved extensive pilot testing on a variety of fruits (Hofstein et al., 1994). A pilot packing line was specifically designed and constructed for testing the efficacy of Aspire™ on citrus and pome fruit. The pilot packing line was used to simulate different commercial treatments practiced in a wide array of packinghouses. Several factors such as: (1) the site of antagonist application, (2) the method of application (dipping, drenching, atomization), and (3) integration of the antagonist with different enhancers were easily evaluated. It is evident from our experience that the positioning of the biocontrol applicators, as well as the timing of the application, can markedly influence the extent of decay control. The pilot packing line was used for evaluation and selection of the most suitable product formulations as well. Aspire™ was further subjected to commercial tests under packinghouse conditions in a number of countries.

Concern has been raised regarding health and safety in relation to the mass introduction of antagonists on our food. Some of the antagonists reported to effectively control postharvest diseases have also been reported to be opportunistic pathogens on immune-compromised humans. Although this might pose an obstacle to public acceptance of this technology, these antagonists are indigenous to agricultural commodities and humans are continuously exposed to them. Even though these antagonists are introduced in large numbers to the surface of a commodity, they survive and grow only in very restricted sites on the fruit surface (surface wounds). After their introduction on the intact fruit surfaces, antagonist populations are reduced to the level of natural epiphytic microflora in a very short period of time. Thus, in spite of the rigorous tests needed to verify their safety to humans and the environment, the use of microbial antagonists to control postharvest diseases of fruits and vegetables is a commercially viable option to the use of synthetic fungicides.

IV. COMMERCIAL APPLICATION: REASONS FOR DECREASING EFFICACY AND INCONSISTENCY

The performance of biocontrol agents cannot by itself be expected to equal that of a synthetic fungicide. Although reports in the literature indicate that some biocontrol agents are as effective as fungicides, in most cases their performance under commercial conditions has been sometimes inconsistent. This may be attributed to factors that have not yet been fully identified. The following factors, however, may have a role in the reduced efficacy of postharvest biocontrol agents under large-scale and commercial conditions.

A. Fermentation and Formulation

When attempting to scale-up the production and develop a commercial formulation of a certain biocontrol agent, it is essential that the microbial cells retain their attributes as colonizers and antagonists on fruit surfaces. Industrial mass production is accomplished under conditions quite different from those in laboratory shake culture. The process must be cost effective and rely on industrial byproducts as nutrients, and fermentation must be completed within 24 to 30 hours (Hofstein et al., 1994). Downstream processing and formulation involve various steps, such as drying, addition of volume materials (inert ingredients), adhesives, emulsifiers, and adjuvants. All these actions may adversely affect the characteristics of the organism that have been identified as important to the biocontrol.

Maintaining cell viability is fundamental in formulating antagonists as commercial products. Biocontrol agents must have a storage life of at least one or two years (Baker and Henis, 1990; Connick et al, 1990). Reproducibility in the performance of a formulated antagonist is the most important requirement of a reliable product. This can be achieved by using strict guidelines for quality assurance (QA). As reported by Hofstein et al. (1994), *in vitro* tests were used for the evaluation of the activity of each fermentation batch of the yeast antagonist *C. oleophila*. This approach, however, though rapid and simple, does not provide crucial information about the biocontrol activity of the antagonist on the fruit. Therefore, *in vivo* tests on fruit were conducted as part of the QA procedure for fermentation batches and formulated products. The main limitation to this approach is time, as it requires one week to obtain results.

B. Delivery Method

The type of delivery system used to apply a biocontrol agent can affect its performance. Before making a decision on the delivery system to be used to apply biocontrol agents on harvested commodities, one should examine the delivery systems currently being used for the application of fungicides. In general, chemical fungicides are applied to a commodity via dip or dump tanks or on-line spray or drench applicators, or as a mixture with coating waxes. Dip tanks are used in many postharvest handling systems to clean and disinfect the produce. In this method, the fruit is submerged for a few minutes in tanks containing saltwater (necessary for floating), and fungicides, to allow fungicidal activity on pathogenic propagules and any infections inflicted during harvesting. In some cases, heated solutions of fungicides are employed to increase the efficacy of disinfection and to reduce the concentration of the fungicide. This method is not very suitable for the application of microbial antagonists because contamination of the dump solution with pathogen propagules can adversely affect the efficacy of a biocontrol agent. If using a dip tank is the only application method available, the fruit should be disinfected with chlorine or some other type of disinfectant prior to treatment with the biocontrol agent. This will help ensure that the dip tank containing the biocontrol agent will be kept as free as possible from a buildup of pathogen propagules.

On-line drenches and sprays are also common delivery methods for fungicides, anti-scald agents, and waxes in most packinghouses (Sommer 1985). Several different applicators are currently in use. Various materials may be sprayed directly onto fruit with cone-type nozzles using either hydraulic or air pressure. Some packinghouses drip materials onto rotating brushes that coat the fruit with the compound. Recently, there has been growing interest in the use of controlled-droplet applicators that create a very fine mist. Devices that place an electronic charge on the droplet have been evaluated in conjunction with controlled-droplet applicators and appear to increase the amount of residue on the fruit. The controlled-droplet systems are non-recirculating and require only small volumes of solution. The on-line drench and spray systems are the most suitable for the application of microbial biocontrol agents. The yeast product, Aspire™, was successfully applied onto citrus fruit using an on-line drench system (Droby et al., 1999). This application system allowed for the combining of the biocontrol agent with reduced concentration of fungicides and/or any other additives needed to enhance biocontrol efficacy.

C. Inoculum Pressure

For postharvest biocontrol, the most important factor determining the efficacy of any microbial antagonist is the implementation of a stringent sanitation program that reduces pathogenic propagules in water systems and on rollers, belts, brushes, and packinghouse floors. As already reported, the performance of biocontrol agents is much more sensitive than that of synthetic chemicals to the effects of pathogen concentration. Chalutz and Wilson (1990), Janisiewicz (1988), and McLaughlin et al. (1990) have demonstrated that, when pathogen spore concentrations increase, biocontrol efficacy decreases. For postharvest biocontrol to be successful, packinghouses must adopt a program to minimize the exposure of fruit to pathogens. Inoculum originating from the orchard may be greatly reduced during and after harvest by minimizing contamination of bins with soil and decayed fruit (Michailides and Spotts, 1986). In the packinghouse, a work area free of decayed fruit and plant remains would also markedly reduce inoculum levels. Contaminated belts, rollers, cups, brush beds, and floors can dramatically affect the levels of fruit decay if contaminated with fungal spores (Spotts, 1994). These surfaces must be sanitized.

D. Physiological Status of the Fruit

A principal factor that impacts the preservation of harvested commodities is the physiological status of the tissue. Once harvested, commodities are senescing rather than developing. Consequently, the susceptibility of fruit tissue to pathogen attack increases due to weakened natural defense mechanisms, as well as partial degradation of cell walls and subsequent increased leakage of solutes. Over-mature fruit is much more susceptible to fungal attack than fruit picked at optimal maturity (Boonyakiat et al., 1987). Roberts (1990) demonstrated that maturity markedly affected biocontrol activity of the yeast antagonists *Cryptococcus laurentii*, *C. flavus,* and *C. albidus* used to control *Mucor piriformis* on pears. Application of one of the above yeast antagonists on surface wounds of freshly harvested d'Anjou pears resulted in complete inhibition of the rot. Similar treatment on ripened pears gave much less disease control. Droby and co-workers (unpublished data) found that biocontrol efficacy and consistency could be improved when biocontrol agents are applied to citrus fruit soon after harvest (within 24 hours). Under commercial conditions, however, treatment of fruits immediately after harvest may not be feasible in certain cases and can be delayed for several days. In this situation, only a combination of a biocontrol agent with low rates of chemical fungicides has been able to provide the level of control needed by the industry.

V. MODE OF ACTION OF YEAST BIOCONTROL AGENTS

A full understanding of the interactions taking place at the wound site between the antagonist cells, the host tissue, and the pathogen is a prerequisite for the development of successful biocontrol strategies. A complex interaction involving host resistance and wound responses, as well as interaction with other microorganisms, has to be taken into consideration. This conceptualization raises some critical questions: (1) What are the effects of antagonists on wound healing and host resistance? (2) How important and widespread are the direct effects of antagonists on pathogens? (3) How do incidental microorganisms or mixtures of antagonists affect pathogen/antagonist interactions? (4) How does nutrient/chemical composition at the wound site affect the antagonist, other microflora, the infection process, and the wound response?

A. Indirect Interactions

To inhibit infection, the antagonist must be present in the wound site prior to arrival of the pathogen or within a short period thereafter. Recent studies on biological control of postharvest diseases of fruits and vegetables have reported the use of antagonistic microorganisms that multiply rapidly, colonize the wound, and outcompete the pathogen for nutrients and space. The yeast *Pichia guilliermondii* (formerly identified as *Debaryomyces hansenii*), *Cryptococcus laurentii*, *Aureobasidium pullulans*, *Candida* spp., *Sporobolomyces roseus*, and the bacteria *Enterobacter cloacae*, *Pseudomonas cepacia*, and *P. syringae* have all been reported to rapidly and extensively colonize the wound site (Droby et al., 1989; Gullino et al., 1991; Janisiewicz et al., 1994; Roberts, 1990a,b; Shefelbine and Roberts, 1990; Smilanick et al., 1992, 1993; Wilson and Wisniewski, 1989; Wisniewski et al., 1989). We observed that the US-7 isolate of *P. guilliermondii* multiplied very rapidly at the wound site and increased in numbers by 1 to 2 orders of magnitude within 24 hours, while the pathogen spores had just started to germinate and grow (Droby et al., 1989).

Several lines of evidence support the assumption that inhibition of pathogen development by the antagonist involves competition for nutrients (Droby et al., 1989). Such competition was demonstrated by *P. guilliermondii* in culture, when both the antagonist and the pathogen were co-cultured in minimal synthetic medium or in wound leachate solutions. The efficacy of the yeast could be markedly reduced by the addition of nutrients to the spore suspension used for inoculation. Similarly,

E. cloacae, a bacterium antagonist, inhibited germination of *Rhizopus stolonifer* spores through nutrient competition (Wisniewski et al., 1989). In both studies, indirect evidence was provided to demonstrate the role of competition for nutrients as part of the mode of action of these two antagonists: (1) inhibition of spore germination or growth of the pathogen during co-culturing with the antagonist was demonstrated; (2) inhibition of the pathogen was dependent on the concentration of the antagonist propagules; and (3) partial or complete reversal of inhibition could be achieved by the addition of exogenous nutrients.

In most reports on biological control of postharvest diseases of fruits and vegetables, a quantitative relationship has been demonstrated between the antagonist concentration and the efficacy of the biocontrol agent. Thus, a delicate balance apparently exists at the wound site between the number of antagonist cells and the pathogen propagules which affects the outcome of the interaction and determines whether or not the wound becomes a site of infection. Manipulation of the initial concentration of the antagonist cells and/or the fungal spores clearly affects infection. On the other hand, we have shown that the number of antagonist cells at the wound site will not always determine its efficacy. Our data suggested that active multiplication and growth of the US-7 yeast cells were required for the yeast to exhibit its biocontrol activity. This was demonstrated by using a mutant of *P. guilliermondii* which lost its biocontrol activity against *P. digitatum* on grapefruit and against *B. cinerea* on apples, even when applied to the wound at concentrations as high as 10^{10} cells/mL (Droby et al., 1991). The cell population of this mutant remained constant at the wound site during the incubation period, while that of the wild type increased 60- to 100-fold within 24 hours. Failure of the mutant to inhibit spore germination of the pathogen in culture on a minimal salt medium suggested that this mutant lost its ability to utilize some nutrients and grow in culture, as well. This could be the reason for its non-efficacy.

B. Direct Interactions

Direct parasitism by the antagonist of a pathogen's vegetative growth and propagules has been reported to play a role in biological control against soilborne and foliar diseases. In this regard, studies have been conducted with *Trichoderma* (Elad et al., 1983). In the postharvest arena, very little information is available on biological control agents that directly parasitize pathogens. Wisniewski et al. (1991), however, have shown that yeast cells of *P. guilliermondii* that were in direct contact with *B. cinerea* hyphae appeared to be lying within a depression of the hyphal cell wall. *P. guilliermondii* similarly attached to hyphae of *P. digitatum.* In addition, the yeast appeared firmly imbedded within these depressions and surrounded by an extracellular matrix. This attachment was blocked when the yeast cells or the pathogen hyphae were exposed to compounds that affect protein integrity, or when respiration was inhibited. This suggested that the proteins involved in attachment are most likely located on the yeast cell surface and/or within the surrounding extracellular matrix. The close contact of the yeast cells with the fungal cell wall would also facilitate the efficient uptake and depletion of nutrients from the immediate vicinity surrounding the fungus.

C. Production of Cell-Wall Hydrolases

P. guilliermondii was also found to exhibit high levels of ß-1,3-glucanase activity when cultured on various carbon sources or on cell walls of several fungal pathogens (Wisniewski et al., 1991). The ability to produce high levels of ß-1,3-glucanase by the yeast has been suggested to be associated with the firm attachment of the yeast cells to fungal hyphae and pitting observed in some areas on fungal mycelium (Wisniewski et al., 1991). In addition, the kinetics of ß-1,3-glucanase production as related to growth may suggest that firm attachment of the yeast cells to fungal hyphae and to plant tissue might be facilitated by extensive production of ß-1,3-glucanase.

Isolate US-7 of *P. guilliermondii* investigated by Wisniewski et al. (1991) appears to produce and secrete high levels of exo-ß-1,3-glucanase into the growth medium. Monitoring the kinetics of an exo-ß-1,3-glucanase production during growth of the yeast revealed that the activity detected in the yeast culture filtrate reached its highest activity after 36 to 48 hours growth at 25°C and correlated with the growth rate of yeast in NYDB medium.

In further studies of the glucanase activity of US-7, an exo-ß-1,3-glucanase with molecular weight of 45 KDa was identified (Avraham, 1994). The exoglucanase activity was confirmed by its ability to hydrolyze various substrates. Characterization of the optimal conditions for enzyme activity has shown that the exo-ß-1,3-glucanase produced by the yeast *P. guilliermondii* exhibited its maximal activity at pH 4.0 to 5.0. Interestingly, however, the exoglucanase retained about 80% of its activity over a wide range of pH (3.0 to 6.0). Low activity was evident at extreme acidic and basic conditions. Optimal temperature for its activity was 50°C. In addition, enzymatic activity was not affected by glucose up to a concentration of 50 m*M* at pH 5. At glucose concentrations of 500 or 1400 m*M* (pH = 5), the activity was reduced by 35 and 64%, respectively.

D. Production of Extracellular Materials

In our early studies, conducted to evaluate antagonist-pathogen interactions at the wound site, extensive production of extracellular materials was found to be present around yeast cells in the wound (Wisniewski et al., 1988). To study the possible role of this extracellular material as part of the mode of action by which *P. guilliermondii* inhibits postharvest pathogens, we have recently observed that an extracellular material extracted from the surface of the yeast cells exhibited antifungal activity against several postharvest pathogens when tested *in vitro* (Droby et al., 1995). Both spore germination and germ tube elongation were inhibited. These results suggested the possible involvement of this extracellular material in the interaction of the yeast with the pathogen. In addition, the extracellular material also inhibited infection and development of green mold decay caused by *P. digitatum* on wounded grapefruits (Droby et al., 1995). Thus, the biocontrol activity of *P. guilliermondii*, and possibly other yeast biocontrol agents, may be dependent not only on its ability to rapidly colonize the wound site and compete for nutrients, but also on its ability to attach firmly to hyphae of the pathogen and to produce extracellular materials, as well as cell-wall-degrading enzymes.

E. Induction of Resistance Mechanisms in the Host

Both physical and biological agents elicit resistance responses in harvested fruits and vegetables (Droby et al., 1993b; Wilson et al., 1993, 1994). Heat treatment, wounding, gamma radiation, ultraviolet-C light, antagonists, attenuated strains, and natural compounds have all been suggested as elicitors of resistance in harvested crops. In this review, we will discuss induction of resistance by antagonists of postharvest diseases. Indeed, some reports have indicated that certain postharvest biocontrol agents may interact with the host tissue, in particular with wounded surfaces, leading to enhanced wound-healing processes. Direct evidence to support this possible mode of action, however, is lacking.

We examined the production of ethylene in yeast-treated tissues. When cell suspensions of the US-7 yeast antagonist were placed on surface wounds of grapefruit, pomelo, table grapes, or carrot root tissue, enhanced ethylene production was evident in all tissues (Droby et al., 1994). When cultured *in vitro*, the yeast cells themselves did not produce ethylene. In addition, induction of ethylene was demonstrated in grapefruit peel discs treated with isolated extracellular material of the yeast *P. guilliermondii*, indicating its possible role as an elicitor of ethylene. Ethylene applied exogenously to the discs or to whole grapefruits also induced resistance to *P. digitatum* infections (Droby et al., 1994). In addition, application of US-7 to lemon wounds enhanced production of the phytoalexin scoparone (Rodov et al., 1994). Thus, the induction of both ethylene and scoparone by fruit tissues in response to the application of a yeast antagonist suggests the involvement of host resistance mechanisms in the biocontrol activity of the yeast. The nature and mechanism of this induction are yet to be elucidated. The yeast antagonist *Candida saitoana* was also found to induce chitinase and cause deposition of papillae along host cell walls in apple tissue (El Ghaouth et al., 1998), providing another example of the induction of host resistance.

VI. ENHANCEMENT OF BIOCONTROL EFFICACY UNDER COMMERCIAL CONDITIONS

Microbial biocontrol agents commercially available for the control of postharvest diseases have been criticized for not providing as consistent a level of disease control as synthetic fungicides. The first generation of biological control agents of postharvest spoilage has relied on the use of single antagonists, and perhaps it is unrealistic to expect disease control comparable to synthetic fungicides. It can be anticipated that the second generation of biologically based controls will utilize means of enhancing and perhaps synergizing control of postharvest diseases with antagonists. The addition of various nutrients has been shown to enhance the biocontrol activity of bacterial and yeast antagonists (El Ghaouth et al., 2000b; Janisiewicz, 1992, 1994).

Recently, a bioactive coating has been developed by Wilson and El-Ghaouth (2000) which consists of the combination of complementary biological approaches for additive and/or synergistic effects. Such combinations may have greater stability and effectiveness than the use of single biocontrol agents alone. Biological control activity of antagonists can also be enhanced by several additives.

A. Antimicrobial Additives

Enhancing the activity of biocontrol agents could be the most important factor in their success in controlling fruit diseases and their ultimate acceptance in commercial disease management. McLauglin et al. (1990) demonstrated that the addition of calcium salts to yeast cell suspensions markedly enhanced the ability of *P. guilliermondii* to control postharvest diseases of apple. This allowed a reduction in the amount of yeast biomass necessary to achieve desirable levels of disease control. Wisniewski et al. (1995) reported that the biocontrol activity of isolate 182 of the yeast *C. oleophila* was enhanced by the addition of 90 or 180 m*M* $CaCl_2$. The combination of sugar analogs, such as 2-deoxy-D-glucose, with the yeast antagonists *Sporobolomyces roseus* or *C. saitoana* enhanced biocontrol against blue mold of apples (El Ghaouth et al., 2000; Janisiewicz, 1994). El Nashawy et al. (1998) reported that the addition of nisin improved biocontrol activity of the yeast *C. oleophila.* Many other additives, such as the GRAS (generally regarded as safe) compounds commonly used in food industry, may enhance activity of biocontrol agents and should be considered as preferred additives.

Results from series of pilot tests on apple and citrus fruit showed that the bioactive coating was significantly more effective in controlling decay than either the antagonist or the polymer chitosan alone (El Ghaouth et al., 2000a). The bioactive coating was effective in controlling postharvest decay caused by *Botrytis cinerea*, *Penicillium expansum*, *P. digitatum*, and *P. italicum* on a variety of fruit in a series of semi-commercial tests conducted in West Virginia, Florida, and California. The bioactive coating was very effective in controlling the natural decay of major apple and citrus varieties. The level of disease control obtained with the bioactive coating on apple and citrus fruit was comparable to that obtained with the recommended fungicides, Thiabendazole and Imazalil. The results obtained from such semi-commercial tests demonstrate the potential of the bioactive coating as a viable alternative to synthetic fungicides (El Ghaouth et al., 2000a).

Synergistic effects from combined biological treatments were also observed with a biocontrol product "bioenhancer" that consisted of an antagonistic yeast with a low dose of an antifungal sugar analog (El Ghaouth et al., 2000b). In large-scale pilot tests on apple and citrus fruit, the bioenhancer displayed greater stability and effectiveness in controlling natural infection than either the antagonist or sugar analog alone. The level of disease control obtained with the bioenhancer was comparable to that obtained with the recommended fungicides (Imazalil and Thiabendazole). The biocontrol activity of the bioenhancer appeared to be due to the synergistic interactions between the antagonistic activity of the yeast and the antifungal property of the sugar analog. This was well illustrated by the curative activity of the bioenhancer. In laboratory tests, the bioenhancer, besides having a protectant effect, displayed a curative activity against major postharvest pathogens in variety of fruit. The results obtained from semi-commercial tests demonstrate the great potential of the bioenhancer as an antifungal preservative for harvested commodities. The complexity of the mode of action displayed by combined alternatives should make the development of pathogen resistance more difficult, present a highly complex disease deterrent barrier, and provide a greater stability and effectiveness than approaches utilizing a single biological agent.

B. Physical Treatments

Other ways to enhance biocontrol efficacy, and possibly ensure consistency, is to integrate the biocontrol agent with physical methods such as curing and heat treatments (Barkai-Golan and Douglas, 1991; Cook et al., 1999), ultraviolet light (Droby et al., 1993a). Another readily applicable way to enhance efficacy is to combine the biocontrol treatment with modified or controlled atmospheres (MA/CA) and cold storage (Sugar et al., 1994). We have recently tested the compatibility of the yeast *C. oleophila* with CA storage against postharvest decay of nectarines (Lurie et al., 1995). In another study, it was demonstrated that the efficacy of *P. guilliermondii* against *P. digitatum* increased when citrus fruit was stored at optimal low storage temperature (Droby and Chalutz, 1994).

Preharvest application of biocontrol agents as a stand-alone treatment or combined with a postharvest application of the biocontrol agent may also prove to be a useful strategy in achieving improved performance against infections. This approach could be used as a tool to manipulate epiphytic populations and change patterns of surface wound colonization. Reports indicate the possibility of reducing postharvest decay caused by *C. gleoesporiodes* on avocado and mangoes by a preharvest spray with *Bacillus subtilis*, an antibiotic-producing bacterium (Korsten et al., 1991). Koomen and Jeffries (1993) have also demonstrated the feasibility of controlling anthracnose on mango fruit with *B. cereus* and *Pseudomonas fluorescens*. Droby et al. (1993c) suggested the possible use of preharvest application of the yeast antagonist *Pichia guilliermondii* to reduce the development of postharvest decay of citrus fruit; a reduction of postharvest *Rhizopus* rot of table grapes was also achieved by a preharvest spray of the yeast *P. guilliermondii* (Ben-Arie et al., 1991). More recently, Benbow and Sugar (1999) showed that preharvest application of the yeasts *Cryptococcus infirmo-ministus*, *C. laurentii,* and *C. oleophila* provided control of postharvest decay on two pear cultivars. *C. infirmo-ministus* gave the most consistent postharvest decay control in fruit treated 3 weeks before harvest. Preharvest introduction of antagonists in conjunction with additional postharvest applications may prove successful in providing acceptable levels of control. To fully explore the potential of this approach, however, obtaining data on the composition of epiphytic populations before and after the introduction of a single antagonist is crucial.

C. Genetic Manipulation of Biocontol Agents

While there are several approaches to improving biocontrol activity of yeast antagonists, one of the most attractive is enhancing genetic traits involved with the ability of the antagonist to inhibit establishment and development of the pathogen at the infection court. A molecular approach would be useful in achieving this goal and would allow full exploitation of the potential of these yeasts. Yeasts selected for their biocontrol activity appear to be very well adapted to growth and the colonization of fruit surface and wounds. This suggests that these yeasts could serve as excellent vehicles to deliver bioactive compounds such as fungal cell-wall-degrading enzymes, or any other antifungal proteins, directly to areas where propagules of the pathogen are most likely to germinate and infect the tissue.

In our efforts to identify genetic traits of the yeast *C. oleophila* and determine its potential to enhance biocontrol activity, we have been studying the yeast's ability to produce fungal cell-wall-degrading enzymes in culture and *in planta*. Both chitinase and ß-1,3-glucanase activities were found in culture filtrate of *C. oleophila* and in fruit wounds treated with yeast cells. These findings imply that the yeast is producing these enzymes constitutively without the effect of exogenous inducers. It is not known, however, if the amounts of these enzymes that are secreted by the yeast are sufficient to play a significant role in inhibiting pathogens. It is also undetermined whether or not the secretion of these enzymes occurs at a time that would have the greatest impact on the infection process. Constitutive expression of high amounts of these enzymes would help ensure that they would play a major role in inhibiting and preventing the pathogen from becoming established. The fact that both enzymes are produced and secreted by the yeast at the wound site demonstrates the feasibility of genetically enhancing the levels of these enzymes to levels that would be inhibitory to the pathogen. This could be accomplished by the over-expression of chitinase and glucanase endogenous genes.

VII. CONCLUSIONS

In view of environmental and health concerns, a determined effort has been made to reduce the use of pesticides. Regulatory agencies have reacted to public pressure and introduced comprehensive legislation to reduce pesticide use. The limited number of viable alternatives to synthetic pesticides, however, is currently the major obstacle in reaching that goal. Microbial control of postharvest diseases has been one of the most extensively studied alternatives and appears to be a viable technology. Research and development of biological control products for postharvest use has been on a fast track. Several commercial products are already available and others will be available in the near future. The opportunities of successful implementation of this technology is growing as information regarding various aspects related to formulation technology, application, microbial ecology, and genetics becomes available. Currently, the main hurdle facing widespread use of postharvest biocontrol strategies is the decreasing efficacy and lack of consistency found when these methodologies are applied as stand-alone treatments under commercial conditions. Unlike the control of tree, field crop, or soilborne diseases, successful commercial control of postharvest diseases of fruits and vegetables must be extremely efficient, in the range of 95 to 98%. As of today, such levels of control can be reached by biofungicides only when supplemented with low levels of synthetic chemical fungicides. However, by employing several biological, chemical, and physical avenues, either singly or in combination, the efficacy of microbial antagonists may be significantly increased. Innovative biological control strategies should take into consideration the growing concern over contamination of the produce with human pathogens, as well as plant pathogens.

The time is ripe to integrate biocontrol agents with one or more physical treatments such as heat treatments, controlled and modified atmospheres, natural biocides, and food-grade preservatives. Such an integrated approach will probably provide adequate control levels comparable to those achieved by chemical fungicides. To achieve this goal, the fruit and vegetable industry should adopt certain changes in packing and sorting lines required for successful implementation of an integrated control strategy.

Molecular approaches may prove useful in developing biocontrol agents with enhanced biocontrol activity and set directions for full exploitation of the genetic potential of these antagonists. Yeast antagonists selected for their biocontrol activity appear to be very well adapted to growth and colonizing fruit surface and especially fruit wound sites. This feature suggests that these yeasts provide an excellent means of delivering bioactive compounds, such as fungal cell-wall-degrading enzymes, directly to areas where pathogen propagules are most likely to germinate and infect the tissue. Combining the ability of these yeasts to rapidly colonize wound sites with enhanced constitutive production of antifungal proteins may prove to be a useful strategy for improving biocontrol effectiveness.

REFERENCES

Aharoni, Y., Coppel, A., and Fallik, E. 1993. Hinokitiol (-thujaplicin) for postharvest decay control on Galia melons. *N.Z. J. Crop Hortic. Sci.,* 21:165.

Avraham, A. 1994. Isolation and Biochemical Characterization of Exo-ß-1,3-Glucanase from *Pichia guilliermondii*, Master's of Science thesis, Faculty of Agriculture, the Hebrew University of Jerusalem, 65 pp.

Baker, C.A. and Henis, J.M.S. 1990. Commercial production and formulation of microbial biocontrol agents. In: *New Directions in Biological Control: Alternatives for Suppressing Agricultural Pests and Diseases,* R.R. Baker and P.E. Dunn, Eds. Allan R. Liss, New York.

Barkai-Golan, R. and Douglas, J.P. 1991. Postharvest heat treatment of fresh fruits and vegetables for decay control. *Plant Dis.,* 75:1085.

Ben-Arie, R., Droby, S., Zutkhi, J., Cohen, L., Weiss, B., Sarig, P., Zeidman, M., Daus, A., and Chalutz, E. 1991. Preharvest and postharvest biocontrol of *Rhizopus* rot of table grapes with yeasts. In: *Biological Control of Postharvest Diseases of Fruits and Vegetables.* USDA-ARS Pub. 92, Proc. Int. Workshop, Shepardstown, WV, pp. 100–113.

Benbow, J.M., and Sugar, D. 1999. Fruit surface colonization and biological control of postharvest diseases of pear by preharvest yeast application. *Plant Dis.,* 83:839–844.

Boonyakiat, D., Chen, P.M., Spotts, R.A., and Richarsdson, D.G. 1987. Effect of harvest maturity on decay and postharvest life of d'Anjou pear. *Scientia Hort.,* 31:131–139.

Chalutz, E. and Wilson, C.L., 1990, Postharvest biocontrol of green and blue mold and sour rot of citrus by *Debaryomyces hansenii. Plant Dis.,* 74:134–137.

Connick, W.J., Jr., Lewis, J.A., and Quimby, P.C., Jr. 1990. Formulation of biocontrol agents for use in plant pathology. In: *New Directions in Biological Control: Alternatives for Suppressing Agricultural Pests and Diseases.* R.R. Baker and P.E. Dunn, Eds. Allan R. Liss, New York.

Cook, D.W.M., Long, P.G., Ganesh, S., and Cheah, L.-H. 1997a. Attachment microbes antagonistic against *Botrytis cinerea*: biological control and scanning electron microscope studies *in vivo. Ann. Appl. Biol.,* 131:503–518.

Cook, D.W.M., Long, P.G., Ganesh, S., and Neilson, H.F. 1997b. Development of an apparatus and protocol for the efficient isolation of bacteria and yeasts that attach to *Botrytis cinerea* germlings. *Ann. Appl. Biol.,* 131:213–226.

Cook, D.W.M., Long, P.G., and Ganesh, S. 1999. The combined effect of delayed application of yeast biocontrol agents and fruit curing for the inhibition of the postharvest pathogen *Botrytis cinerea* in Kiwifruit. *Postharvest Biol. Technol.,* 16:233–243.

Droby, S. and Chalutz, E. 1994a. Successful biocontrol of postharvest pathogens of fruits and vegetables. In: *Proc. Brighton Crop Protection Conference — Pests and Diseases,* pp. 1265–1272.

Droby, S. and Chalutz, E. 1994b. Mode of action of biocontrol agents for postharvest diseases. In: *Biological Control of Postharvest Diseases of Fruits and Vegetables: Theory and Practice,* C.L. Wilson and M.E. Wisniewski, Eds., CRC Press, Boca Raton, FL, pp. 63–75.

Droby, S., Chalutz, E.,Wilson, C.L., and Wisniewski, M.E. 1989. Characterization of the biocontrol activity of *Debaryomyces hansenii* in the control of *Penicillium digitatum* on grapefruit. *Can. J. Microbiol.,* 35:794–800.

Droby, S., Chalutz, E., and Wilson, C.L. 1991. Antagonistic microorganisms as biological control agents of postharvest diseases of fruits and vegetables. *Postharvest News Inform.,* 2:169–173.

Droby, S., Chalutz, E., Horev, B., Cohen L., Gaba, V., Wilson, C.L., and Wisniewski, M.E. 1993a. Factors affecting UV-induced resistance in grapefruit against the green mold decay caused by *Penicillium digitatum. Plant Pathol.,* 42:418–424.

Droby, S., Chalutz, E., Wilson, C.L., and Wisniewski, M.E. 1993b. Biological control of postharvest diseases: a promising alternative to the use of synthetic fungicides. *Phytoparasitica,* 20:149–153.

Droby, S., Hofstein, R., Wilson, C.L., Wisniewski, M., Fridlender, B., Cohen, L., Weiss, B., Daus, A., Timar, D., and Chalutz, E. 1993c. Pilot testing of *Pichia guilliermondii*: a biocontrol agent of postharvest diseases of citrus fruit. *Biological Control,* 3:47–52.

Droby, S., Robin, D., Chalutz, E., and Chet, I. 1993d. Possible role of glucanase and extracellular polymers in the mode of action of yeast antagonists of postharvest diseases. *Phytoparasitica,* 2:167.

Droby, S., Chalupovicz, L., Chalutz, E., Wisniewski, M.E., and Wilson, C.L. 1995. Inhibitory activity of yeast cell wall materials against postharvest fungal pathogens. *Phytopathology,* 85:1123.

Eckert, J.W. and Ogawa, J.M. 1988. The chemical control of postharvest diseases: deciduous fruits, berries, vegetables, and root/tuber crops. *Ann. Rev. Phytopathol.,* 26:433–469.

Elad, Y., Chet, I., Boyle, P., and Henis, Y. 1983. Parasitism of *Trichoderma* spp. on *Rhizoctonia solani* and *Sclerotium rolfsii* — scanning electron microscopy and fluorescence microscopy. *Phytopathology,* 73:85–88.

El-Ghaouth, A., Arul, J., Grenier, J., and Asselin, A. 1992. Antifungal activity of chitosan on two postharvest pathogens of strawberry fruits. *Phytopathology,* 82:398–402.

El-Ghaouth, A., Wilson, C.L., and Wisniewski, M. 1998. Ultrastructural and cytochemical aspects of the biological control of *Botrytis cinerea* by *Candida saitoana* in apple fruit. *Phytopathology,* 88:282–291.

El-Ghaouth, A., Smilanick, J., Eldon, G., Brown, E., Ippolito, A., and Wilson, C.L. 2000a. Application of *Candida saitoana* and glycolchitosan for the control of postharvest diseases of apple and citrus fruit under semi-commercial conditions. *Plant Dis.,* 84:243–248.

El-Ghaouth, A., Smilanick, J., Wisniewski, M., and Wilson, C.L. 2000b. Improved control of apple and citrus fruit decay with a combination of *Candida saitoana* with 2-deoxy-D-glucose. *Plant Dis.,* 84:249–253.

Fokkema, N.J. 1992. The phyllosphere as an ecological neglected milieu: a plant pathologists point of view. In: *Microbial Ecology of Leaves,* J.H. Andrews and S. Hirano, Eds., Springer-Verlag, New York, pp. 3–18.

Fokkema, N.J. and van den Heuvel, J. 1986. *Microbiology of the Phyllosphere,* Cambridge University Press, London, 392 pp.

Gullino, M.L., Aloi, C., Palitto, M., Benzi, D., and Garibaldi, A. 1991. Attempts at biocontrol of postharvest diseases of apple. *Med. Fac. Landbouw. Rijksuiv. Gent,* 56:195.

Hofstein, R., Fridlender, B., Chalutz, E., and Droby, S. 1994. Large scale production and pilot testing of biocontrol agents of postharvest diseases. In: *Biological Control of Postharvest Diseases of Fruits and Vegetables: Theory and Practice,* C.L. Wilson and M.E. Wisniewski, Eds., CRC Press, Boca Raton FL, 182 pp.

Janisiewicz, W.J. 1988. Biological control of diseases of fruit. In: *Biocontrol of Plant Diseases,* Vol. II, K.G. Mukergi and K.L. Grag, Eds., CRC Press, Boca Raton, FL, pp. 153–165.

Janisiewicz, W.J. 1994. Enhancement of biocontrol of blue mold with nutrient analog 2-deoxy-D-glucose on apples and pears. *Appl. Environ. Microbiol.,* 60:2671–2676.

Janisiewicz, W.J. 1998. Biocontrol of postharvest diseases of temperate fruits: challenges and opportunities. In: *Plant-MicrobeInteractions and Biological Control,* J. Boland and L. D. Kuykendall, Eds., Marcel Dekker, New York, pp. 171–198.

Janisiewicz, W. and Roitman, J. 1988. Biological control of blue mold and gray mold on apple and pear with *Pseudomonas cepacia. Phytopathology,* 78:1697–1700.

Janisiewicz, W.J., Peterson, D.L., and Bors, R. 1994. Control of storage decay of apples with *Sporobolomyces roseus. Plant Dis.,* 78:466–470.

Kable, P.F. 1971. Significance of short-term latent infections in the control of brown rot in peach fruits. *Phytopathol. Z.,* 70:173.

Koomen, I. and Jeffries, P. 1993. Effect of antagonistic microorganisms on the postharvest development of *Colletotrichum gloeosporioides* on mango. *Plant Pathol.,* 42:230–237.

Korsten, L., Villiers, E.E., Jager, E.S., Cook, N., and Kotze, J.M. 1991. Biological control of avocado postharvest diseases. *South African Avocado Growers' Association Yearbook,* 14:57–59.

Leibinger, W., Breuker, B., Hahn, M., and Mendegen, K. 1997. Control of postharvest pathogens and colonization of apple surface by antagonistic microorganisms in the field. *Phytopathology,* 87:1103–1110.

Lurie, S., Droby, S., Chalupowicz, L., and Chalutz, E. 1995. Efficacy of *Candida oleophila* strain 182 in preventing *Penicillium expansum* infection of nectarine fruits. *Phytoparasitica,* 23:231–234.

McLaughlin, R.J., Wisniewski, M.E., Wilson, C.L., and Chalutz, E. 1990. Effect of inoculum concentration and salt solutions on biological control of postharvest diseases of apple with *Candida* sp. *Phytopathology,* 80:456–461.

Michailides, T.J. and Spotts, R.A. 1986. Factors affecting dispersal of *Mucor piriformis* in pear orchards and into packinghouse. *Plant Dis.,* 70:1060–1063.

Piano, S., Neyrotii, V., Migheli, Q., and Gullino, L. 1997. Biocontrol capability of *Metschnikowia pulcherrima* against *Botrytis* postharvest rot of apple. *Postharvest Biol. Technol.,* 11:131–140.

Pusey, P.L., Hotchkiss, M.W., Dulmage, H.T., Baumgardner, R.A., Zehr, E.I., and Wilson, C.L. 1988. Pilot tests for commercial production and application of *Bacillus subtilis* (B-3) for postharvest control of peach brown rot. *Plant Dis.,* 72:622–626.

Roberts, R.G. 1990a. Biological control of gray mold of apple by *Cryptococcus laurentii. Phytopathology,* 80:526–530.

Roberts, R.G. 1990b. Biological control of mucor rot of pear by *Cryptococcus laurentii, C. flavus,* and *C. albidus. Phytopathology,* 80:1051.

Rodov, V., Ben-Yehoshua, S., Díhallewin, G., Castia, T., and Farg, D. 1994. Accumulation of phytoalexins scoparone and scopoletin in citrus fruit subjected to various postharvest treatments. *Acta Hort.,* 381:517–523.

Shefelbine, P.A. and Roberts, R.G. 1990. Population dynamics of *Cryptococcus laurentii* in wounds in apple and pear fruit stored under ambient or controlled atmospheric conditions. *Phytopathology,* 80:1020.

Smilanick, J.L. and Dennis-Arrue, R. 1992. Control of green mold of lemons with *Pseudomonas* species. *Plant Dis.,* 76:481–485.

Smilanick, J.L., Denis-Arrue, R., Bosch, J.R., Gonzales, A.R., Henson, D.J., and Janisiewicz, W.J. 1993. Biocontrol of postharvest brown rot of nectarines and peaches by *Pseudomonas* species. *Crop Protection,* 12:513–520.

Snowdon, A.L. 1990. *A Color Atlas of Post-Harvest Diseases and Disorders of Fruits and Vegetables,* Vols. 1 and 22, CRC Press, Boca Raton, FL.

Sommer, N.F. 1982. Postharvest handling practices and postharvest diseases of fruits. *Plant Dis.,* 66:357–364.

Sommer, N.E. 1985. Strategies for control of postharvest diseases of selected commodities. In: *Postharvest Technology of Horticultural Crops,* A.A. Kader, R.F. Kasmire, G.F. Mitchel, M.S. Reid, N.F. Sommer, and J.F. Thompson, Eds., Cooperative Extension, University of California, Davis, pp. 83–99.

Stretch, A.W. 1989. Biological control of blueberry and cranberry fruit rots (*Vaccinium corymbosu* L. and *Vaccinium macrocarpon* Ait. *Acta Hort.,* 241:301–306.

Sugar, D., Roberts, R.G., Hilton, R.J., Reghetti, T.L., and Sanchez, E.E. 1994. Integration of cultural methods with yeast treatment for control of postharvest decay in pear. *Plant Dis.,* 78:791–795.

Teixidoí, N., Vinas, I., Usal, J., and Magan, N. 1998. Control of blue mold of apples by preharvest application of *Candida sake* grown in media with different water activity. *Phytopathology,* 88:960–964.

Whitesides, S.K., Daoust, R.S., and Gouger, R.J. 1994. Commercialization of biological control agents: an industry perspective. In: *Biological Control of Postharvest Diseases of Fruits and Vegetables: Theory and Practice,* C.L. Wilson and M.E. Wisniewski, Eds. CRC Press, Boca Raton FL, 182 pp.

Wilson, C.L. and Chalutz, E. 1989. Postharvest biological control of *Penicillium* rots of citrus with antagonistic yeasts and bacteria. *Sci. Hort.,* 40:105–112.

Wilson, C.L. and El-Ghaouth, A. 1993. Multifaceted biological control of postharvest diseases of fruits and vegetables. In: *Pest Management: Biologically Based Technologies,* R.D. Lumsden and J.L. Vaughn, Eds., American Chemical Society Press, Washington, D.C., pp. 181–185.

Wilson, C.L. and Pusey, P.L. 1985. Potential for biological control of postharvest plant diseases. *Plant Dis.,* 69:375–378.

Wilson, C.L. and Wisniewski, M.E. 1989. Biological control of postharvest diseases of fruits and vegetables: an emerging technology. *Ann. Rev. Phytopathol.,* 27:425–441.

Wilson, C.L. and Wisniewski, M.E. 1994. *Biological Control of Postharvest Diseases of Fruits and Vegetables: Theory and Practice*. CRC Press, Boca Raton, FL, 182 pp.

Wilson, C.L., Wisniewski, M.E., Droby, S., and Chalutz, E. 1993. A selection strategy for microbial antagonists to control postharvest diseases of fruits and vegetables. *Sci. Hort.,* 40:105–112.

Wilson, C.L., El Ghaouth, A., Chalutz, E., Droby, S., Stevens, C., Lu, J.Y., Khan, V., and Arul, J. 1994. Potential of induced resistance to control postharvest diseases of fruits and vegetables. *Plant Dis.,* 78:837–844.

Wisniewski, M.E., Wilson, C.L., and Chalutz, E. 1988. Biological control of postharvest diseases of fruit: inhibition of *Botrytis* rot on apple by an antagonistic yeast. *Proc. Ann. Meet. Electron Microsc. Soc. Am.,* 46:290–291.

Wisniewski, M., Biles, C., Droby, S., McLaughlin, R., Wilson, C., and Chalutz, E. 1991. Mode of action of the postharvest biocontrol yeast, *Pichia guilliermondii*. I. Characterization of the attachment to *Botrytis cinerea*. *Physiol. Molec. Plant Pathol.,* 39:245–258.

Wisniewski, M., Wilson, C., and Hershberger, W. 1989. Characterization of inhibition of *Rhizopus stolonifer* germination and growth by *Enterobacter cloacae*. *Can. J. Botany,* 67:2317–2323.

Wisniewski, M.E., Droby, S., Chalutz, E., and Eilam, Y. 1995. Effects of Ca^+ and Mg^+ on *Botrytis cinerea* and *Penicilium expansum in vitro* and on the biocontrol activity of *Candida oleophila*. *Plant Pathol.,* 44:1016–1024.

14 Genetic and Biological Control of Aflatoxigenic Fungi

D. Bhatnagar, P.J. Cotty, and T.E. Cleveland

CONTENTS

I. INTRODUCTION

Aflatoxins, metabolites produced by several members of *Aspergillus* section *Flavi*, particularly, the fungi *Aspergillus flavus* and *A. parasiticus*, occur in food and feed crops before harvest and during storage. These compounds are toxic and extremely carcinogenic when introduced into animal systems; therefore, many investigations have focused upon identifying technologies to inhibit fungal growth and/or aflatoxin synthesis in crops to prevent contamination and subsequent consumption of these toxic compounds by animals and humans.

A "gene cluster" (over 70 kilobases in size) in *A. flavus* and *A. parasiticus* has recently been identified upon which reside almost all the genes involved in aflatoxin biosynthesis. A few of these genes appear to be "environmentally responsive," and their expression may be regulated by several factors that affect aflatoxin production, such as temperature, pH, plant metabolites and carbon, and nitrogen sources. The molecular regulation of this gene cluster has been characterized in some detail, including the identification of a regulatory gene and a common regulatory relationship between fungal development and toxin synthesis. This critical information on aflatoxin biosynthesis provides us with an opportunity to target specific sites to interfere with aflatoxin formation in plants.

Aspergillus flavus is the most common causal agent of aflatoxin contamination. Communities of fungi that fall within the species *Aspergillus flavus* are highly diverse. Strains of *A. flavus* may belong to different genetically isolated groups that have widely varying abilities to produce aflatoxins. Some natural *A. flavus* strains produce no aflatoxins. These atoxigenic strains have been "seeded" into agricultural fields so that the atoxigenic strain predominates the fungal communities. The increase in atoxigenic strain occurrence results in reduced vulnerability of crops to contamination without an increase in the overall quantity of *A. flavus* in the field. Identification of critical genes governing aflatoxin formation could also lead to the conversion of any ecologically successful *A. flavus* or *A. parasiticus* strain, through gene disruption, into a designer non-aflatoxigenic biocontrol strain that could be used in aflatoxin management programs directed at competitively excluding aflatoxigenic strains in specific environments.

Efforts are also underway to inhibit fungal growth and/or aflatoxin formation through enhancement of host-plant resistance. A great deal of knowledge about the mechanisms involved in natural resistance in certain corn inbreds has been produced through inoculation of kernels with reporter gene-containing *A. flavus* (tester strains). By monitoring the degree of fungal infection and toxin production, resistance mechanisms have been identified on kernel surfaces, beneath the kernel pericarp, and in the embryo. Certain kernel physical components and antifungal proteins have also been shown to correlate with high levels of resistance in corn varieties. Additional studies have led to the identification of a variety of antifungal (to *A. flavus* infection) peptides, proteins, and genes from several host and non-host species. Knowledge gained from studies that establish plant-*Aspergillus* interactions, host resistance mechanisms, and consequent aflatoxin formation are being used in plant breeding and genetic engineering strategies to suppress aflatoxin production in crops.

II. AFLATOXIN-PRODUCING FUNGI

All aflatoxin-producing fungi may be assigned taxonomically to *Aspergillus* section *Flavi*. These include *Aspergillus flavus* and *Aspergillus parasiticus*, as well as several less common taxa including *Aspergillus nomius*.[1] The aflatoxin-producing group is considerably more complex than previously thought; for example, strains of *Aspergillus tamarii* have recently been shown to produce aflatoxins,[2] and new taxa may soon be described as important aflatoxin producers, historically only found in the previously named taxa.[2–4]

A. flavus, a species that is ubiquitous in warm tropical and desert environments,[1,5] is a highly diverse asexual species that can be divided on the basis of physiological, morphological, and genetic criteria; however, most crop contamination with aflatoxins is apparently caused by either the S or the L strains of *A. flavus*.[6] The S strain can be separated on the basis of sclerotial morphology and habitat. The S-strain isolates produce, on average, significantly higher levels of aflatoxins than typical or L-strain isolates. In some agricultural regions, S-strain isolates dominate and are responsible for most of the aflatoxin-producing potential of the resident *A. flavus* communities.[7–9] However, S-strain isolates frequently vary in virulence, with certain isolates failing to produce the primary pectinase needed to ramify through host tissues.[10,11] Thus, isolates with the greatest aflatoxin-producing potential do not always have high virulence. Conversely, isolates that do not produce aflatoxins in crops may be very effective in colonizing and ramifying through plant host tissue.[6]

Aspergillus flavus populations within agricultural fields are complex, with members of the population further divided genetically by vegetative compatibility which limits gene flow between dissimilar individuals.[12] Both S-strain and L-strain isolates are composed of many vegetative compatibility groups (VCGs).[13,14] Genetically distinct *A. flavus* strains frequently interact during dispersal, growth, and crop infection.[1,13] Multiple VCGs commonly reside within a gram of soil or infect a single seed.[13] The influence of these individuals on each other during crop infection may vary widely.[15] VCGs evolve largely when distinct clonal lineages gradually diverge as they spread spatially. This gradual divergence among VCGs can be measured by distinct random amplified polymorphic DNA (RAPD) markers and isozyme profiles.[16] VCGs may also differ both physiologically and morphologically,[14,17] and the characteristic most frequently examined is the aflatoxin-producing ability.

The aflatoxin-producing ability of *Aspergillus flavus* isolates varies widely, as well. A significant proportion of *A. flavus* populations may not produce aflatoxins (atoxigenic),[7] whereas others have the potential to contaminate infected seed with over a million parts per billion (ppb). Isolates within a VCG tend to have similar aflatoxin-producing potentials, and certain VCGs have no known members that produce aflatoxins.[14,17] The lack of aflatoxin-producing ability has been most thoroughly studied in isolates of *A. flavus* and *A. parasiticus* and other aspergilli used in food production.[18] These latter isolates, typically named *A. oryzae* and *A. sojae*, are domesticated strains of *A. flavus* and *A. parasiticus*, respectively, and have been used to produce soy sauce and other fermented foods for centuries.[4,19]

III. ECONOMIC SIGNIFICANCE OF AFLATOXIN CONTAMINATION

Aflatoxins have been found in many foods of animal and plant origin, including corn meal, peanuts, cottonseed, spices, cassava, pistachio nuts, rice, cocoa, bread, macaroni, copra, Brazil nuts, oilseeds, pumpkin seeds, meat pies, milk, cheese, sausage, and cooked meat. In the U.S., frequent preharvest contamination of corn, cotton, peanuts, and tree nuts is of major concern because of the economic impact of destroying contaminated crops.

Since their discovery in 1960, aflatoxins have been implicated in carcinogenicity, mutagenicity, teratogenicity, hepatotoxicity, and aflatoxicosis. Currently, 18 different aflatoxins are known. The most important members of this family of toxins are B_1, B_2, G_1, G_2 (see Figure 14.1), M_1, and M_2. Of these, aflatoxin B_1 (AFB_1) is the most common and also the most carcinogenic. AFM_1 is a contaminant in the milk of cows fed with AFB_1-contaminated feed and is considerably less toxic than AFB_1.

Epidemiological studies have provided evidence of the carcinogenicity of aflatoxin B_1 to humans (for several reviews, see reference 20). The liver is the primary target organ in many animal species; however, tumors in other organs have also been observed in aflatoxin-treated species. Ironically, this effect arises as a result of the detoxification response in animals. Highly reactive aflatoxin derivatives (8,9-epoxy-aflatoxin B_1) can intercolate DNA and form DNA-aflatoxin adducts. Subsequent cellular repair of the adducts often leads to G-to-T transversion in the coding region of genes, particularly that of the tumor suppressor gene, p53. A very high incidence (67%) of liver carcinomas in Senegal, China, Swaziland, and Mozambique bear the characteristics of aflatoxin-induced mutation of the p53 tumor suppressor gene. This mutation has also been associated with liver cancer in Mexico. The binding of AFB_1 to DNA also leads to the formation of single-stranded gaps. As a result, it inhibits DNA polymerase activity at DNA binding sites. This stimulates an error-prone repair system that may induce mutation. Furthermore, it has been suggested that AFB_1 is teratogenic due to its prenatal effects on certain animals. Its inhibitory effect on protein synthesis of eukaryotic cells can impair differentiation in sensitive primordial cells. According to epidemiological studies, raising the permissible limits of aflatoxins in foods in the U.S. would not greatly increase the incidence of liver cancer; however, there is a worldwide epidemic of hepatitis C, and hepatitis C patients with impaired liver function may be much more susceptible to aflatoxins than healthy people.

The national economy would be affected adversely both by the losses incurred by crop and livestock producers when aflatoxin-contaminated crops are destroyed due to regulatory restrictions and by the multiplier effect this would have on other industries as a result of the reduced spending power of producers.[21,22] Additionally, the costs of chemical analyses, quality control and regulatory programs, research and development, extension services, law suits, and human illness must all be borne by the national economy. The direct cost of aflatoxin contamination in corn in 1980 to all of the southeastern states was estimated to be greater than $237 million.[23]

While the short-run costs are substantial for the individual and for society, they may be greater in the long run if recurrent aflatoxin (for example, in corn) cannot be eliminated or detoxified. Farmers who are unable to market their corn will ultimately shift acreage to other crops such as soybeans or grain sorghum, which have less year-to-year risks and considerably less net returns. Thus, the growers in affected areas will have fewer cropping options and be forced to bear market forces on the few crops they can grow.

The economics of aflatoxin contamination are an issue primarily relevant to developed countries where food is in ample supply. In developing countries, where food is sometimes in short supply, long-term health implications of aflatoxin contamination are commonly overlooked. However, economies of developing countries could be seriously affected when the presence of even the smallest amount of toxin in export commodities is rejected by countries that strictly adhere to regulatory guidelines for levels of toxin in agricultural products for human or animal consumption.

IV. CONTROL OF PREHARVEST AFLATOXIN CONTAMINATION

Aflatoxin formation before or after crop harvest cannot be prevented thus far, but it can be reduced by appropriate management practices. Due to the human and animal health implications, intense efforts worldwide are underway to remove aflatoxin from food and feed supplies. Attention has been focused on the preharvest control of aflatoxin contamination, because that is when the fungi first colonize host tissues. This emphasis would obviate the need to detoxify large quantities of contaminated materials and avoid the uncertainties of gaining approval from regulatory agencies for the use of detoxified seeds for animal feed or human food. However, control strategies should also include methods for detoxifying contaminated products resulting from prevention measures that are not always completely successful. Detoxification can prevent total loss of valuable foodstuffs and reduce the burden of contamination at the farm gate (for reviews on detoxification procedures see references 20, 24–26).

A. Conventional Methods

Several agronomic practices have been shown to reduce preharvest aflatoxin contamination in certain crops,[27] including the use of pesticides (fungicides and insecticides), altered cultural practices (such as irrigation), and the use of resistant varieties. However, such procedures have only a limited potential for reducing aflatoxin levels in the field, especially in years when environmental conditions are particularly favorable to the contamination process (for review, see reference 28).

B. Use of Natural Product Inhibitors To Control Aflatoxin Contamination

There are several plant-derived inhibitors of aflatoxin synthesis, and this subject has been reviewed extensively.[29] Inhibitors with unknown modes of action have been discovered in our laboratory[30,31] that could be directly applied to crops in the field. Examples of natural products that may have potential in augmenting host-plant resistance against *A. flavus* infection are certain plant-derived volatile compounds.[30,32–34]

C. Control of Aflatoxin Contamination through Biotechnology

Because conventional methods are only partially effective and are not expected to achieve the extremely low or negligible levels of aflatoxin required to meet regulatory guidelines for the sale and export of commercial food and feed, there is an increasing need to develop new technology to reduce and eventually to eliminate preharvest aflatoxin contamination. Three biotechnological approaches are being developed to exclude toxigenic fungi from their environmental niches and to regulate fungal growth or aflatoxin biosynthesis in crops:

1. Inhibit biosynthetic or secretory processes responsible for aflatoxin accumulation.
2. Replace aflatoxigenic strains with non-aflatoxigenic (biocompetitive) strains in the field.
3. Enhance host resistance by marker-assisted plant breeding or by genetic engineering of plant varieties to specifically express antifungal agents in the susceptible plant tissues (e.g., infected seed tissues).

V. AFLATOXIN BIOSYNTHETIC PATHWAY

Elimination of preharvest aflatoxin contamination through plant-induced inhibition of biosynthetic or secretory processes responsible for toxin production[35–37] would significantly benefit from additional knowledge about the fundamental molecular and biological mechanisms that regulate the synthesis of aflatoxin by the fungus. Previous studies have determined that aflatoxins are synthesized by the polyketide metabolic pathway (for reviews, see references 38–40). The generally accepted scheme for aflatoxin biosynthesis is acetate → polyketide precursor → norsolorinic acid, NOR → averantin, AVN → 5′-hydroxyaverantin, HAVN → averufanin, AVNN → averufin, AVF → versiconal hemiacetal acetate, VHA → versiconal, VAL → versicolorin B, VERB → versicolorin A, VERA → demethylsterigmatocystin, DMST → sterigmatocystin, ST → *O*-methylsterigmatocystin, OMST → aflatoxin B_1, AFB_1 (Figure 14.1).

Specific enzyme activities, including those of reductase, dehydrogenase, cyclase, desaturase, P-450 monooxygenase, and *O*-methyltransferase, have been associated with precursor conversions in the aflatoxin pathway (for reviews, see references 38–41). Some of these enzymes have been partially purified, whereas others have been purified to homogeneity (for reviews, see references 38–41 and references therein). The genes encoding most of these enzymes have been cloned (for reviews, see references 36, 37, 42) (Figure 14.1). Alternate pathways may exist at several steps in the aflatoxin pathway;[43] therefore, more than one enzyme may catalyze the same reaction — for example, the reductase/dehydrogenase encoded by *nor*1, *norA,* and *norB.*[44,45] Also, independent reactions and different chemical precursors involved in AFB_1 and AFB_2 syntheses are catalyzed by common enzyme systems — that is, *O*-methyltransferses encoded by *omtA* and *omtB*[43,46–50] and the P-450 oxidoreductase encoded by *ordA*.[51,52]

Genetic studies of *A. flavus* and *A. parasiticus* were hampered by the lack of a sexual stage in these fungal species. Nonetheless, by means of parasexual cycle analysis, over 30 genes have been mapped to eight linkage groups.[53,54] Pulsed-field gel electrophoresis has helped resolve karyotypes and define genetic maps of these imperfect fungi.[55,56] Karyotyping of several *A. flavus* and *A. parasiticus* strains shows that there are six to eight chromosomes ranging in size from approximately 3 to ≥7 Mb.[55] The aflatoxin genes were mapped to linkage group VII in *A. flavus.*[56]

Genetic complementation has been a valuable tool in the cloning of aflatoxin biosynthesis genes.[57] Chromosomal walking and cross-hybridization studies have established that genes for aflatoxin synthesis in *A. parasiticus* and *A. flavus*,[58–60] as well as sterigmatocystin synthesis in *A. nidulans,* are clustered.[61] One of these genes, *aflR*, involved in the transcriptional regulation of aflatoxin and sterigmatocystin biosynthesis,[62–64] has been characterized from *A. parasiticus, A. flavus,* and *A. nidulans*. The function of AFLR proteins is conserved among the three aspergilli.[64] Expression of genes in the AF/ST clusters is co-regulated by AFLR, and AFLR binds to the promoters of aflatoxin biosynthesis genes.[65–67] AFLR also appears to regulate its own expression.[62] The transcription activation domain in *A. parasiticus* AFLR has been localized to its carboxy-terminal region.[68] Recently, a regulatory association between aflatoxin biosynthesis and fungal development has also been suggested.[69–71] Regulatory factors other than AFLR that play a role in aflatoxin gene expression as well as in fungal survival must be identified to provide effective manipulation of these toxigenic fungi.

Expression "probes" based on cloned aflatoxin genes and the use of reporter gene technology could be used to identify phytological agents that naturally inhibit aflatoxin biosynthesis.[72–74] This knowledge could then be used to improve resistance to aflatoxin contamination through marker-assisted plant breeding or genetic engineering procedures. Plant chemicals from crops vulnerable to aflatoxin that modulate aflatoxin biosynthesis have been reported;[30,33,34,75–77] however, additional knowledge is needed with regard to the identity, synthesis, and tissue sites of these plant components that influence aflatoxin synthesis during plant-fungus interaction.

The complete characterization of aflatoxin genes and their regulation not only has been extremely beneficial in our understanding of how the toxin is produced by the fungus when it invades a crop, but has also aided in the success of other projects seeking to develop non-aflatoxigenic biocompetitive fungi or to monitor crop resistance to fungal growth and aflatoxin formation.

VI. STRAIN INTERACTIONS AND BIOLOGICAL CONTROL

Not only may individual isolates, strains, and VCGs differ in aflatoxin-producing potential, but communities of fungi found in different areas may also have different aflatoxin-producing potentials.[7,8] The latter may contribute to variability in crop vulnerability to aflatoxin contamination in different regions and different fields. The lack of correlation between the ability to produce aflatoxins and a producing strain's ability to colonize and infect developing cotton bolls[6] suggested that atoxigenic strains of *A. flavus* could exclude aflatoxin-producing strains through competition during infection of developing crops and thereby prevent aflatoxin contamination.[6,7,78,79] In both greenhouse and field experiments, simultaneous wound inoculation of developing cotton bolls and corn ears with toxigenic and atoxigenic strains resulted in reductions in aflatoxin contamination of the developing crop parts as compared with controls inoculated with only the toxigenic strains.[80,81] Although not all atoxigenic strains reduce contamination by aflatoxin-producing strains during co-

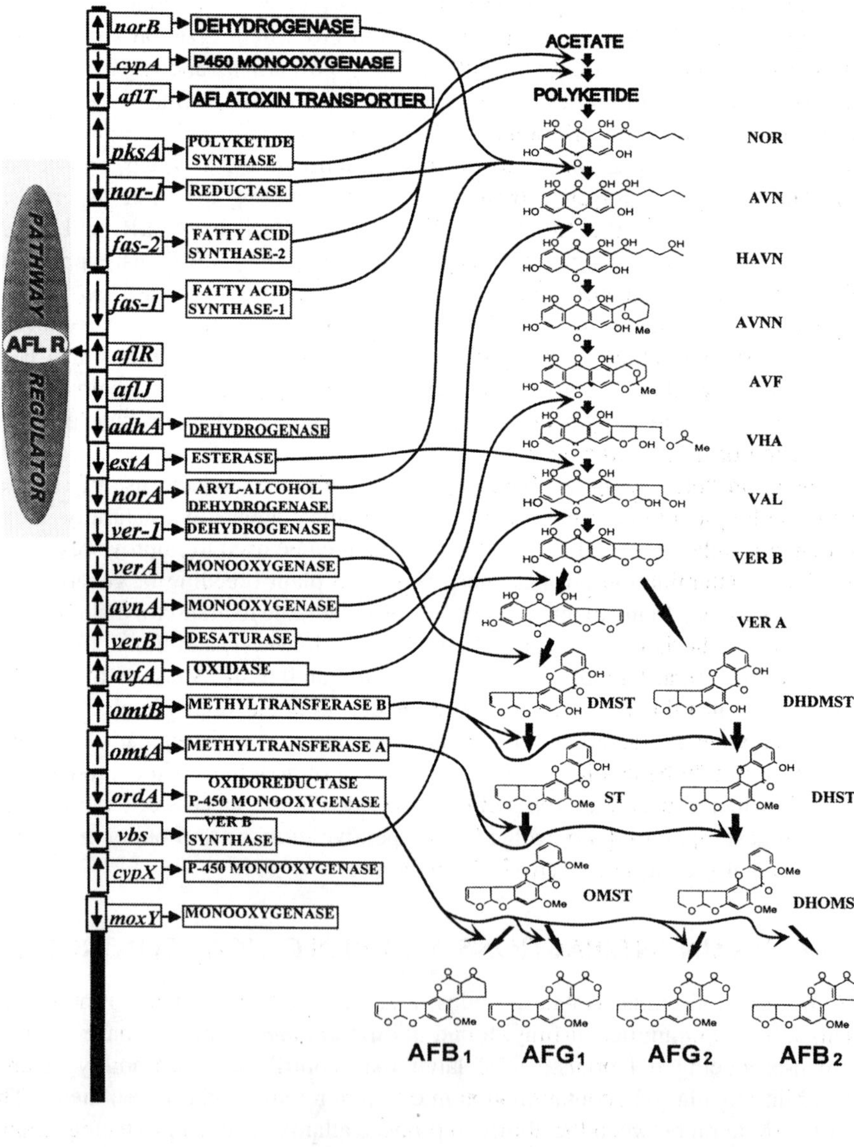

infection of crops, certain atoxigenic strains consistently cause reductions of 90% or more.[15,81]

Greenhouse and laboratory tests indicate that the mode of action of the atoxigenic strains is primarily through competitive exclusion.[82] Application of atoxigenic strains in the field has also been shown to be effective at preventing postharvest aflatoxin contamination both when the crop is infected by aflatoxin producers naturally in the field and when inoculated after harvest.[80] Taken together, the greenhouse and laboratory tests suggest that competitive exclusion of aflatoxin-producing strains of *A. flavus* with atoxigenic strains of the same fungal species may provide an efficient method for reducing aflatoxin accumulation throughout crop production, storage, and utilization.[6,78,81,83] These tests, however, rely on direct application of the atoxigenic strains to either fresh wounds in the crop or to harvested kernels. To circumvent this drawback, Cotty et al.[84] have developed a practical method for applying the atoxigenic strains in an agricultural setting. Efforts to use atoxigenic strains to prevent contamination is limited to three crops: peanuts and corn in southeastern

---

FIGURE 14.1 Summary of the cluster of aflatoxin pathway genes, corresponding biosynthetic enzymes, and precursor intermediates involved in the aflatoxin B_1 and B_2 synthesis. The generally accepted aflatoxin B_1 and B_2 biosynthetic pathway in *A. parasiticus* and *A. flavus* and the identified enzymes for some specific conversion steps and cloned genes are schematically presented. The regulatory gene, *aflR*, coding for the pathway regulatory factor (AFLR protein), controls the expression of the structural genes at the transcriptional level. The *fas*1, *fas*2, and *pksA* gene products, fatty acid synthase, and polyketide synthase, respectively, are involved in the conversion steps between the initial acetate unit to the synthesis of the decaketide backbone in aflatoxin synthesis. The *nor*1 gene encodes a reductase for the conversion of NOR to AVN. The *avnA* gene encodes a P-450 monooxygenase for the conversion of AVN to HAVN. The *aflJ* gene has also been demonstrated to be involved in the regulation of aflatoxin biosynthesis, but the role is under investigation. The *adhA* (homology to an alcohol dehydrogenase), *norA* (homology to an aryl-alcohol dehydrogenase), *ver*1 (encoding a dehydrogenase), *avfA*, and *cyp*450 gene products have been demonstrated to be functioning at various stages of the pathway, but their exact enzymatic role has not been fully characterized and is under investigation. The *omtA* gene encodes an *O*-methyltransferase for the conversion of ST to OMST and DHST to DHOMST. The *vbs* gene encodes a Ver B synthase (cyclase), which has been reported to be involved in the conversion of VHA to Ver B. The *ordA* gene encodes an oxidoreductase for the conversion of OMST to AFB_1 and DHOMST to AFB_2 and is also involved in the conversion of OMST to AFG_1 and DHOMST to AFG_2. The *estA* gene encodes an esterase involved in the conversion of VHA to VAL. The vertical bar on the left represents at least a 75-kb aflatoxin pathway gene cluster with identified genes shown in the open boxes. The names of the individual genes are labeled next to the open boxes. Arrows inside the open boxes indicate the direction of transcription. Arrows indicate the relationships from the genes to the enzymes they encode, from the enzymes to the bioconversion steps they are involved in, and from the intermediates to products in the aflatoxin bioconversion steps. Abbreviations: NOR, norsolorinic acid; AVN, averantin; HAVN, 5′-hydroxyaverantin; AVNN, averufanin; AVF, averufin; VHA, versiconal hemiacetal acetate; VAL, versiconal; Ver B, versicolorin B; Ver A, versicolorin A; DHST, dihydrosterigmatocystin; ST, sterigmatocystin; OMST, *O*-methylsterigmatocystin; DHOMST, dihydro-*O*-methylsterigmatocystin; AFB_1 aflatoxin B_1; AFB_2, aflatoxin B_2.

FIGURE 14.2 Cover of the 1998 Cotton Foundation annual report shows wheat colonized by an atoxigenic strain of *A. flavus* as it is applied (right) and after the fungus has grown out (left). The cotton industry has been enthusiastic about the potential for atoxigenic strain technology to prevent aflatoxin contamination.

U.S. and cottonseed in Arizona.[80,83,85–88] Emphasis at the Southern Regional Research Center has been on the application of atoxigenic strain technology to control aflatoxin contamination of cottonseed in regions of Arizona. In the U.S., aflatoxin contamination of cottonseed is most consistent and severe in the irrigated western desert.[87] Cottonseed produced in these valleys has a relatively high value per acre due to both high yields and high demand for cottonseed within the area.

A. Developing Atoxigenic Strain Technology

An aflatoxin-prevention technology based on atoxigenic strains of *A. flavus* was developed (Figure 14.2) for use in the region of Arizona with the most frequent and severe aflatoxin contamination of cottonseed.[84] For use in Arizona, a solid formulation of atoxigenic *A. flavus* has been developed in which whole wheat seeds are sterilized and colonized with the strain of choice.[83,88] This formulation has multiple year stability and tolerates exposures up to 70°C.[88] The fungus, in this formulation, can withstand both the severe conditions of on-farm storage during the summer months of the desert and direct exposure to Arizona's severe summer conditions after application.

For atoxigenic strains of *A. flavus* to be useful during crop production, they must be applied at a time and in a manner that allows them to compete successfully with aflatoxin-producing strains. In theory, application of an atoxigenic *A. flavus* strain when overall *A. flavus* levels are low should give the atoxigenic strain preferential exposure to the developing crop and thus the advantage in competing for crop resources during infection and during *A. flavus* population increases associated with crop production.[1,83] The atoxigenic strains are routinely applied at 10 lb per acre, but it should be emphasized that it is the timing of applications that dictates success.

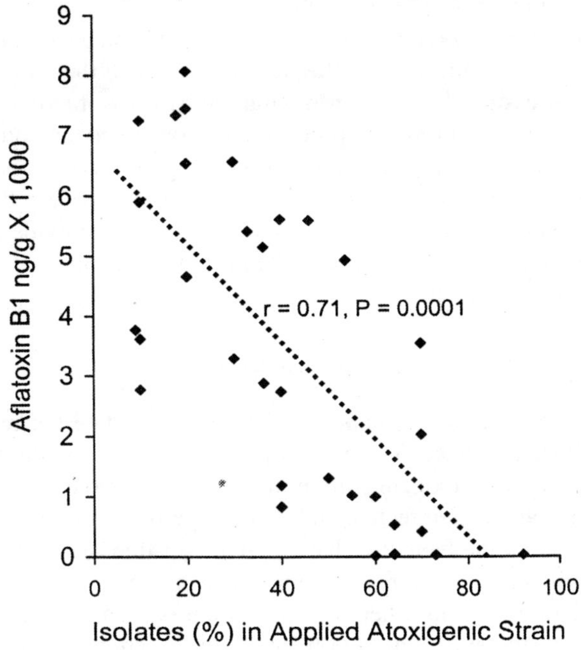

FIGURE 14.3 Relationship between the incidence of an applied atoxigenic strain in the infected portion of the crop and the quantity of aflatoxins in that crop. (Redrawn from Cotty, P.J., *Phytopathology*, 84, 1270, 1994.)

During seasons when aflatoxin contamination is severe, *A. flavus* populations increase as the cotton crop is produced.[89]

Strains are seeded into cotton fields immediately prior to first bloom. As in greenhouse tests,[15] it was found that strains differ in efficacy in the field, and selection of the proper strain and the inoculum rate may potentially determine the extent of success of a biological control strain.[90] The strains are applied to the soil surface under the crop canopy in the form of colonized sterile wheat seeds. When the crop is subsequently irrigated, the atoxigenic strain utilizes the resources in the colonized wheat seed, sporulates, and disperses to the crop. Wheat seed colonized by atoxigenic strain *Aspergillus flavus* AF36 has been evaluated in small-scale test plots since 1989.[83] Strain seeding has caused large and significant changes in the *A. flavus* population on the crop and in the soil. Applications have resulted in the applied atoxigenic strain becoming dominant in the field and aflatoxin-producing strains becoming less frequent.[83] These changes in the *A. flavus* populations have been associated with very significant reductions (75 to 99%) in aflatoxin contamination.[83] An inverse relationship between the incidence of the applied strain on seed from treated and control plots and the concentrations of aflatoxins in the seed has been well established (Figure 14.3).

A. flavus typically becomes associated with crops in the field during crop development. Field plot tests indicate that atoxigenic strain applications do not increase the amount of *A. flavus* on the crop at maturity or the percent of the cottonseed crop infected by *A. flavus*.[83] Both toxigenic and atoxigenic *A. flavus* that become associated with the crop in the field remain with the crop during harvest, storage, and processing.[91] Thus, crop vulnerability to aflatoxin contamination remains until the crop is ultimately processed or consumed, and atoxigenic strains seeded into agricultural fields prior to crop development continue to remain with the crop, providing long-term postharvest protection.[91] Atoxigenic strains applied both prior to harvest and after harvest have been shown to provide protection from aflatoxin contamination in corn,[80] even when toxigenic strains are associated with the crop prior to application.

B. Suitability of Biological Control Strategy

Economics of aflatoxin contamination will probably dictate the regions in which atoxigenic strains are utilized. The current projected cost to produce materials by a grower cooperative for atoxigenic strain applications is expected to be \$5/acre or less. If treatments are 70% effective and an average of 40 to 70% of seed is above 20 ppb aflatoxins and the benefit of having aflatoxin-free seed is \$20 to \$40 a ton, then growers will gain an average return above an initial \$5/acre investment of \$0.60–\$14.60/acre in regions where one ton of cottonseed is produced per acre. In severe aflatoxin years, even with a 90% reduction in contamination, growers may achieve no economic gain because the resulting crop will still contain over 20 ppb aflatoxin B_1. In certain regions, cottonseed crops containing over 1000 ppb are not uncommon.[92] Benefits could also arise from the applied atoxigenic strains remaining with the crop until use and thus preventing increased contamination during transit and in storage at dairies and reducing litigation. Economics might be improved by both long-term and cumulative benefits resulting from the strain's ability to remain in fields until the next crops are planted. Further, field-plot tests have demonstrated that atoxigenic strain applications have a long-term influence on *A. flavus* populations resident in agricultural fields.[93,94] Thus, atoxigenic strain applications have benefits over multiple seasons, and long-term, area-wide changes in the aflatoxin-producing potential of *A. flavus* populations can be achieved.

Just as dust doesn't stay in the field where it was raised, fungi do not stay in the field to which they were applied. Over time, therefore, applications could reduce contamination in an area as a whole. This should facilitate the development of either gin-wide or community-wide management programs. In areas where multiple crops are affected by contamination (e.g., corn, cotton, and peanuts), treatments to one crop could possibly benefit all crops. Nonetheless, a better understanding of the long-term benefits of atoxigenic strain applications are needed prior to development of area-wide management programs based on this technology.

Tests to evaluate the longevity of changes to *A. flavus* communities induced by atoxigenic strain applications to commercial cotton fields in Arizona have been conducted since 1996. The experimental program included treatments over a 3-year period (1120 acres total) and monitoring the *A. flavus* community from 1996 through 1999.[84,93] Both a scaled-up laboratory procedure for producing inoculum[88] and a quality control program were approved by the Environmental Protection Agency

(EPA) for use under Experimental Use Permit 69224-1. During this period, over 11,000 lb of colonized wheat were produced in the laboratory. The product was shipped without refrigeration to growers in food-grade, 5-gallon polyethylene buckets and was stored on-farm without special care until use. Different treatment regimes were applied to different fields, with some fields receiving treatment only in a single year and others receiving treatments in multiple years.[94]

Sterile wheat seed colonized by an atoxigenic strain was applied to 22 fields ranging in size from 10 to 160 acres from 1996 to 1998. The material was applied either by air or on the ground at the rate of 10 lb per acre. In order to monitor changes to the composition of *A. flavus* populations, soil samples were collected prior to application each year. From 1996 through 1999, over 8000 isolates of *A. flavus* were cultivated from soil samples taken from the treatment areas. Isolates were characterized by strain, and those assigned to the L strain of *A. flavus* were further characterized by vegetative compatibility analysis in order to determine applied strain distribution.[94]

One year after application, atoxigenic strain incidence was greatly increased, and incidence of the highly toxigenic S strain was greatly decreased in treated and adjacent fields.[93] The applied strain incidence gradually declined by the second year after application; however, even with this decline, the atoxigenic strain remained in treated fields at levels significantly higher than prior to treatment. The incidence of the applied strain in fields adjacent to treated fields 1 and 2 years after application was variable.[93] Thus, when planning area-wide management, possible directional movement of the applied strains from treated to untreated fields should be considered. Crop and crop stage, not only in treated fields but also in nearby untreated fields, at application are important determinants of the extent to which long-term and area-wide benefits are achieved.

Results of initial field studies have allowed the Arizona Cotton Growers Association to pursue research and development of an area-wide aflatoxin management program utilizing atoxigenic strains of *A. flavus* as a central component.[95] The Arizona Cotton Research and Protection Council has taken the lead in partnership with the Agricultural Research Service to develop an effective management program and to establish a grower-owned facility for the production of commercially useful quantities of atoxigenic strain inoculum. The Arizona growers applied in 1998 to the EPA for an atoxigenic strain registration that would permit treatment of all cotton in Arizona. In 1999, an expansion of the Experimental Use Permit was received that allowed treatment of up to 20,000 acres per annum.[96] Over 10,000 acres were treated in 1999. Improvements to the manufacturing facility were expected before treatment of the year 2000 crop. Once the Arizona growers have a facility that can produce the quantity and quality of material required, they will have to undergo a development series during which the agronomic practices associated with atoxigenic strain use are optimized to achieve both single-season benefit and long-term, area-wide reduction in the aflatoxin-producing potential of fungi resident in Arizona's agricultural areas.

Selection and seeding of fungal strains could, ultimately, become a useful tool in reducing the vulnerability of all crops grown in treated areas to aflatoxin contamination. Such selection and seeding of fungal strains may also be useful for other fungi that either cause contamination problems or provide needed activities in the agroecosystem. Some management of or modification to the local fungal strains may be required to enhance effectiveness in different geographical regions. Development of methodologies to specifically genetically modify *A. flavus* communities offers the opportunity to exert an unprecedented control over the safety and benefits of the filamentous fungi in the agricultural environment. This extends beyond the use of our extensive knowledge of the regulation of aflatoxin biosynthesis to construct specific gene-disrupted strains without the ability to produce aflatoxins. This potential extends to the development of "designer microbes" ideally suited to the environment in which they must compete and with safety features that may preclude animal pathogenesis or allergenicity.

VII. ENHANCEMENT OF HOST-PLANT RESISTANCE TO THE AFLATOXIN CONTAMINATION PROCESS

Traditional breeding or genetic engineering of plants with genes expressing either resistance against fungal infection or inhibition of aflatoxin biosynthesis most likely would achieve significant control of aflatoxin contamination. This is especially true for corn, which possesses a large amount of natural genetic diversity with respect to fungal infection. Naturally resistant crop germplasm, if identified, provides us not only with a source of resistance but also nature's lesson as to the specific requirements of resistance (e.g., antifungal compounds, regulation of these compounds, and physiological conditions for bioactivity). Studies in which various crops have been inoculated with aflatoxin-producing fungi have shown clearly (particularly in corn) that resistance mechanisms exist; however, our knowledge about resistance is incomplete, and little is known about the specific genetic and/or biochemical traits required for the expression of significant resistance against the aflatoxin contamination process.

Research efforts focusing on the development of crop germplasm with resistance to insect injury and effects of drought, acting along with resistance against fungal growth and aflatoxin production, could prove beneficial. In corn, the focus has included: (1) indirect protection of developing kernels by husk cover and anti-nutritional substances in silks, (2) direct protection provided by kernel compounds that block fungal development, and (3) direct protection provided by external integuments of developing kernels.[97] Investigations have also centered on identification of resistance to drought stress and insect damage as useful and obtainable traits.[97] Insect damage is often positively correlated with aflatoxin contamination; however, insects probably play a more important role in the infection process of *A. flavus* when conditions are less favorable for the fungus.[98] Under conditions more favorable to *A. flavus* (high temperatures, drought stress), the role of insect injury in aflatoxin contamination probably diminishes.[98] *A. flavus* may even display limited parasitic abilities under such conditions stressful to the plant.[99] In cotton, pink bollworm damage has been closely associated with aflatoxin contamination, and resistance to pink bollworm was thought to be a potential solution to the cottonseed aflatoxin contamination problem.[100] Early field-plot results with transgenic Bt cottons indicated that these cottons were both highly resistant to pink bollworm and to aflatoxin contamination.[101] When commercial crops were evaluated, the transgenic Bt cultivars continued to express very strong, sustained resistance to the pink bollworm; however, Bt cottonseed lots highly contaminated with aflatoxins were rapidly identified,[101] and in some locations differences in aflatoxin contamination between transgenic Bt and conventional cultivars were not seen in the commercial crop.[91,92,101] This may be true for corn, as well.[102]

Current research efforts are primarily focused on kernel/seed morphologic and chemical (e.g., antifungal proteins) resistance to fungal infection. The resistant genotypes generally inhibit aflatoxin production indirectly, through the inhibition of fungal growth.[73,103,104]

A. Current Progress in Plant Breeding Strategies

1. Screening Technologies

Screening crops for resistance to kernel or seed infection by *A. flavus* or for resistance to aflatoxin production is a more difficult task than most disease screening. Successful screening in the past[105] has been hindered by the lack of: (1) a resistant control, (2) inoculation methods that yield infection/aflatoxin levels high enough to differentiate among genotypes (natural infection is undependable), (3) repeatability across different locations and years, and (4) rapid and inexpensive methods for assessment of fungal infection and aflatoxin levels. Several plant inoculation methods for assessing crops such as corn and cotton for resistance to *A. flavus* invasion and subsequent aflatoxin contamination have been tried with varying degrees of success — for example, the pinbar inoculation technique for inoculating corn kernels through husks with the fungal conidia, the silk inoculation technique in corn, infesting corn ears with insect larvae infected with *A. flavus* conidia, wound inoculation of cotton bolls or tree nuts to mimic the exit holes of insects, etc.[106–109] Amending soils containing developing peanuts has also been examined for assessing resistance to *A. flavus* infection.[110,111]

a. Corn

Screening studies to evaluate various crop genotypes for resistance to aflatoxin contamination have demonstrated differential levels of resistance in the crops, but the genotypes all showed only partial resistance. However, two resistant corn inbreds (Mp420 and Mp313E) were discovered and tested in field trials at different locations and released as sources of "resistant" germplasm.[112,113] The pinbar inoculation technique (a precision kernel-wounding technique) was one of the methods employed in the initial trials, and contributed towards the separation of partially resistant from susceptible lines.[112] Several other corn inbreds, demonstrating partial resistance to aflatoxin contamination in Illinois field trials (employing a modified pinbar technique), also were discovered.[114] Another source of partially resistant germplasm is the corn-breeding population GT-MAS:gk, which was derived from visibly classified segregating kernels obtained from a single fungus-infected hybrid ear.[115] It tested resistant in trials conducted over a 5-year period, where a kernel knife inoculation technique was employed. These discoveries of partially resistant germplasm may have been facilitated by the use of inoculation techniques capable of repeatedly providing high infection/aflatoxin levels for genotype separation to occur. These corn lines do not generally possess commercially acceptable agronomic traits; however, they may be sources of resistance genes and, as such, provide a basis for the rapid development of host resistance strategies to eliminate aflatoxin contamination.

b. Peanuts

Several sources of resistant peanut germplasm have also been identified from a core collection representing the entire peanut germplasm collection.[110] Over 95% of this core has been preliminarily screened in a single environment; 16 genotypes tested over 3 years in two environments still display low levels of aflatoxins. A possible link between low linoleic acid content in peanuts and low preharvest aflatoxin production has been indicated.[110] Significant correlations have also been observed between leaf temperature and aflatoxin levels and/or visual stress ratings and aflatoxin levels. The preliminary screening of peanut genotypes using either or both of these traits could greatly reduce expenses involved in developing resistant cultivars. The promising germplasm, however, has less than acceptable agronomic characteristics and is being hybridized with those with commercially acceptable features. Resistant lines are also being crossed to pool resistances to aflatoxin production. To date, some success has been achieved in identifying resistant peanut germplasm, and field studies are being conducted by various researchers to verify this trait. Methods to improve screening of peanuts for resistance to *A. flavus* have been developed. A system of evaluating peanuts in the field through the manipulation of drought stress was successfully tested.[116] Also, an *in vitro* seed culture system demonstrating water stress responses in peanuts, similar to field responses, and variations in peanut phytoalexins and aflatoxin levels appears potentially useful.[111,117]

c. Tree Nuts

Among tree nuts, strategies for controlling preharvest aflatoxin formation by breeding for host resistance have been mostly studied in almonds and walnuts.[118–120] The approach employed in this effort is to integrate multiple genetic mechanisms for control of *Aspergillus* spp. as well as Navel orangeworm (*Paramyelois transitella* Walk), which appears important for initial fungal infection. Resistance to fungal colonization by incorporating seed coat resistance to infection is being pursued; however, genotypes demonstrating inhibition to fungal infection in seed tissues have been inconsistent over different environments.[118] Studies have also been conducted with figs and pistachios to identify the mode of infection of these crops by *A. flavus*. Once this parameter is clearly understood, strategies could be developed to identify germplasm with agronomically desirable characteristics and resistance to fungal infection.[120,121]

Generally speaking, crop varieties showing reduced levels of aflatoxin contamination have been produced by plant breeding; however, unacceptable levels of toxin still resulted when the plants were exposed to severe environmental pressures. Hindrances to resistance screening, discussed above, certainly played key roles in this lack of success.[112] Screening for resistance can produce useful results, but specific resistance traits or markers must be measured before appropriate breeding and selection techniques can be properly exploited.[122] Plating kernels or seed to determine the frequency of fungal infection or examining corn kernels for emission of a bright greenish-yellow fluorescence (BGYF) are methods that have been used for assessing *A. flavus* infection.[105] While these methods can indicate the presence of *A. flavus* in seed, neither can provide accurate quantitative or tissue-localization data useful for effective resistance breeding. Several other protocols have been developed and used for separation and relatively accurate quantitation of aflatoxins,[123,124] but fungal growth quantitation has not been possible until recently, with the genetically engineered fungi containing a reporter gene as a growth indicator.

2. Novel Screening Methods To Better Assess Fungal Infection and Growth

A laboratory kernel screening assay (KSA) has been developed and used to study resistance to aflatoxin production in GT-MAS:gk corn kernels.[125] KSA is an inexpensive humidity chamber technique in which kernels are screened at 100% humidity and 31°C, conditions that favor *A. flavus* growth and aflatoxin production.[73] Data from KSA experiments can be obtained 2 weeks after experiments are initiated. KSA experiments have confirmed GT-MAS:gk resistance to aflatoxin production and have demonstrated that the resistance is maintained even when the pericarp barrier, in otherwise viable kernels, is breached by wounding.[125] The wounding experiment facilitates both differentiation between different resistance mechanisms and comparison with other traits (e.g., fungal growth, protein induction). Apparently, there are two levels of resistance: one at the pericarp and another at the subpericarp level. KSA studies have demonstrated that the pericarp wax composition influences the kernel resistance phenotype[103,104,126] and that there are quantitative and qualitative differences in pericarp wax between GT-MAS:gk and susceptible genotypes.[127]

Kernel screening assays have also confirmed sources of resistance among inbreds tested in Illinois field trials.[73,114] When selected resistant Illinois inbreds (MI82, CI2, and T115) were examined by the modified KSA, which included an *A. flavus*-GUS transformant (a strain genetically engineered with a gene construct consisting of a ß-glucuronidase reporter gene linked to an *A. flavus* ß-tubulin gene promoter for monitoring fungal growth), a positive relationship between the degree of fungal infection and aflatoxin levels was established.[73,128,129] Moreover, kernel resistance to fungal infection in nonwounded and wounded kernels was clearly demonstrated visually and quantitatively.[129] Thus, it is now possible to accurately assess fungal infection levels and to determine if a correlation exists between infection and aflatoxin levels in the same kernels. *A. flavus* GUS transformants with the reporter gene linked to an aflatoxin biosynthetic pathway gene could also provide a quick and economical way to indirectly measure aflatoxin levels.[129,130]

The kernel screening assay has several advantages to complement traditional breeding techniques:[114,125] (1) it can be performed and repeated several times throughout the year and outside of the growing season; (2) it requires few kernels; (3) it can detect/identify different kernel resistance mechanisms; (4) it can dispute or confirm field evaluations (e.g., identify escapes); and (5) it can correlate findings and inoculations in the field. Field trials are irreplaceable for confirmation of resistance; however, the KSA may eliminate many preliminary field screenings and facilitate an in-depth investigation of kernel responses to fungal infection and aflatoxin production.

Using these new technologies, the fungus can now be "tracked" during its invasion process in various kernel/seed compartments. Studies employing KSA,[73] as well as other techniques,[131] have demonstrated that kernel embryos are colonized before endosperm tissue is invaded by aflatoxin-producing fungi. Embryo viability has also been shown to be necessary for the expression of kernel resistance.[125] It is possible that resistance, especially subpericarp, is a function of the kernel's ability to limit fungal colonization to a small area after wounding. Limiting fungal ingress may help prevent fungal spread through the kernel and interruption of whole-kernel expression of embryo-based resistance mechanisms, thus denying easy access of the fungus to a substrate most conducive to aflatoxin production.[128,129] The high levels of aflatoxins detected in susceptible kernels have often been considered the primary result of fungal metabolic activity on an embryonic substrate; however, there is evidence that these high levels of aflatoxin production may result from later fungal activity in the endosperm.[132]

The reporter gene constructs, when placed under the control of specific aflatoxin pathway gene promoter, can be utilized to identify environmental and nutritional signals in the plants and their sites of synthesis in specific plant tissues, which in turn are important in governing the degree of aflatoxin production. Further experiments can now be conducted to: (1) elucidate how environmental factors (fungal growth substrates, host plants, etc.) influence genetic regulation of aflatoxin biosynthesis; (2) utilize reporter gene assays to assess the influence of plant biochemicals on aflatoxin gene expression during the host plant *A. flavus* interaction; and (3) determine the effect of selected plant volatiles derived from the plant lipoxygenase pathway on fungal development, reproduction, and sporulation, processes critical to fungal survival and sharing genetic connection with the aflatoxin biosynthetic process.

3. Identification of Resistance Markers and Their Functions in Crops Vulnerable to Aflatoxin

A great preponderance of resistance "markers" potentially of value in plant breeding for resistance to the aflatoxin contamination process have been discovered in corn. In other crops, resistance markers that can be utilized in breeding to select for resistant progeny have proven more difficult to identify. The possible link between low linoleic acid content in peanuts and low preharvest aflatoxin production was suggested, but after further analysis the nature of this relationship has proven to be elusive. Progress has been made in identifying chromosome regions in corn associated with resistance to *A. flavus* infections and inhibition of aflatoxin production in corn using RFLP analysis in three resistant lines (R001, LB31, and Tex6), after mapping populations were developed using B73 and/or Mo17 elite inbreds.[133,134] In some cases, chromosomal regions were associated with resistance to *Aspergillus* ear rot but not with aflatoxin inhibition, and vice versa, whereas other chromosomal regions were found to be associated with both traits. This suggests that these two traits may be at least partially under separate genetic control. Also, it was observed that variation can exist in the chromosomal regions associated with *Aspergillus* ear rot and aflatoxin inhibition in different mapping populations, suggesting the presence of different resistance genes in different resistant germplasm. RFLP may provide the basis for employing a successful strategy of pyramiding different types of resistances into commercially viable germplasm, while avoiding the introduction of undesirable traits. However, there is a real need to identify specific traits at the gene level associated with the above RFLP markers and assign a specific biochemical or other function to the trait. This information is needed in order to make decisions on which traits to include to pyramid resistance through traditional plant breeding or transfer specific resistance traits into crops vulnerable to aflatoxin contamination by genetic engineering.

Studies demonstrating subpericarp (wounded-kernel) resistance in corn kernels have led to research with the aim of identifying subpericarp resistance mechanisms. When kernels of susceptible genotypes were allowed to imbibe water at 100% humidity at 31°C for 3 days prior to being subjected to the KSA protocol, aflatoxin levels were drastically and significantly reduced compared to unimbibed controls.[103,104] Kernel proteins induced during imbibition may have inhibited growth and/or fungal elaboration of aflatoxins. Examinations of kernel proteins of several genotypes revealed several differences between resistant and susceptible genotypes.[135] In imbibed susceptible kernels, decreased aflatoxin levels were associated with germination-induced synthesis of ribosome inactivating protein (RIP) and zeamatin.[136] Both zeamatin and RIP have been demonstrated to inhibit *A. flavus* growth *in vitro*.[136] These studies implicate proteins as potentially involved in kernel resistance to *A. flavus* infection and aflatoxin production.

Two kernel proteins have been identified from resistant inbred Tex6 which may contribute to resistance to aflatoxin production.[137] The 28-kDa protein inhibits *A. flavus* growth, while a 100-kDa protein inhibits toxin formation with little effect on fungal growth. With the recent elucidation of aflatoxin biosynthetic pathway genes and enzymes and regulatory mechanisms (see earlier section on aflatoxin biosynthesis),[138] technology could be developed with the potential for identifying seed/kernel resistance mechanisms that directly inhibit aflatoxin biosynthetic activity. An examination of kernel protein profiles of 13 corn genotypes revealed that a 14-kDa trypsin inhibitor protein is present at relatively high concentrations in seven resistant genotypes, but is present only in low concentrations in six susceptible ones.[139] This protein exhibits strong bioactivity against the growth of *A. flavus, A. parasiticus,* and a morphologically diverse group of other fungi.[128,129,140,141] Thus, comparisons of kernel protein profiles between susceptible and resistant genotypes may shorten the time it takes to identify resistance-associated proteins to be used in marker-assisted breeding.

4. Genetic Engineering Strategies

Several prerequisites must be met in order to employ genetic engineering as a means of developing host resistance against aflatoxin-contamination in crops: (1) resistance genes, native or foreign, should be identified that express inhibitory activity against

A. flavus; (2) gene promoters also must be selected that will allow appropriate expression of antifungal genes, at a desired time, in the candidate crop; and (3) genetic transformation needs to be adapted to each specific crop.[142] Resistance to *A. flavus* infection in plants could consist of an interaction of multiple components and biochemical changes that are either preformed or induced upon past invasion. Stimulation by elicitors may result in changes in gene expression and induction of pathogenicity-related (PR) proteins.[143–145] Multiple genes governing constitutive and inducible metabolic factors/mechanisms may exist and be expressed optimally at different stages of seed/kernel maturation. Development of optimal host-plant resistance through genetic engineering will require a sound understanding of the multiple factors that endow a plant with resistance to fungal attack.

B. Candidate Antifungal Compounds

Identifying resistance (e.g., in corn) makes it possible to correlate resistance with many endogenous small molecular weight compounds and biomacromolecules in kernel tissues already implicated as antifungal at various stages of kernel development in grain crops.[146–152] However, compounds with activity against other fungal species are ineffective against *A. flavus*, thus it is important to select the best candidate genes for these inhibitory compounds before plant genetic engineering procedures are initiated. A list of candidate antifungal compounds includes RIPs, lectins, relatively small molecular weight polypeptides, cell-surface glycoproteins, hydrolases, and certain basic proteins.[153] For example, the 14-kDa trypsin inhibitor,[139] shown to be correlated with kernel resistance to *A. flavus* infection of corn, when expressed in transgenic tobacco greatly enhances resistance to the tobacco pathogen, *Colletotrichum destructivum* (see details below).[154,155] A putative peptide also has been partially purified from aqueous kernel extracts of resistant inbred, Tex6, which demonstrated antifungal activity against *A. flavus*.[156] The corn kernel pathogenesis-related (PR) proteins appear to have a function during the normal process of seed germination;[144,147,148,150] however, they are induced to accumulate in response to fungal infection, and their expression is tissue-specific.[144,157] A further investigation of the kernel PR proteins using resistant and susceptible genotypes to examine specific tissue expression of these proteins under varying kernel physiology may facilitate the isolation of factors responsible for subpericarp resistance.

A number of potentially useful antifungal enzymes/proteins are produced either constitutively or in response to fungal attack in plants. These include chitinases and ß-1,3-glucanases,[152,158–160] osmotins,[161] protease inhibitors,[162] and polygalacturonase inhibitor proteins (PGIPs).[163] In addition, small molecular weight peptides have been isolated from organisms other than plants that also show promise as antifungal agents — for example, the cecropins[164] and magainins[165] of insect and amphibian origins, respectively, and their synthetic analogs.[166,167]

Several recent studies have suggested the potential of manipulating/inducing the lipoxygenase (LOX) pathway in plants to ward off fungal attack.[168] *A. flavus* exhibits strong lipolytic activity during infection on oilseeds[169] and at times causes substantial deterioration of the crop seeds and oils it has contaminated. Lipase activity originating from fungal degradation of host-plant membrane tissues releases fatty acids from bound triglycerides and triggers the LOX-hydroperoxide lyase (HPLS) enzyme pathway, converting linoleic and linolenic acids into hexanal and *cis*-3-hexenal, respectively.[170] Then, *cis*-3-hexenal is usually isomerized to *trans*-2-hexenal, both enzymatically and nonenzymatically.[170] Recently, it has been reported that specific LOX decay products, such as 13S-hydroperoxy fatty acid, jasmonic acid, and C_6–C_{10} alkenals and alkanals, may function as important signal molecules in host-pathogen interactions.[32,76,117,171,172] Jasmonic acid has also been shown to inhibit aflatoxin production and delay spore germination of *A. flavus*.[173] The antifungal properties of small chain alkanals and alkenals (derived from the LOX pathway) produced by cotton leaves have been demonstrated in solid and liquid cultures of aflatoxigenic *Aspergillus* spp.[30,75,174] Because of the mode of activation of these volatile aldehydes and because of the significant antifungal activity they exhibit, these compounds could function as "gaseous phytoalexins" in the cotton plant.[75]

There are only a few candidate genes whose expression products demonstrate convincing inhibitory activity against *A. flavus* and show promise for transformation of plants to reduce infection of seed by this particular fungal species. Included among these antifungal products are certain small lytic peptides. It is relatively easy to chemically synthesize genes encoding small peptides (using an oligonucleotide synthesizer) for transformation of plants, as only relatively small coding regions are required for their complete synthesis. Cecropins, for example, are lytic peptides of 22 to 23 amino acids in linear arrays that comprise antimicrobial systems found in insects and pig intestine.[164,175] The broad antibacterial activities of cecropins are due to the formation of large pores in the cell membrane.[176,177] They apparently do not lyse erythrocytes or other higher eucaryotic cells[178] but do inhibit growth of *A. flavus* mycelia.[179] Rajasekaran et al.[155] reported that a synthetic lytic peptide when transformed into tobacco greatly enhances resistance to *Colletotrichum destructivum*. In addition to lytic peptide genes, a variety of other candidate antifungal genes from bacterial, plant, and mammalian sources have a good potential to be active against *A. flavus* upon transformation into plants. Genes encoding LOX are available from plant sources.[180,181] The LOX products such as 13-hydroperoxylinoleic acid and its breakdown products/volatiles, such as hexenal and hexanal, are antifungal and interfere with the aflatoxin pathway. Genes encoding for haloperoxidases are also available for possible genetic engineering of plants for antifungal resistance.[182,183] In bioassays using *A. flavus* as the test organism, addition of a myeloperoxidase greatly enhanced (90-fold) the lethality of H_2O_2 by catalyzing its conversion to sodium hypochlorite.[184] A bacterial chloroperoxidase also greatly reduced the viability of *A. flavus* conidiospores.[183] H_2O_2 is induced in plants by wounding or injuring of plant tissues, a process often associated with pest attack, thus the substrate for these unique peroxidases should be available in the specific host-plant tissues under attack.[185]

C. Gene Promoters

Promoter elements that allow constitutive, wound-inducible, or tissue-specific expression of antifungal genes in plants have been identified. Characterized promoter elements that are useful in obtaining optimum expression of antifungal genes in plants include the CaMV 35S[186] and ubiquitin 3 promoter elements (constitutive),[187] the protease inhibitor II promoter (wound-inducible),[188] and storage protein gene promoters (seed-specific expression).[189] Peanuts have successfully been transformed with a wound-inducible promoter from soybean vegetative storage protein (*vsp*).[190] When the *vsp*-promoter/GUS gene fusion is inserted into peanut, expression of the GUS gene follows temporal and spatial patterns as would be predicted from soybean.[190] The *E. coli* ß-glucuronidase reporter gene has been used to assess the level of gene expression obtained under the control of some of the above promoters in transformed cotton.[142,191]

D. Transformation Methods

Antifungal genes in suitable gene expression vectors have been used in the transformation of plants by a variety of methods. The two most common methods include *Agrobacterium*-mediated gene transfer[192,193] and biolistic particle delivery or "gene gun" technology.[142,155,167,191,194] After gene transfer, transformed tissues are identified by growth on selective medium, and whole plants are regenerated from the selected, transformed cells. Cotton transformation has been accomplished using the *Agrobacterium*-based system and cotton hypocotyl sections, although the subsequent regeneration procedure in this system is not necessarily straightforward and can be lengthy. Cotton regeneration from transformed hypocotyl tissue involves the development of transformed embryogenic cell lines, embryoid formation, dissection, desiccation of embryos, and subsequent germination of the embryos.[195,196] Coker cultivars 201 and 312 of cotton have been transformed with *Agrobacterium*-mediated systems and regenerated.[197,198] Transformation of other commercially important cotton varieties has proven difficult due to the inability to generate embryogenic cell lines; however, *Agrobacterium*-mediated transformation has been successfully employed on elite Acala and Coker cultivars.[199] Problems, nevertheless, remain regarding the efficiency of this method of transformation and its adaptability to a wide range of germplasm. To circumvent the problem of cultivar-dependent regeneration, investigators have used the biolistic approach to transform cotton.[142,191,200,201] Peanut is also currently being transformed with antifungal genes by microprojectile bombardment of embryogenic tissues.[202] Walnut has been successfully transformed with barley lectin and nettle lectin antifungal genes.[203] The somatic embryo is the targeted tissue in walnut, and a new technique for cryopreservation of walnut somatic embryos is being used for long-term storage of embryo lines to prevent somaclonal variation and loss of lines to contamination.[203]

VIII. CONCLUSION

Several approaches are being explored and developed using new methods in biotechnology to eliminate pre-harvest aflatoxin contamination of food and feed. These approaches have resulted from recently acquired information about: (1) the ecology and epidemiology of aflatoxin-producing fungi, (2) molecular mechanisms governing aflatoxin biosynthesis, and (3) plant-derived metabolites that inhibit aflatoxin biosynthesis. Experience in our laboratory suggests a combined approach utilizing both host defense augmentation and biological control will be necessary to complement existing conventional methods in the eventual elimination of aflatoxin from the food and feed supply.

REFERENCES

1. Cotty, P.J., Bayman, P., Egel, D.S., and Elias, K.S., Agriculture, aflatoxins, and *Aspergillus*, in *The Genus Aspergillus: From Taxonomy and Genetics to Industrial Applications*, Powell, K.A., Renwick, A., and Peberdy, J.F., Eds., Plenum Press, New York, 1994, p. 1.
2. Goto, T., Wicklow, D.T., and Ito, Y., Aflatoxin and cyclopiazonic acid production by a sclerotium-producing *Aspergillus tamarii* strain, *Appl. Environ. Microbiol.*, 62, 4036, 1996.
3. Cotty, P.J. and Cardwell, K.F., Divergence of West African and North American communities of *Aspergillus* section *Flavi*, *Appl. Environ. Microbiol.*, 65, 2264, 1999.
4. Egel, E.S., Cotty, P.J., and Elias, K.S., Relationships among isolates of *Aspergillus* sect. *Flavi* that vary in aflatoxin production, *Phytopathology*, 84, 906, 1994.
5. Boyd, M.L. and Cotty, P.J., Spatiotemporal distribution and density of *Aspergillus* section *Flavi* propagules in Sonoran desert habitats [abstract], *Phytopathology*, 88, S10, 1998.
6. Cotty, P.J., Virulence and cultural characteristics of two *Aspergillus flavus* strains pathogenic to cotton, *Phytopathology*, 79, 814, 1989.
7. Cotty, P.J., Aflatoxin-producing potential of communities of *Aspergillus* section *Flavi* from cotton producing areas in the United States, *Mycol. Res.*, 101, 698, 1997.
8. Horn, B.W. and Dorner, J.W., Soil populations of *Aspergillus* species from section *Flavi* along a transect through peanut-growing regions of the United States, *Mycologia*, 90, 767, 1998.
9. Orum, T.V., Bigelow, D.M., Cotty, P.J., and Nelson, M.R., Using predictions based on geostatistics to monitor trends in *Aspergillus flavus* strain composition, *Phytopathology*, 89, 761, 1999.
10. Cotty, P.J., Cleveland, T.E., Brown, R.L., and Mellon, J.E., Variation in polygalacturonase production among *Aspergillus flavus* isolates, *Appl. Environ. Microbiol.*, 56, 3885, 1990.
11. Cleveland, T.E. and Cotty, P.J., Invasiveness of *Aspergillus flavus* in wounded cotton bolls is associated with production of a specific fungal polygalacturonase, *Phytopathology*, 81, 155, 1991.
12. Leslie, J.F., Fungal vegetative compatibility, *Ann. Rev. Phytopathol.*, 31, 127, 1993.
13. Bayman, P. and Cotty, P.J.,Vegetative compatibility and genetic variation in the *Aspergillus flavus* population of a single field., *Can. J. Bot.*, 69, 1707, 1991.
14. Horn, B.W. and Greene, R.L., Vegetative compatibility within populations of *Aspergillus flavus*, *A. parasiticus*, and *A. tamarii* from a peanut field, *Mycologia*, 87, 324, 1995.
15. Cotty, P.J. and Bhatnagar, D., Variability among atoxigenic *Aspergillus flavus* strains in ability to prevent aflatoxin contamination and production of aflatoxin biosynthetic pathway enzymes, *Appl. Environ. Microbiol.*, 60, 2248, 1994.
16. Elias, K.S. and Cotty, P.J., Use of esterase isozymes to distinguish *Aspergillus flavus* group species, strains, and vegetative compatibility groups, *Phytopathology*, 84, 1150, 1994.

17. Bayman, P. and Cotty, P.J., Genetic diversity in *Aspergillus flavus*: association with aflatoxin production and morphology, *Can. J. Bot.*, 71, 23, 1993.
18. Beuchat, L.R., Traditional fermented food products, in *Food and Beverage Mycology*, Beuchat, L.R., Ed., AUI, Westport, 1978, p. 224.
19. Geiser, D.M., Pitt, J.I., and Taylor, J.W., Cryptic speciation and recombination in the aflatoxin-producing fungus *Aspergillus flavus*, *Proc. Natl. Acad. Sci.*, 95, 388, 1998.
20. Eaton, D.L. and Groopman, J.D., Eds., *The Toxicology of Aflatoxins: Human Health, Veterinary and Agricultural Significance,* Academic Press, San Diego, CA, 1994, p. 554.
21. Charmley, L.L., Trenholm, H.L., Preluskey, D.B., and Rosenberg, A., Economic losses and decontamination, *Natural Toxins*, 3, 199, 1995.
22. Shane, S.M., Economic issues associated with aflatoxins, in *The Toxicology of Aflatoxins: Human Health, Veterinary, and Agricultural Significance*, Eaton, D.L. and Groopman, J.D., Eds., Academic Press, London, 1994, p. 513.
23. Nicholas, Jr., T.E., Economic impact of aflatoxin in corn, in *Aflatoxin and Aspergillus flavus in Corn*, Diener, U.L., Asguith, R.L., and Dickens, J.W., Eds., Auburn University, AL, 1983, p. 67.
24. Bhatnagar, D., Lillehoj, E.B., and Bennett, J.W., Biological detoxification of mycotoxins, in *Mycotoxins and Animal Foods*, Smith, J.E. and Henderson, R.S., Eds., CRC Press, Boca Raton, FL, 1991, p. 815.
25. Park, D.L., Lee, L.S., Price, R.L., and Pohland, A.E., Review of decontamination by ammoniation: current status and regulation, *J. Am. Oil Chemists' Soc.*, 71, 685, 1988.
26. Sinha, K.K., Detoxification of mycotoxins and food safety, in *Mycotoxins in Agriculture and Food Safety*, Sinha, K.K. and Bhatnagar, D., Eds., Marcel Dekker, New York, 1998, p. 381.
27. Jones, R.K., The influence of cultural practices on minimizing the development of aflatoxin in field maize, in *Aflatoxin in Maize: Proceedings of the Workshop*, Zuber, M.S., Lillehoj, E.B., and Renfro, B.L., Eds., Cimmyt, Mexico, 1987, p. 136.
28. Cleveland, T.E. and Bhatnagar, D., Molecular strategies for reducing aflatoxin levels in crops before harvest, in *Molecular Approaches to Improving Food Quality and Safety*, Bhatnagar, D. and Cleveland, T.E., Eds., Van Nostrand Reinhold, New York, 1992, p. 205.
29. Zaika, L.L. and Buchanan, R.L., Review of compounds affecting biosynthesis or bioregulation of aflatoxins, *J. Food Prot.*, 50, 691, 1987.
30. Zeringue, Jr., H.J. and McCormick, S.R., Relationships between cotton leaf-derived volatiles and growth of *Aspergillus flavus*, *J. AOCS*, 66, 581, 1989.
31. Bhatnagar, D. and McCormick, S.P., The inhibitory effect of neem (*Azadirachta indica*) leaf extracts on aflatoxin synthesis in *Aspergillus parasiticus*, *J. Am. Oil Chemists' Soc.,* 65, 1166, 1988.
32. Zeringue, Jr., H.J., Effects of C_6–C_{10} alkenals and alkanals on eliciting a defense response in the developing cotton boll, *Phytochemistry,* 31, 2305, 1992.
33. Greene-McDowelle, D.M., Ingber, B., Wright, M.S., Zeringue, Jr., H.J., Bhatnagar, D., and Cleveland, T.E., The effects of selected cotton-leaf volatiles on growth, development and aflatoxin production of *Aspergillus parasiticus*, *Toxicon*, 37, 883, 1999.
34. Wright, M., Greene-McDowelle, D.M., Zeringue, Jr., H.J., Bhatnagar, D., and Cleveland, T.E., Effect of volatile aldehydes from aflatoxin-resistant varieties of corn on *Aspergillus parasiticus* growth and aflatoxin biosynthesis, *Toxicon*, 38, 1215, 2000.

35. Bhatnagar, D., Cleveland, T.E., Brown, R.L., Cary, J.W., Yu, J., and Chang, P.-K., Preharvest aflatoxin contamination: elimination through biotechnology, in *Ecological Agriculture and Sustainable Development.* Vol. 1. *Indian Ecological Society and Center for Research in Rural and Industrial Development*, Dhaliwal, G.S., Randhawa, N.S., Arora, R., and Dhawan, A.K., Eds., Chaman Enterprises, New Delhi, 1998, p. 110.
36. Payne, G.A. and Brown, M.P., Genetics and physiology of aflatoxin biosynthesis, *Ann. Rev. Phytopathol.*, 36, 329, 1998.
37. Brown, M.P., Brown-Jenco, C.S., and Payne, G.A., Genetic and molecular analysis of aflatoxin biosynthesis, *Fungal Genet. Biol.*, 26, 81, 1999.
38. Dutton, M.F., Enzymes and aflatoxin biosynthesis, *Microbiol. Rev.*, 52, 274, 1988.
39. Bhatnagar, D., Ehrlich, K.C., and Cleveland, T.E., Oxidation-reduction reactions in biosynthesis of secondary metabolites, in *Mycotoxins in Ecological Systems*, Bhatnagar, D., Lillehoj, E.B., and Arora, D.K., Eds., Marcel Dekker, New York, 1991, p. 255.
40. Minto R.E. and Townsend C.A., Enzymology and molecular biology of aflatoxin biosynthesis, *Chem. Rev.*, 97, 2537, 1997.
41. Anderson, J.A., Enzymes in aflatoxin biosynthesis, *World J. Micro. Biotechnol.,* 8, 96, 1992.
42. Keller, N.P. and Hohn, T.M., Metabolic pathway gene clusters in filamentous fungi, *Fungal Genet. Biol.*, 21, 17, 1997.
43. Bhatnagar, D., Cleveland, T.E., and Kingston, D.G.I., Enzymological evidence for separate pathways for aflatoxin B_1 and B_2 biosynthesis, *Biochemistry*, 30, 4343, 1991.
44. Trail, F., Chang, P.-K., Cary, J., and Linz, J.E., Structural and functional analysis of the *nor*-1 gene involved in the biosynthesis of aflatoxin by *Aspergillus parasiticus*, *Appl. Environ. Microbiol.*, 60, 4078, 1994.
45. Cary, J.W., Wright, M., Bhatnagar, D., Lee, R., and Chu, F.S., Molecular characterization of an *Aspergillus parasiticus* dehydrogenase gene, *norA*, located on the aflatoxin biosynthesis gene cluster, *Appl. Environ. Microbiol.*, 62, 360, 1996.
46. Yabe, K., Ando, Y., and Hamasaki, T., Biosynthetic relationship among aflatoxins B_1, B_2, G_1, G_2, *Appl. Environ. Microbiol.*, 54, 101, 1988.
47. Keller, N.P., Dischinger, H.C., Bhatnagar, D., Cleveland, T.E., and Ullah, A.H.J., Purification of a 40-kilodalton methyltransferase active in the aflatoxin biosynthetic pathway, *Appl. Environ. Microbiol.*, 59, 479, 1993.
48. Motomura, M., Chihaya, N., Shinozawa, T., Hamasaki, T., and Yabe, K., Cloning and characterization of the *O*-methyltransferase I gene (*dmtA*) from *Aspergillus parasiticus* associated with the conversions of demethylsterigmatocystin to sterigmatocystin and dihydrodemethylsterigmatocystin to dihydrosterigmatocystin in aflatoxin biosynthesis, *Appl. Environ. Microbiol.*, 65, 11, 4987, 1999.
49. Yu, J., Chang, P.-K., Payne, G.A., Cary, J.W., Bhatnagar, D., and Cleveland, T.E., Comparison of the *omtA* genes encoding *O*-methyltransferases involved in aflatoxin biosynthesis from *Aspergillus parasiticus* and *A. flavus*, *Gene*, 163, 121, 1995.
50. Yabe, K., Nakamura, M., and Hamasaki, T., Enzymatic formation of G-group aflatoxins and biosynthetic relationship between G- and B-group aflatoxins, *Appl. Environ. Microbiol.*, 65(9), 3867, 1999.
51. Prieto, R. and Woloshuk, C.P., *ord*l, an oxidoreductase gene responsible for conversion of *O*-methylsterigmatocystin to aflatoxin in *Aspergillus flavus*, *Appl. Environ. Microbiol.*, 63, 1661, 1997.

52. Yu, J., Chang, P.-K., Ehrlich, K.C., Cary, J.W., Montalbano, B., Dyer, J.M., Bhatnagar, D., and Cleveland, T.E., Characterization of critical amino acids of an *Aspergillus parasiticus* cytochrome P-450 monooxygenase encoded by *ordA* that is involved in biosynthesis of aflatoxins B_1, G_1, B_2, and G_2, *Appl. Environ. Microbiol.*, 64, 4834, 1998.
53. Bennett, J.W. and Papa, K.E.,The aflatoxigenic *Aspergillus* spp., in *Advances in Plant Pathology: Genetics of Plant Pathogenic Fungi,* Vol. 6, Sidhu, G.S., Ed., Academic Press, New York, 1988, p. 263.
54. Payne, G.A. and Woloshuk, C.P., The transformation of *Aspergillus flavus* to study aflatoxin biosynthesis, *Mycopathologia*, 107, 139, 1989.
55. Keller, N.P., Cleveland, T.E., and Bhatnagar, D., Variable electrophoretic karyotypes of members of *Aspergillus* section *Flavi*, *Curr. Genet.*, 21, 371, 1992.
56. Foutz, K.R., Woloshuk, C.P., and Payne, G.A., Cloning and assignment of linkage group loci to a karyotypic map of the filamentous fungus *Aspergillus flavus*, *Mycologia*, 87, 787, 1995.
57. Skory, C.D., Chang, P.-K., Cary, J.W., and Linz, J.E., Isolation and characterization of a gene from *Aspergillus parasiticus* associated with conversion of versicolorin A to sterigmatocystin in aflatoxin biosynthesis, *Appl. Environ. Microbiol.*, 58, 3527, 1992.
58. Yu, J., Chang, P.-K., Cary, J.W., Wright, M., Bhatnagar, D., Cleveland, T.E., Payne, G.A., and Linz, J.E., Comparative mapping of aflatoxin pathway gene clusters in *Aspergillus parasiticus* and *Aspergillus flavus*, *Appl. Environ. Microbiol.*, 61, 2365, 1995.
59. Trail, F., Mahanti, N., Rarick, M., Mehigh, R., Liang, S.-H., Zhou, R., and Linz, J.E., A physical and transcriptional map of an aflatoxin gene cluster in *Aspergillus parasiticus* and the functional disruption of a gene involved in the aflatoxin pathway, *Appl. Environ. Microbiol.*, 61, 2665, 1995.
60. Silva, J.C., Minto, R.E., Barry, III, C.E., Holland, K.A., and Townsend, C.A., Isolation and characterization of the versicolorin B synthase gene from *Aspergillus parasiticus*: expansion of the gene cluster, *J. Biol. Chem.*, 271, 13600, 1996.
61. Brown, D.W., Yu, J.H., Kelkar, H.S., Fernandes, M., Nesbitt, T.C., Keller, N.P., Adams, T.H., and Leonard, T.J., Twenty-five coregulated transcripts define a sterigmatocystin gene cluster in *Aspergillus nidulans*, *Proc. Natl. Acad. Sci. USA*, 93, 1418, 1996.
62. Chang, P.-K., Ehrlich, K.C., Yu, J., Bhatnagar, D., and Cleveland, T.E., Increased expression of *Aspergillus parasiticus aflR*, encoding a sequence-specific DNA-binding protein, relieves nitrate inhibition of aflatoxin biosynthesis, *Appl. Environ. Microbiol.*, 61, 2372, 1995.
63. Flaherty, J.E. and Payne, G.A., Overexpression of *aflR* leads to upregulation of pathway gene transcription and increased aflatoxin production in *Aspergillus flavus*, *Appl. Environ. Microbiol.*, 63, 3995, 1997.
64. Yu, J.-H., Butchko, R.A.E., Fernandes, M., Keller, N.P., Leonard, T.J., and Adams, T.H., Conservation of structure and function of the aflatoxin regulatory gene *aflR* from *Aspergillus nidulans* and *A. flavus*, *Curr. Genet.*, 29, 549, 1996.
65. Fernandes, M., Keller, N.P., and Adams, T.H., Sequence-specific binding by *Aspergillus nidulans* AFLR, a C6 zinc cluster protein regulating mycotoxin biosynthesis, *Molec. Microbiol.*, 28, 1355, 1998.
66. Ehrlich, K.C., Montalbano, B.G., and Cary, J.W., Binding of the C6-zinc cluster protein, AFLR, to the promoters of aflatoxin pathway biosynthesis genes in *Aspergillus parasiticus*, *Gene,* 230, 249, 1999.

67. Cary, J.W., Barnaby, N., Ehrlich, K.C., and Bhatnagar, D., Isolation and characterization of experimentally induced, aflatoxin biosynthetic pathway deletion mutants of *Aspergillus parasiticus*, *Appl. Microbiol. Biotechnol.*, 51, 808, 1999.
68. Chang, P.-K., Yu, J., Bhatnagar, D., and Cleveland, T.E., The carboxy-terminal portion of the aflatoxin pathway regulatory protein AFLR of *Aspergillus parasiticus* activates *GAL1::lacZ* gene expression in *Saccharomyces cerevisiae*, *Appl. Environ. Microbiol.*, 65, 2508, 1999.
69. Kale, S.P., Cary, J.W., Bhatnagar, D., and Bennett, J.W., Characterization of experimentally induced nonaflatoxigenic variant strains of *Aspergillus parasiticus*., *Appl. Environ. Microbiol.*, 62, 3399, 1996.
70. Guzman-de-Pena, D. and Ruiz-Hirrera, J., Relationship between aflatoxin biosynthesis and sporulation in *Aspergillus parasiticus*, *Fungal Genet. Biol.*, 21, 198, 1997.
71. Hicks, J.K., Yu, J.-H., Keller, N.P., and Adams, T.H., *Aspergillus* sporulation and mycotoxin production both require inactivation of the FadA G protein-dependent signaling pathway, *EMBO J.*, 16, 4916, 1997.
72. Flaherty, J.E., Weaver, M.A., Payne, G.A., and Woloshuk, C.P., A beta-glucuronidase reporter gene construct for monitoring aflatoxin biosynthesis in *Aspergillus flavus*, *Appl. Environ. Microbiol.*, 61, 2482, 1995.
73. Brown, R.L., Cleveland, T.E., Payne, G.A., Woloshuk, C.P., Campbell, K.W., and White, D.G., Determination of resistance to aflatoxin production in maize kernels and detection of fungal colonization using an *Aspergillus flavus* transformant expressing *Escherichia coli* ß-glucuronidase, *Phytopathology*, 85, 983, 1995.
74. Cleveland, T.E., Cary, J.W., Bhatnagar, D., Yu, J., Chang, P-K., Chlan, C.A., and Rajasekaran, K., Use of biotechnology to eliminate aflatoxin in preharvest crops, *Bull. Inst. Compr. Agr. Sci.*, 5, 75, 1997.
75. Zeringue, Jr., H.J. and McCormick, S.P., Aflatoxin production in cultures of *Aspergillus flavus* incubated in atmospheres containing cotton leaf-derived volatiles, *Toxicon*, 28, 445, 1990.
76. Zeringue, Jr., H.J, Brown, R.L., Neucere, J.N., and Cleveland, T.E., Relationship between C_6–C_{12} alkenal volatile contents and resistance of maize genotypes to *Aspergillus flavus* and aflatoxin production, *J. Agri. Food Chem.*, 44, 403, 1996.
77. Calvo, A.M., Hinze L.L., Gardner H.W., and Keller N.P., Sporogenic effect of polyunsaturated fatty acids on development of *Aspergillus* spp., *Appl. Environ. Microbiol.*, 65, 3668, 1999.
78. Cole, R.J. and Cotty, P.J., Biocontrol of aflatoxin production by using biocompetitive agents, in *A Perspective on Aflatoxin in Field Crops and Animal Food Products in the United States*, Robens, J.F., Ed., Agricultural Research Service, Beltsville, MD, 1990, p. 62.
79. Cotty, P.J., Use of Native *Aspergillus flavus* Strains To Prevent Aflatoxin Contamination, U.S. Patent No. 5,171,686, 1992.
80. Brown, R.L., Cotty, P.J., and Cleveland, T.E., Reduction in aflatoxin content of maize by atoxigenic strains of *Aspergillus flavus*, *J. Food Prot.*, 54, 623, 1991.
81. Cotty, P.J., Effect of atoxigenic strains of *Aspergillus flavus* on aflatoxin contamination of developing cottonseed, *Plant Dis.*, 74, 233, 1990.
82. Cotty, P.J. and Bayman, P., Competitive exclusion of a toxigenic strain of *Aspergillus flavus* by an atoxigenic strain, *Phytopathology*, 93, 1283, 1993.
83. Cotty, P.J., Influence of field application of an atoxigenic strain of *Aspergillus flavus* on the populations of *A. flavus* infecting cotton bolls and on the aflatoxin content of cottonseed, *Phytopathology*, 84, 1270, 1994.

84. Cotty, P.J., Harwell, D.R., and Sobek, E.A., The EPA approved experimental use program to *Aspergillus flavus* AF36, in *Proceedings of the 1996 USDA/ARS Aflatoxin Elimination Workshop*, Fresno, CA, 1996, p. 3.
85. Dorner, J.W., Cole, R.J., and Blankenship, P.D., Use of a biocompetitive agent to control preharvest aflatoxin in drought stressed peanuts, *J. Food Prot.*, 55, 888, 1992.
86. Dorner, J.W., Cole, R.J., and Wicklow, D.T., Aflatoxin reduction in corn through field application of competitive fungi, *J. Food Prot.*, 62, 650, 1999.
87. Cotty, P.J., Aflatoxin contamination: variability and management, in *Cotton: A College of Agriculture Report*, Series P-87, Silvertooth, J. and Bantlin, M., Eds., University of Arizona, Tucson, 1991, p. 114.
88. Bock, C.H. and Cotty, P.J., Wheat seed colonized with atoxigenic *Aspergillus flavus*: characterization and production of a biopesticide for aflatoxin control, *Biocontrol Sci. Technol.*, 9, 529, 1999.
89. Lee, L.S., Lee, L.V., and Russell, T.E., Aflatoxin in Arizona cottonseed: field inoculation of bolls by *Aspergillus flavus* spores in wind-driven soil, *J. Am. Oil Chem. Soc.*, 63, 530, 1986.
90. Dorner, J.W., Cole, R.J., and Blankenship, P.D., Effect of inoculum rate of biological control agents on preharvest aflatoxin contamination of peanuts, *Biological Control*, 12, 171, 1998.
91. Cotty, P.J., Update on methods to prevent aflatoxin formation, *Oil Mill Gazetteer*, 104, 8, 1997.
92. Bock, C.H. and Cotty, P.J., The relationship of gin date to aflatoxin contamination of cottonseed in Arizona, *Plant Dis.*, 83, 279, 1999.
93. Cotty, P.J., Atoxigenic strains of *Aspergillus flavus* have been applied to commercial cotton fields for three years, *Proc. 1999 Beltwide Cotton Conf.*, 1, 108, 1999.
94. Cotty, P.J., Long-term influences of atoxigenic strains on *Aspergillus flavus* communities in commercial agriculture, in *Proceedings of the 1999 USDA/ARS Aflatoxin Elimination Workshop*, Atlanta, GA, 2000, p. 82.
95. Antilla, L. and Cotty, P.J., Production of commercially useful quantities of atoxigenic strain inoculum, in *Proceedings of the 12th Annual USDA/ARS Aflatoxin Elimination Workshop*, 2000, p. 83.
96. Anon., *Aspergillus flavus* AF36: pesticide tolerance exemption, *Fed. Regist.*, 64, 28371, 1999.
97. Lisker, N. and Lillehoj, E.B., Prevention of mycotoxin contamination (principally aflatoxins and *Fusarium* toxins) at the preharvest stage, in *Mycotoxins and Animal Foods*, Smith, J.E. and Henderson, R.S., Eds., CRC Press, Boca Raton, FL, 1991, p. 689.
98. Payne, G.A., Process of contamination by aflatoxin-producing fungi and their impact on crops, in *Mycotoxins in Agriculture and Food Safety*, Sinha, K.K. and Bhatnagar, D., Eds., Marcel Dekker, New York, 1998, p. 279.
99. Payne, G.A., Aflatoxin in maize, *CRC Crit. Rev. Plant Sci.*, 10, 423, 1992.
100. Cotty, P.J. and Lee, L.S., Aflatoxin contamination of cottonseed: comparison of pink bollworm damaged and undamaged bolls, *Trop. Sci.*, 29, 273, 1989.
101. Cotty, P.J., Howell, D.R., Bock, C., and Tellez, A., Aflatoxin contamination of commercially grown transgenic Bt cottonseed, in *Proceedings of the Beltwide Cotton Production Research Conferences*, National Cotton Council of America, Memphis, TN, 1997, p. 108.
102. Wilson, D.M., Wright, D.L., Wiatrak, P., Herzog, D., Buntin, G.D., and Lee, D., Comparison of preharvest aflatoxin accumulation in Bt and non-Bt corn in Florida and Georgia, in *Proceeding of the 1999 USDA/ARS Aflatoxin Elimination Workshop*, Atlanta, GA, 2000, p. 32.

103. Guo, B.Z., Russin, J.S., Brown, R.L., Cleveland, T.E., and Widstrom, N.W., Resistance to aflatoxin contamination in corn as influenced by relative humidity and kernel germination, *J. Food Prot.*, 59, 276, 1996.
104. Guo, B.Z., Russin, J.S., Cleveland, T.E., Brown, R.L., and Damann, K.E., Evidence for cutinase production by *Aspergillus flavus* and its possible role in infection of corn kernels, *Phytopathology*, 86, 824, 1996.
105. King, S.B. and Wallin, J.R., Methods for screening corn for resistance to kernel infection and aflatoxin production by *Aspergillus flavus*, in *Aflatoxin and Aspergillus flavus in Corn,* Southern Cooperative Bull. 279, Diener, U.L., Asquith, R.L, and Dickens, J.W., Eds., Auburn University, AL, 1983, p. 77.
106. Cotty, P.J., Effects of cultivar and boll age on aflatoxin in cottonseed after inoculation with *Aspergillus flavus* at simulated exit holes of the pink bollworm, *Plant Dis.*, 73, 489, 1989.
107. King, S.B. and Scott, G.E., Field inoculation techniques to evaluate maize for reaction to kernel infection by *Aspergillus flavus*, *Phytopathology*, 72, 782, 1982.
108. Tucker, Jr., D.H., Trevathan, L.E., King, S.B., and Scott, G.E., Effect of four inoculation techniques on infection and aflatoxin concentration of resistant and susceptible corn hybrids inoculated with *Aspergillus flavus*, *Phytopathology*, 76, 290, 1986.
109. Mahoney, N. and Molyneux, R.J., Contamination of tree nuts by aflatoxigenic fungi: aflatoxin content of closed-shell pistachios, *J. Agric. Food Chem.*, 46, 1906, 1998.
110. Holbrook, C.C., Wilson, D.M., and Matheron, M.E., An update on breeding peanut for resistance to preharvest aflatoxin contamination, in *Proceedings of the USDA-ARS Aflatoxin Elimination Workshop*, Atlanta, GA, 1995, p. 3.
111. Dorner, J., Cole, R., Sanders, T., and Blankenship, P., Inter-relationship of kernel water activity, soil temperature, maturity, and phytoalexin production in preharvest aflatoxin contamination of drought-stressed peanuts, *Mycopathologia*, 105(2), 117, 1989.
112. Scott, G.E. and Zummo, N., Sources of resistance in maize to kernel infection by *Aspergillus flavus* in the field, *Crop Sci.*, 28, 505, 1988.
113. Windham, G.L. and Williams, W.P., *Aspergillus flavus* infection and aflatoxin accumulation in resistant and susceptible maize hybrids, *Plant Dis.*, 82, 281, 1998.
114. Campbell, K.W. and White, D.G., Evaluation of corn genotypes for resistance to *Aspergillus* ear rot, kernel infection, and aflatoxin production, *Plant Dis.*, 79, 1039, 1995.
115. Widstrom, N.W., McMillan, W.W., and Wilson, D., Segregation for resistance to aflatoxin contamination among seeds on an ear of hybrid maize, *Crop Sci.*, 27, 961, 1987.
116. Mehan, V.K., Rao, R.C., McDonald, D., and Williams, J.H., Management of drought stress to improve field screening of peanuts for resistance to *Aspergillus flavus*, *Phytopathology*, 78, 659, 1988.
117. Basha, S.M., Cole, R.J., and Pancholy, S.K., A phytoalexin and aflatoxin-producing peanut seed culture system, *Peanut Sci.,* 21, 130, 1994.
118. Gradziel, T.M. and Dandekar, A., Integrated fungal/insect resistance to aflatoxin contamination of almond appears durable in field trails despite differential environmental and genotype by environment responses of individual components, in *Proceedings of the USDA-ARS Aflatoxin Elimination Workshop*, St. Louis, MO, 1998, p. 5.
119. Gradziel, T.M. and Wang, D., Susceptibility of California almond cultivars to aflatoxigenic *Aspergillus flavus*, *Hort. Sci.*, 29, 33, 1994.
120. Doster, M.A., Michailides, T.J., and Morgan, D.P., Aflatoxin control in pistachio, walnut, and figs: identification and separation of contaminated nuts and figs, ecological relationships, and agronomic practices, in *Proceedings of the USDA-ARS Aflatoxin Elimination Workshop*, Atlanta, GA, 1995, p. 63.

121. Mahoney, N.E. and Rodriguez, S.B., Aflatoxin variability in pistachios, *Appl. Environ. Microbiol.*, 62(4), 1197, 1996.
122. Brown, R.L., Chen, Z.-Y., Cleveland, T.E., and Russin, J.S., Advances in the development of host resistance in corn to aflatoxin contamination by *Aspergillus flavus*, *Phytopathology*, 89,113, 1999.
123. Shotwell, O., Aflatoxin detection and determination in corn, in *Aflatoxin and Aspergillus flavus in Corn,* Southern Cooperative Bull. 279, Diener, U.L., Asquith, R.L., and Dickens, J.W., Eds., Auburn University, AL, 1983, p. 38.
124. Wilson, D.M., Sydenham, E.W., Lombaert, G.A., Trucksess, M.W., Abramson, D., and Bennett, G.A., Mycotoxin analytical techniques, in *Mycotoxins in Agriculture and Food Safety*, Sinha, K.K. and Bhatnagar, D., Eds., Marcel Dekker, New York, 1998, p. 135.
125. Brown, R.L., Cotty, P.J., Cleveland, T.E., and Widstrom, N.W., Living maize embryo influences accumulation of aflatoxin in maize kernels, *J. Food Prot.,* 56, 967, 1993.
126. Guo, B.Z., Russin, J.S., Cleveland, T.E., Brown, R.L., and Widstrom, N.W., Wax and cutin layers in maize kernels associated with resistance to aflatoxin production by *Aspergillus flavus*, *J. Food Prot.*, 58, 296,1995.
127. Russin, J.S., Guo, B.Z., Tubajika, K.M., Brown, R.L., Cleveland, T.E., and Widstrom, N.W., Comparison of kernel wax from corn genotypes resistant or susceptible to *Aspergillus flavus*, *Phytopathology*, 87, 529, 1997.
128. Brown, R.L., Chen, Z.-Y., Lax, A.R., Cary, J.W., Cleveland, T.E., Russin, J.S., Guo, B.Z., Williams, W.P., Davis, G., Windham, G.L., and Payne, G.A., Determination of maize kernel biochemical resistance to aflatoxin elaboration: mechanisms and biotechnological tools, in *Proceedings of the USDA-ARS Aflatoxin Elimination Workshop*, Memphis, TN, 1997, p. 58.
129. Brown, R.L., Cleveland, T.E., Payne, G.A., Woloshuk, C.P., and White, D.G., Growth of an *Aspergillus flavus* transformant expressing *Escherichia coli* ß-glucuronidase in maize kernels resistant to aflatoxin production, *J. Food Prot.*, 60, 84, 1997.
130. Payne, G.A., Characterization of inhibitors from corn seeds and the use of a new reporter construct to select corn genotypes resistant to aflatoxin accumulation, in *Proceedings of the USDA-ARS Aflatoxin Elimination Workshop*, Memphis, TN, 1997, p. 66.
131. Keller, N.P., Butchko, R.A.E., Sarr, B., and Phillips, T.D., A visual pattern of mycotoxin production in maize kernels by *Aspergillus* spp., *Phytopathology*, 84, 483, 1994.
132. Woloshuk, C.P., Cavaletto, J.R., and Cleveland, T.E., Inducers of aflatoxin biosynthesis from colonized maize kernels are generated by an amylase activity from *Aspergillus flavus*, *Phytopathology*, 87, 164, 1997.
133. White, D.G., Rocheford, T.R., Kaufman, B., and Hamblin, A.M., Further genetic studies and progress on resistance to aflatoxin production in corn, in *Proceedings of the USDA-ARS Aflatoxin Elimination Workshop*, Atlanta, GA, 1995, p. 7.
134. White, D.G., Rocheford, T.R., Kaufman, B., and Hamblin, A.M., Chromosome regions associated with resistance to *Aspergillus flavus* and inhibition of aflatoxin production in maize, in *Proceedings of the USDA-ARS Aflatoxin Elimination Workshop*, Atlanta, GA, 1995, p. 8.
135. Guo, B.Z., Brown, R.L., Lax, A.R., Cleveland, T.E., Russin, J.S., and Widstrom, N.W., Protein profiles and antifungal activities of kernel extracts from corn genotypes resistant and susceptible to *Aspergillus flavus*, *J. Food Prot.*, 61, 98, 1998.

136. Guo, B.Z., Chen, Z.-Y., Brown, R.L., Lax, A.R., Cleveland, T.E., Russin, J.S., Mehta, A.D., Selitrennikoff, C.P., and Widstrom, N.W., Germination induces accumulation of specific proteins and antifungal activities in corn kernels, *Phytopathology*, 87, 1174, 1997.
137. Huang, Z., White, D.G., and Payne, G.A., Corn seed proteins inhibitory to *Aspergillus flavus* and aflatoxin biosynthesis, *Phytopathology*, 87, 622, 1997.
138. Brown, R.L., Bhatnagar, D., Cleveland, T.E., and Cary, J.W., Recent advances in preharvest prevention of mycotoxin contamination, in *Mycotoxins in Agriculture and Food Safety*, Sinha, K.K. and Bhatnagar, D., Eds., Marcel Dekker, New York, 1998, p. 351.
139. Chen, Z.-Y., Brown, R.L., Lax, A.R., Guo, B.Z., Cleveland, T.E., and Russin, J.S., Resistance to *Aspergillus flavus* in corn kernels is associated with a 14-kDa protein, *Phytopathology*, 88, 276, 1998.
140. Chen, Z.-Y., Brown, R.L., Lax, A.R., Guo, B.Z., Cleveland, T.E., and Russin, J.S., A maize kernel trypsin inhibitor is associated with resistance to *Aspergillus flavus* infection, in *Proceedings of the USDA-ARS Aflatoxin Elimination Workshop*, Memphis, TN, 1997, p. 34.
141. Chen, Z.-Y., Brown, R.L., Lax, A.R., Cleveland, T.E., and Russin, J.S., Inhibition of plant pathogenic fungi by a corn trypsin inhibitor over-expressed in *Escherichia coli*, *Appl. Environ. Microbiol.*, 65, 1320, 1999.
142. Chlan, C.A., Lin, J., Cary, J.W., and Cleveland, T.E., A procedure for biolistic transformation and regeneration of transgenic cotton from meristematic tissue, *Plant Molec. Biol. Rep.*, 13, 31, 1995.
143. Farmer, E.E. and Ryan, C.A., Octadecanoid precursors of jasmonic acid activate the synthesis of wound-inducible proteinase inhibitors, *Plant Cell*, 4, 129, 1992.
144. Cordero, M.J., Raventos, D., and San Segundo, B., Induction of PR proteins in germinating maize seeds infected with the fungus *Fusarium moniliforme*, *Physiol. Molec. Plant Pathol.*, 41, 189, 1992.
145. Linthorst, H.J.M., Pathogenesis-related proteins of plants, *Crit. Rev. Plant Sci.*, 10, 123, 1991.
146. Vigers, A.J., Roberts, W.K., and Selitrennikoff, C.P., A new family of plant antifungal proteins, *Molec. Plant-Microbe Interactions*, 4, 315, 1991.
147. Huynh, Q.K., Borgmeyer, J.R., and Zobel, J.F., Isolation and characterization of a 22-kDa protein with antifungal properties from maize seeds, *Biochem. Biophys. Res. Comm.*, 182, 1, 1992.
148. Huynh, Q.K., Hironaka, C.M., Levine, E.B., Smith, C.E., Borgmeyer, J.R., and Shah, D.M., Antifungal proteins from plants, *J. Biol. Chem.*, 267, 6635, 1992.
149. Kumari, S.R. and Chandrashekar, A., Proteins in developing sorghum endosperm that may be involved in resistance to grain moulds, *J. Sci. Food Agric.*, 60, 275, 1992.
150. Darnetty, L.J.F., Muthukrishnan, S., Swegle, M., Vigers, A.J., and Selitrennikoff, C.P., Variability in antifungal proteins in the grains of maize, sorghum and wheat, *Physiologia Plantarum*, 88, 339, 1992.
151. Neucere, J.N., Cleveland, T.E., and Dischinger, C., Existence of chitinase activity in mature corn kernels (*Zea mays* L.), *J. Agric. Food Chem.*, 39, 1326, 1991.
152. Neucere, J.N., Brown, R.L., and Cleveland, T.E., Correlation of antifungal properties and ß-1,3-glucanases in aqueous extracts of kernels from several varieties of corn, *J. Agric. Food Chem.* 43, 275, 1995.

153. Cleveland, T.E., Cary, J.W., Brown, R.L., Bhatnagar, D., Delucca, A.J., Yu, J., Chang, P.-K., and Rajasekaran, K., Use of biotechnology to eliminate aflatoxin in preharvest crops, in *Seminar on Mycotoxins Bulletin*, Kinki University, Nara, Japan, 5, 75, 1997.
154. Cary, J.W., Rajasekaran, K., Delucca, A.J., Jacks, T.J., Lax, A.R., Cleveland, T.E., Chlan, C., and Jaynes, J., Transformation and analysis of cotton and tobacco tissues expressing antifungal proteins and peptides, in *Proceedings of the USDA-ARS Aflatoxin Elimination Workshop*, Memphis, TN, 1997, p. 55.
155. Rajasekaran, K., Cary, J.W., Delucca, A.J., Jacks, T.J., Lax, A.R., Cleveland, T.E., Chen, Z., Chlan, C., and Jaynes, J., Agrobacterium-mediated transformation and analysis of cotton expressing antifungal peptides, in *Proceedings of the USDA-ARS Aflatoxin Elimination Workshop*, Memphis, TN, 1999, p. 66.
156. Huang, Z.-Y., White, D.G., and Payne, G.A., Characterization of inhibitory compounds to aflatoxin biosynthesis in the corn inbred line Tex6, in *Proceedings of the USDA-ARS Aflatoxin Elimination Workshop*, Atlanta, GA, 1995, p. 27.
157. Casacuberta, J.M., Raventos, D., Puigdomenech, P., and San Segundo, B., Expression of the gene encoding the PR-like protein PRms in germinating maize embryos, *Molec. Gen. Genet.,* 234, 97, 1992.
158. Cornelissen, B.J.C. and Melchers, L.S., Strategies for the control of fungal diseases with transgenic plants, *Plant Physiol.,* 101, 709, 1993.
159. Broglie, K., Chet, I., Holliday, M., Cressman, R., Biddle, P., Knowlton, S., Mauvais, C.J., and Broglie, R., Transgenic plants with enhanced resistance to the fungal pathogen *Rhizoctonia solani, Science*, 254, 1194, 1991.
160. Meins, F., Neuhaus, J.-M., Sperisen, C., and Ryals, J., The primary structure of plant pathogenesis-related glucanohydrolases and their genes, in *Genes Involved in Plant Defense*, Boller, T. and Meins, F., Eds., Springer-Verlag, New York, 1992, p. 245.
161. Singh, N.K., Nelson, D.E., Kuhn, D., Hasegawa, P.M., and Bressan, R.A., Molecular cloning of osmotin and regulation of its expression by ABA and adaption to low water potential, *Plant Physiol.*, 90, 1096, 1989.
162. Ryan, C.A., Protease inhibitors in plants: genes for improving defenses against insects and pathogens, *Ann. Rev. Phytopathol.*, 28, 425, 1990.
163. Toubart, P., Desiderio, A., Salvi, G., Cervone, F., Daroda, L., DeLorenzo, G., Bergmann, C., Darvill, A.G., and Albersheim, P., Cloning and characterization of the gene encoding the endopolygalacturonase-inhibiting protein (PGIP) of *Phaseolus vulgaris* L, *Plant J.*, 2, 367, 1992.
164. Bowman, H.G. and Hultmark D., Cell-free immunity in insects, *Ann. Rev. Microbiol.*, 31, 103, 1987.
165. Zasloff, M., Magainins, a class of antimicrobial peptides from *Xenopus* skin, *Proc. Natl. Acad. Sci. USA*, 84, 5449, 1987.
166. Rajaskaran, K., Hudspeth, R.L., Cary, J.W., Anderson, D.M., and Cleveland, T.E., High frequency stable transformation of cotton (*Gossypium hirsutum* L.) by particle bombardment of embryogenic cell suspension cultures, *Plant Cell Rep.*, 19, 539, 2000.
167. Rajasekaran, K., Chlan, C., and Cleveland, T.E., Tissue culture and genetic transformation of cotton, in *Emerging Technologies in Cotton Breeding*, Jenkins, J.J. and Saha, S., Eds., Oxford University Press, New York, 2000 (in press).
168. Doehlert, D.C., Wicklow, D.T., and Gardner, H.W., Evidence implicating the lipoxygenase pathway in providing resistance to soybeans against *Aspergillus flavus*, *Phytopathology*, 83, 1473, 1993.
169. Raper, K.B. and Fennell, D.I., The *Aspergillus flavus* group, in *The Genus Aspergillus*, Powell, K.E., Renwick, J.F., and Peberdy, J.F., Eds., Williams & Wilkins, Baltimore, MD, 1965, p. 393.

170. Vick, B.A. and Zimmerman, D.C., Oxidative systems for modification of fatty acids: the lipoxygenase pathway, in *The Biochemistry of Plants, A Comprehensive Treatise.* Vol. 9. *Lipids: Structure and Function*, Stumpf, P.K., Ed., Academic Press, New York, 1987, p. 53.
171. Burow, G.B., Nesbitt, T.C., Dunlap, J., and Keller, N.P., Seed lipoxygenase products modulate *Aspergillus* mycotoxin biosynthesis, *Am. Phytopathol. Soc.*, 10(3), 380, 1997.
172. Gundlach, H., Miller, M.J., Kutchan, T.M., and Zenk, M.H., Jasmonic acid is a signal transducer in elicitor-induced plant cell cultures, *Proc. Natl. Acad. Sci. USA*, 89, 2389, 1992.
173. Goodrich, T.M., Mahoney, N.E., and Rodriguez S.B., The plant growth regulator methyl jasmonate inhibits aflatoxin production by *Aspergillus flavus*, *Microbiology,* 141, 2831, 1995.
174. Gueldner, R.C., Wilson, D.M., and Heidt, A.R., Volatile compounds inhibiting *Aspergillus flavus*, *J. Agric. Food Chem.*, 33, 411, 1985.
175. Lee, J.-Y., Boman, A., Chuanxin, S., Anderson, M., Mutt, H., Jornvall, H., Mutt, V., and Bowman, H.G., Antibacterial peptides from pig intestine: isolation of a mammalian cecropin, *Proc. Natl. Sci. Acad. USA*, 86, 9159, 1989.
176. Christensen, B., Fink, J., Merrifield, R.B., and Mauzerall, D., Channel-forming properties of cecropins and related model compounds incorporated into planar lipid membranes, *Proc. Natl. Acad. Sci. USA*, 85, 5072, 1988.
177. Kagan, B.L., Selsted, M.E., Ganz, T., and Lehrer, R.I., Antimicrobial defensin peptides form voltage-dependent ion-permeable channels in planar lipid membranes, *Proc. Natl. Acad. Sci. USA*, 87, 210, 1990.
178. Steiner, H., Hultmark, D., Engstrom, A., Bennich, H., Boman, H.G., Sequence and specificity of two antibacterial proteins involved in insect immunity, *Nature*, 292, 246, 1981.
179. DeLucca, A.J., Jacks, T.J., and Brogden, K.A., Binding between lipopolysaccharide and Cecropin A, *Molec. Cell. Biochem.*, 151, 141, 1995.
180. Bell, E. and Mullet, J.E., Lipoxygenase gene expression is modulated in plants by water deficit, wounding, and methyl jasmonate, *Molec. Gen. Genet.,* 230, 456, 1991.
181. Melan, M., Dong, X., Endara, M.E., Davis, K.R., Ausubel, F.M., and Peterman, T.K., An *Arabidopsis thaliana* lipoxygenase gene can be induced by pathogens, abscissic acid, and methyl jasmonate, *Plant Physiol.*, 101, 441, 1993.
182. Wolffram, C., van Pee, K.-H., and Lingens, F., Cloning and high-level expression of a chloroperoxidase gene from *Pseudomonas pyrrocinia*, *FEBS Lett.*, 238, 325, 1988.
183. Jacks, T.J., Delucca, A.J., and Morris, N.M., Effects of chloroperoxidase and hydrogen peroxide on the viabilities of *Aspergillus flavus* conidiospores, *Molec. Cell Biochem.*, 195, 169, 1999.
184. Jacks, T.J., Cotty, P.J., and Hinojosa, O., Potential of animal myeloperoxidase to protect plants from pathogens, *Biochem. Biophys. Res. Commun.*, 178, 1202, 1991.
185. Jacks, T.J. and Hinojosa, O., Superoxide radicals in intact tissues and in dimethyl sulfoxide-based extracts, *Phytochemistry,* 33, 563, 1993.
186. Schardl, C.L., Byrd, A.D., Benzion, G., Altschuler, M.A., Hildebrand, D.F., and Hunt, A.G., Design and construction of a versatile system for the expression of foreign genes in plants, *Gene*, 61, 1, 1987.
187. Garbarino, J.E. and Belknap, W.R., Isolation of a ubiquitin-ribosomal protein gene (*ubi*3) from potato and expression of its promoter in transgenic plants, *Plant Molec. Biol.,* 24, 119, 1994.

188. Thornburg, R.W., An, G., Cleveland, T.E., Johnson, R., and Ryan, C.A., Wound-inducible expression of potato inhibitor II — chloramphenicol acetyltransferase gene fusion in transgenic tobacco plants, *Proc. Natl. Acad. Sci. USA*, 84, 744, 1987.
189. Burrow, M., Sen, P., Chlan, C., and Murai, M., Two temporal and three spatial regulatory elements are responsible for developmental control of the bean seed storage protein beta-phaseolin, *Plant J.*, 2, 537, 1992.
190. Ozias-Akins, P., Fan, H., and Wang, A., Genetic engineering of peanut with Bt and function of a soybean promoter in peanut, in *Proceedings of the USDA-ARS Aflatoxin Elimination Workshop*, Atlanta, GA, 1995, p. 19.
191. Chlan, C., Rajasekaran, K., and Cleveland, T.E., Transgenic cotton, in *Biotechnology in Agriculture and Forestry Series*, Vol. 46, Bajaj, Y.P.S., Ed., Springer-Verlag, Heidelberg, 1995, p. 283.
192. Horsch, R.B., Fry, J.B., Hoffman, N., Eichholtz, L.D., Rogers, S.G., and Fraley, R.T., A simple and general method for transferring genes into plants, *Science*, 227, 1229, 1985.
193. Umbeck, P., Johnson, G., Barton, K., and Swain, W., Genetically transformed cotton (*Gossypium hirsutum* L.) plants, *Bio/Technology*, 5, 263, 1987.
194. Bio-Rad Biolistic PDS-1000/He Particle Delivery System, Bull. 1700, 1996.
195. Davidonis, G. and Hamilton, R.H., Plant regeneration from callus tissue of *Gossypium hirsutum* L, *Plant Sci. Lett.*, 32, 89, 1983.
196. Trolinder, N. and Goodin, J.R., Somatic embryogenesis and plant regeneration in cotton (*Gossypium hirsutum* L.), *Plant Cell Rep.*, 6, 231, 1987.
197. Firoozabady, E., DeBoer, D., Merlo, D., Halk, E.L., Amerson, L., Rashka, E., and Murray, E.E., Transformation of cotton (*Gossypium hirsutum* L.) by *Agrobacterium tumefaciens* and regeneration of transgenic plants, *Plant Molec. Biol.*, 10, 105, 1987.
198. Bayley, C., Trolinder, N., Ray, C., Morgan, M., Quisenberry, J., and Ow, D.W., Engineering 2,4-D resistance into cotton, *Theoret. Appl. Genetics,* 83, 645, 1992.
199. Rajasekaran, K., Grula, J.W., Hudspeth, R.L., Pofelis, S., and Anderson, D.M., Herbicide-resistant Acala and Coker cottons transformed with a native gene encoding mutant forms of acetohydroxyacid synthase, *Molec. Breeding*, 2, 307, 1996.
200. Finer, J.J. and McMullen, M.D., Transformation of cotton (*Gossypium hirsutum* L.) via particle bombardment, *Plant Cell Rep.*, 8, 586, 1990.
201. McCabe, D.E. and Martinelli, B.J., Transformation of elite cotton cultivars via particle bombardment of meristems, *Bio/Technology*, 11, 596, 1993.
202. Weissinger, A., Cade, R., and Urban, L., Transformation of Peanut cv. 'NC 7' with genes encoding defensive peptides, in *Proceedings of the USDA-ARS Aflatoxin Elimination Workshop*, Atlanta, GA, 1995, p. 20.
203. Mendum, M., McGranahan, G., Dandekar, A., and Uratsu, S., Progress in engineering walnuts for resistance to *Aspergillus flavus*, in *Proceedings of the USDA-ARS Aflatoxin Elimination Workshop*, Atlanta, GA, 1995, p. 21.

Microbial Food Contamination and International Trade

15 The Impact of Foodborne Infections and Toxins on International Trade: Israeli Examples

Isaac Klinger

ABSTRACT

The trend of governments worldwide is to protect the interests of local producers as much as possible from the import of agricultural products (AP) from other countries. Such trends are controlled today by regulations issued by the World Trade Organization (WTO) limiting the restrictions imposed on imported agricultural products to only those of a sanitary and phytosanitary (SPS) nature. In the same agreement, it is also mentioned that the utilization of non-tariff barriers (NTB) concerning SPS conditions should be minimized. Some countries in the world enjoy such agricultural advantages as minimal limitations in water supply, agricultural land, and mild weather, allowing them benefits in offering APs to the world market. Israel signed the WTO agreement, and APs are imported as far as commercial interest allows.

The natural conditions of Israel's agriculture prescribe the need for importing some essential APs, such as red meat, grains, fish and seafood, dried fruits, and other food products, as well as some feed components. Israel is self-supporting in regard to most APs; however, there is no way to supply the population with protein of animal origin, in particular red meat and fish. The health, veterinary, and plant protection authorities face the difficult position of protecting the Israeli consumer from health hazards and protecting local agriculture from animal and plant health problems associated with imported APs.

Israel was the first country in the world to apply regulations aimed at preventing the entrance of APs that might introduce bovine spongiform encephalopathy (BSE) into the country in both food and feed. Countries reported to have beef herds infected with BSE are not allowed to export beef to Israel, and importation of feeding materials of animal origin is restricted to non-mammalian origin only. Every shipment of feed of animal origin is examined for the presence of mammalian protein. Those found positive are rejected, and the exporting company is refused for further shipments.

Every shipment of red meat is examined visually in the port of entry by the port veterinarian, and samples are taken for laboratory examinations, including bacteriological examinations; tests for the presence of drugs, pesticides, and substances with hormonal activity; and detection of environmental pollutants, such as heavy metals, PCBs, and others.

Similar examinations are performed with imported fish and seafood. Although located along the seashore of the Mediterranean, Israel suffers from a shortage of food from the sea. The creation of the Assuan Dam on the Nile disturbed the natural balance of organic materials in this part of the Mediterranean Sea, and as a result the seashores of Israel are poor in fish. Most of the fishery products produced in the country originate from aquaculture. Thus, fish are also imported from those parts of the world carrying exotic human diseases. Imported fish are examined visually and in the laboratory in a manner similar to examinations of red meat, but they are performed specifically for the problems associated with products from the sea.

16 Use of Risk Assessment as a Tool for Evaluating Microbial Food Contaminants

Michael D. McElvaine

CONTENTS

I. INTRODUCTION

Various methods of risk assessment and analysis have been used by engineers, economists, and other scientist since the 1930s.[1] In the 1950s and 1960s, risk assessment was applied to chemical contaminants that may be in the environment or our food and water.[2] Later, the U.S. Food and Drug Administration (FDA) developed safety assessments, a variation of risk assessment, to evaluate food additives and chemical contaminants.[3] Application of formal risk assessment to biological (multiplying) organisms was not attempted until the 1980s and 1990s. Even now, the field of risk assessment for microbial contamination of food is still in its infancy.

When applying risk assessment methodology to microbes, there are two related but very different issues. First, we want to know whether a product is contaminated or not, or the probability that each unit may be contaminated. Second, we want to know how many microbes there are. While chemical contaminants are inert for the most part, microbes may increase, decrease, or stay at the same numbers. This is the area of focus for the field of microbial risk assessment. From a human health aspect, we want to use the risk assessment to evaluate the dose-response of a population and the range of health impacts from these exposures. The current focus on food safety and international trade has increased the interest in developing better information about microbial food contaminants in order to improve food safety regulations and to facilitate international trade.

II. BACKGROUND

I currently work with the U.S. Department of Agriculture's Office of Risk Assessment and Cost Benefit and Analysis (ORACBA). This unit was created by the Agriculture Reorganization Act of 1994[4] and began operations in October 1995. There is no other group in any U.S. Cabinet office that has our specific statutory functions. We are charged with reviewing major regulations ($100 million or greater economic impact and that affect human health, safety, or the environment) to assure that they are based on sound scientific and economic analysis. We also provide education and training, coordination of risk assessment activities and resources, and guidance in the development of risk assessments and other associated analyses. When I was given the suggested title for this article, I had some idea of what might be of interest to the readers. After reviewing the titles of the presentations for this meeting and listening to many of the presentations, I came to the conclusion that, after hours of presentation of hard-core science, a discussion of microbial risk assessment and policy implications might help to provide an overall context to this meeting.

III. WHAT IS RISK ASSESSMENT?

Risk assessment in itself is not hard science. It is the art of taking scientific information and transforming it into information that is useful to provide the basis for a decision-maker to act upon. There are currently only a few universities in the U.S. that have degree programs in risk assessment. Most everyone who practices the art of risk assessment is a "retread" from other fields. It is a new field of interest that is being adopted by an increasing number of agencies and companies to better inform their decisions.

While many readers may have an opinion on the question of what risk assessment actually is, I doubt that many understand the concept of risk assessment in the context to which I am referring, as there have been many uses of risk assessment for other purposes (chemical contaminants, economic impacts, catastrophic events), so in the following discussion I will try to clarify this particular application of risk assessment.

In the general sense, risk assessment is a methodology for evaluating a given scenario (situation) to determine:[5]

1. What can go wrong?
2. How likely is it?
3. What is the magnitude of the impact?

The general use of risk assessment is not new. In essence, it is the kind of evaluation that our prehistoric ancestors used to determine whether it was worthwhile to face the risk of being eaten by predators in order to hunt for food or face starvation. We all make less dramatic decisions every day that are in essence risk analysis. Many people drive to work every day. Do they consciously think about the risk of this activity, or do they take it for granted that the value of earning a salary is greater than the risk of an automobile accident or other corporal risk?

In the context of biological (i.e., microbial) risk assessment, the methodology is an organized framework for presenting information about the risks of a situation, assessing the probability of negative events, and assessing the magnitude of these negative events. The next step, the economic part of the equation, is to compare the magnitude of bad outcomes with the costs of interventions necessary to reduce this magnitude of bad outcomes (we usually translate these magnitudes into dollar values to provide transparency to the decision-makers). This is where probability and magnitude provide insight for decision-makers.

IV. BRIEF RETROSPECTIVE OF RISK ASSESSMENT

Risk assessment is an innate part of being human (as far as our being able to reason) and is a basic premise in our lives that guides our actions. It was only in the 20th century that the art of risk assessment began to be presented in a formalized manner to evaluate scientific and economic questions. In the earlier years of the USDA, various forms of risk assessment methods were used to guide regulatory and programmatic decisions. Only recently has the USDA begun to accept the organized methods of risk analysis to guide current regulatory decisions. The modern use of formal quantitative risk assessment can be traced back to the advent of nuclear power plants in the U.S.,[6] when quantitative methods using Monte Carlo analysis were used to evaluate new issues arising from their development. Monte Carlo analysis was a significant step in the advancement of quantitative risk assessment methods. This methodology was soon adapted by physicists, economists, and others to evaluate situations where probability and magnitude of impact were factors in decisions. If you have financial advisors who guide you in your investments for retirement, they are probably (pun intended) using these methods to guide their advice.

V. CHEMICAL RISK ASSESSMENT

In the 1950s and 1960s, there was an awakening as to the toxic effects of chemicals in our environment which stimulated the development of the chemical risk assessment paradigm. This paradigm was the central focus of the 1983 National Research Council report, *Risk Assessment in the Federal Government: Managing the Process*.[7]

This book is commonly referred to as the "red book," and it is in this work where the four steps of risk assessment were first laid out:

1. Hazard identification
2. Exposure assessment
3. Dose-response assessment
4. Risk characterization

The FDA also has a long history of development of risk assessment methods for both contaminants of foods and, later, food additives. For both of these issues, the FDA was evaluating what was considered "safe" in the context of food consumption. The concept of "safety" is a value-laden term that can mean very different things in different contexts and to different people, even among those who speak English. When we attempt to translate this concept into other languages, things get even more confused. This increases the challenge for negotiating international agreements related to issues that affect human or environmental safety.

VI. RISK ASSESSMENT FOR REPLICATING ORGANISMS

For chemical assessments, we are dealing with inert compounds, which (usually) do not change in concentration. The critical issues here are assessing the exposure dose and the response to that dose. When dealing with microbes, they often are not static, as the number of microbes can change over time. The USDA Agricultural Research Service (ARS) and the Animal and Plant Health Inspection Service (APHIS) both tackled this problem early in the 1990s. ARS constructed a microbial risk assessment modeling program to estimate the levels of microorganisms subjected to various conditions and mitigations.[8] At about the same time, APHIS evaluated the issue of replicating organisms that cause animal diseases. APHIS is charged with protecting the U.S. from the incursion of foreign animal and plant diseases. The agency is thus concerned with whether the pest/microbe is present and whether it could become established in the U.S. Because APHIS had traditionally operated under the (unofficial) motto of "If in doubt, keep it out," they have an excellent record of preventing foreign disease outbreaks. This "zero risk" policy has proven useful but is no longer tenable in a modern world guided by the North American Free Trade Agreement (NAFTA) and the General Agreement on Tariffs and Trade (GATT).

APHIS began exploring the use of risk assessment in the early 1990s, using both quantitative and qualitative methods as were appropriate for the specific issues.[9] In recent years, APHIS has used risk assessment as the basis for regionalization to address the risk of foreign animal diseases.[9] APHIS recognition of a country or region is often viewed as the gold standard for the world to guide trade with other countries or regions. Given this large responsibility, APHIS is still struggling with identifying standards to be used in evaluating the risks of importing foreign animal diseases from various countries and regions.

VII. FOODBORNE PATHOGENS

It should be obvious at this point that we should apply the methods of risk analysis to microbial pathogens in foods. Unfortunately, this is a relatively new area without well-established methods and analyses. The Sanitary and Phytosanitary Agreements of the GATT[10] point to risk assessment to guide the international trade of animal and plant products. The exact guidance for risk assessments is not contained in the GATT documents but they do refer to various international organizations as the standard setters. These international organizations are still in the process of developing standards for risk assessment. WTO named the Codex Alimentarius Commission (CAC) as the standard-setting organization for microbial risk assessment. The CAC has developed guidelines for risk assessment of foods which are currently being reviewed by the Codex Committee on Food Hygiene. Information about the development of these standards is available from the CAC website: http://www.fao.org/es/esn/codex/default.htm.

VIII. MICROBIAL RISK ASSESSMENT ISSUES

When we focus on the issue of risk assessment of microbes in food we are faced with a collection of issues. The first question is whether the microbe is present. When a regulatory body sets a policy prohibiting the presence of certain bacteria in food by classifying the bacteria as an adulterant, it has truncated the range of mitigation options. With a regulatory standard of *no* bacteria of a certain species, we have limited our management options to address the broader issue of public health impact. An obvious example of this is the current FDA and USDA Food Safety and Inspection Service (FSIS) policy for zero tolerance of *Listeria monocytogenes* on food. The FDA Center for Food Safety and Applied Nutrition (CFSAN) has begun a risk analysis project to evaluate the public health impact of *L. monocytogenes* contamination of foods.[11] This is a groundbreaking effort in that it will open the playing field for the use of risk assessment to establish international standards.

In the context of international trade, a myriad of issues can be discussed. What is an acceptable level of risk? What is an acceptable level of safety? How do we proceed under GATT? These are all important concerns that need to be addressed within the context of risk and acceptable level of risk. The WTO will deal with these issues in the future. Questions that scientific risk assessors need to ask in regard to regulatory issues, particularly international ones, include:

1. Is the product contaminated?
2. At what level is the product contaminated?
3. What is the public health impact of this level of contamination?
4. How do we assess equivalency for trade issues?

A. Is the Product Contaminated?

When we assess a contamination that is infrequent and/or unacceptable, we want to know if the contaminating pathogen is present. Our goal is to have every unit of product free of pathogens. We estimate the occurrence of contamination (yes/no) and manage the process by not allowing contaminated product to proceed in the product flow. This is the kind of program that the USDA has for *Escherichia coli* O157:H7 in hamburger. This pathogen has been identified as an adulterant by regulation, and no product that is thus contaminated can be offered for sale.[12] It should be noted that currently *E. coli* O157:H7 is only classified as an adulterant in the final ground product; however, the USDA has proposed new guidelines that would classify *E. coli* O157:H7 as an adulterant if found further up in the product stream.

B. At What Level Is the Product Contaminated?

It is almost impossible to assure that all meat and poultry is free of all pathogens such as *Salmonella* and *Campylobacter* and others; therefore, the risk assessor wants to determine what levels of pathogen may be present and what mitigations may be needed to reduce these pathogens to a reasonable level. The risk assessor also needs to know what the dose of exposure will be in order to evaluate the potential health impacts.

C. What Is the Public Health Impact from This Level of Contamination?

For risk assessors and risk managers, this is usually the prime concern, as disease and death are the ultimate negative impacts of contaminated foods. There are multiple factors that have an impact on this area of microbial risk assessment, such as:

1. *What is the range of doses?* Predictive microbial risk assessment as well as data on quantity of foods consumed by different population groups provide information to estimate the quantity of pathogens in the exposure dose.
2. *What is the dose/response for the average population?* Many studies have been done over the last 50 years to evaluate the infectious dose of various pathogenic enteric organisms. Unfortunately (for dose/response modelers), these studies were done several decades ago and are difficult to repeat in modern times due to current human studies restrictions. Microbial risk assessors must rely on the old studies, animal models, or information supplied from epidemiologic investigation of foodborne disease outbreaks.
3. *What is the dose response for the most sensitive populations?* The most sensitive populations may include very young, very old, or immunocompromised people; people undergoing chemotherapy; transplant patients; and people with other chronic debilitating diseases. These populations have been identified as the target populations to evaluate the safety of microbial pathogens as well as chemicals in our food.

D. How Do We Assess Equivalency for Trade Issues?

This issue has gained importance, as all GATT-participating countries are still learning how to interpret the language of the Sanitary and Phytosanitary Agreement.

Microbial risk assessment will be the tool to address whether a food product processed a different way than another country can be treated as equivalent. A pertinent example of this issue is European soft cheeses that are made from unpasteurized milk. The U.S. requires that the milk used in cheese production be pasteurized. Is the European standard using non-pasteurized milk to produce cheese equivalent to the U.S. standards, or is this a technical barrier to trade? Microbial risk assessors must face this and other issues in the near future.

IX. CONCLUSIONS

While risk assessment is currently recognized as a desirable tool for setting policy both at the national and international level, the methodology of risk assessment is still being formalized, especially for issues of microbial contaminants. I have presented several issues and examples that focus on the challenges we face in an era of international free trade. Clearly, there is much further work to be done in the areas of microbial risk assessment and development of international sanitary policy. Microbial risk assessment has been identified as the best tool at this point for evaluating these issues.

NOTE

Statements and opinions in this paper do not necessarily represent the opinions of the USDA or the Office of Risk Assessment and Cost-Benefit Analysis.

REFERENCES

1. Morgan, M.G., Risk analysis and management, *Scientific Am.,* July:32–41, 1993.
2. Ruckleshaus, W.D., Risk, science, and democracy, in *Readings in Risk*, Glickman, T.S. and Gough M., Eds., Resources for the Future, Washington, D.C., 1990, pp. 105–119.
3. Wilson, J.D., Of what use are dose-response relationships for human pathogens?, *ORACBA News*, 4(1):1–5, 1999.
4. Federal Crop Insurance Reform and Department of Agriculture Reorganization Act of 1994, Public Law 103-354, Title III, Section 304, Oct. 13, 1994.
5. Kaplan, S. and Garrick B.J., On the quantitative definition of risk, *Risk Analysis,* 1(1):4–22, 1981.
6. Rasmussen, N.C., The application of probabilistic risk assessment techniques to energy technologies, *Ann. Rev. Energy*, 6:123–138, 1981.
7. National Research Council, *Risk Assessment in the Federal Government: Managing the Process*, National Academy Press, Washington, D.C., 1983.
8. U.S. Department of Agriculture, Agricultural Research Service, Predictive Microbial Risk Assessment Program, http://www.arserrc.gov/mfs/pathogen.htm, May 17, 1999.
9. Chang, L.L. Miller, L., Ahl, A. and McElvaine, M., Eds., What APHIS *Is Doing in Risk Assessment*, Animal and Plant Health Inspection Service, U.S. Department of Agriculture, Riverdale, MD, 1994.

10. World Trade Organization Agreement on the Application of Sanitary and Phytosanitary Measures, Foreign Agriculture Service, U.S. Department of Agriculture website: http://www.fas.usda.gov/itp/policy/gatt/sps_txt.html, May 17, 1999.
11. USDA, Risk assessment of the public health impact of foodborne *Listeria monocytogenes*, *Fed. Regist.*, 64(88): 24661–24663, 1999.
12. USDA, Ground beef processing guidance material, *Fed. Regist.*, 64(11):2872–2873, 1999.

17 The Codex Alimentarius: What It Is and Why It Is Important

H. Michael Wehr

CONTENTS

I. INTRODUCTION

The Codex Alimentarius (meaning *food code*) Commission, or Codex as the organization is usually termed, is an international food standards organization whose importance has increased substantially since the signing of the General Agreement on Tariffs and Trade (GATT) Uruguay Round Trade Agreements and implementation of the World Trade Organization (WTO). With the formal recognition given to Codex in the WTO Agreement on the Application of Sanitary and Phytosanitary Measures (SPS Agreement), Codex has moved to the forefront of international science-based, standards-setting bodies associated with food safety and quality. Its standards, guidelines, and recommendations have substantial impact on food products that are internationally traded. The work of Codex has historically included substantial work in the field of food hygiene, including the establishment of general principles for the hygienic production of food and the development of commodity-specific codes of hygienic practice. More recently, Codex has expanded its work in the food hygiene area to include such areas as the important fields of microbiological risk assessment, microbiological risk management, and antibiotic resistance in bacteria in food.

This presentation describes the history and organization of Codex, why it has become increasingly important, and how it develops and adopts food safety and quality standards. The policies Codex has established to undertake its scientific standards-setting activities are described, as are the proposed Codex principles for risk analysis and its component parts: risk assessment, risk management, and communication. The presentation will also discuss the work of Codex in the field of food hygiene, focusing on the Codex Committee on Food Hygiene (CCFH). The types of food hygiene codes of practice and other food hygiene texts that are developed by Codex are reviewed, and the work of CCFH in the developing fields of microbiological risk assessment and microbiological risk management is discussed. Brief mention will also be made of other ongoing work in Codex.

II. WHAT IS CODEX?

The Codex Alimentarius Commission (CAC) is an international intergovernmental body that develops science-based food safety and commodity standards, guidelines, and recommendations to promote consumer protection and facilitate world trade.[1,2] Codex is a subsidiary body of two United Nations organizations, the Food and Agriculture Organization (FAO) and the World Health Organization (WHO). Currently, Codex has 165 member countries.

Codex was established in 1962 when the FAO and WHO recognized the need for international food standards to protect the health of consumers and to guide the world's growing food industry in producing quality food. Under the "General Principles of the Codex Alimentarius,"[1] Codex is charged with developing food standards for adoption and use by member countries. The Codex Alimentarius itself is a collection (14 volumes) of international food standards adopted by the Codex Alimentarius Commission and presented in a uniform manner. The purpose of these standards is to protect the health of consumers and facilitate fair practices in food trade.[1] Codex texts are in the form of specific food standards, codes of practice, and recommendations.

The scope of the Codex Alimentarius includes standards for all principle foods, whether processed, semi-processed, or raw, for distribution to the consumer. Materials to be used for further processing into foods are included in a standard to the extent necessary to achieve the purpose of the standard. The Codex Alimentarius includes provisions for food hygiene, food additives, pesticide residues, contaminants, labeling and presentation, and methods of analysis and sampling. It also includes provisions of an advisory nature in the form of codes of practice, guidelines, and recommended measures. Since its inception, Codex has adopted more than 235 commodity standards, 3200 maximum residue limits (MRLs) for pesticides, 50 MRLs for veterinary drugs, and 25 guidelines for contaminants.[3] Codex has evaluated over 1000 food additives and has established over 40 hygienic and technological codes of practice. The work and activities of Codex are available on the Internet through the FAO home page.[1]

III. WHY IS CODEX IMPORTANT?

While Codex has been recognized within the international food scientific and regulatory communities since its inception, only limited adoption of Codex standards and texts by countries has occurred. This situation can be expected to change because of recent multilateral trade agreements. In 1994, the GATT Uruguay Round of International Trade Negotiations established the WTO and several trade agreements, including the Agreement on the Application of Sanitary and Phytosanitary Measures (SPS Agreement) and the Agreement on Technical Barriers to Trade (TBT Agreement).[4] Article 3.1 of the SPS Agreement requires countries to base their sanitary and phytosanitary measures on international standards, guidelines, or recommendations unless they can scientifically justify a more stringent standard. Section 3(a) of Annex A of the SPS Agreement states that, for food safety, certain standards, guidelines, and recommendations established by the Codex Alimentarius Commission shall be recognized as international standards under the Agreement. These include those relating to food additives, veterinary drug and pesticide residues, contaminants, methods of analysis and sampling, and codes and guidelines of hygienic practice. Article 3.2 of the SPS Agreement states that a country's sanitary and phytosanitary measures which conform to international standards, guidelines, or recommendations shall be presumed to be consistent with the relevant provisions of the SPS Agreement and of General Agreements on Tariff and Trade established in 1994 (GATT, 1994). Thus, while Codex standards are voluntary, countries have an obligation to base their own food safety standards on the work of Codex. Additionally, products produced in accordance with Codex standards essentially have safe harborage in international trade.

Codex is also important within the context of the TBT Agreement, which deals with product specifications not related to safety, including commodity composition, packaging, and certain labeling standards. The TBT Agreement also requires countries to use international standards. While Codex is not specifically referenced in the TBT Agreement, the WTO has recognized Codex standards as those to be used preferentially in resolving technical trade disputes. Codex thus becomes a true international focal point for food safety and quality, with major impact on international trade and, potentially, on domestic food regulations. Its importance is therefore significantly greater than it was previously.

IV. THE STRUCTURE OF CODEX

Codex undertakes its work through the Codex Alimentarius Commission and a series of horizontal general subject committees, vertical commodity committees, *ad hoc* task forces, and regional coordinating committees. Horizontal committees are those committees whose work applies to multiple commodities, such as the Codex Committee on Food Hygiene. Vertical committees are those that deal with a single commodity type or single commodity grouping (e.g., milk and milk products) and whose work encompasses all aspects associated with that commodity (e.g., composition, food additives, pesticide residues, food hygiene, product labeling, methods of analysis). An Executive Committee with international geographic representation acts for the Commission between its meetings; the Commission meets once every 2 years. The structure of Codex is shown in Figure 17.1.

Eight horizontal committees establish Codex standards, codes of practice, or guidelines that apply across all commodities. Included in this category is the work of Codex committees that involve pesticide residue MRLs, veterinary drug MRLs, and maximum permitted use levels for food additives. Also included in this category is the work of the Codex committees on Food Labeling, Food Import and Export Inspection and Certification Systems, General Principles (which is the initial forum for Codex policy and procedural matters), Methods of Analysis and Sampling, and Nutrition and Foods for Special Dietary Uses. The work of the final general subject committee, the Codex Committee on Food Hygiene, is discussed later in the presentation. Readers are referred to the Codex Procedures Manual[1] for specific terms of reference for all Codex committees.

A series of Codex commodity committees develops standards specific for various food products. These standards provide essential composition information for product types as well as information relating to residues and contaminants (pesticides and other compounds as appropriate), food additives, food hygiene, labeling, packaging, and methods of analysis. For these latter areas, the commodity standards reference general standards developed by the horizontal committees. Among the most active of commodity committees are those involved with fish and fishery products, milk and milk products, nutrition and foods for special dietary uses (a category which, as noted above, is also considered a horizontal committee), fresh fruits and vegetables, and processed fruits and vegetables. Each of these horizontal and vertical committees is hosted by a country (refer to countries indicated under the committee in Figure 17.1). The host country is responsible for arranging and securing the meeting site, pays for the meeting, and provides the chairperson for the committee. A series of regional coordinating committees is also maintained by Codex. These committees at one time were primarily involved with the establishment of standards applicable only to the region. Today, they are primarily involved with reviewing and analyzing the broader Codex issues as they may affect the region. Regional coordinating committees also serve as a venue for establishing regional positions on proposed international standards. Most recently, Codex established the category of *ad hoc* task forces to deal with specific subjects on a time-limited basis. Current *ad hoc* task forces include those for foods derived through biotechnology, fruit and vegetable juices, and animal feeding.

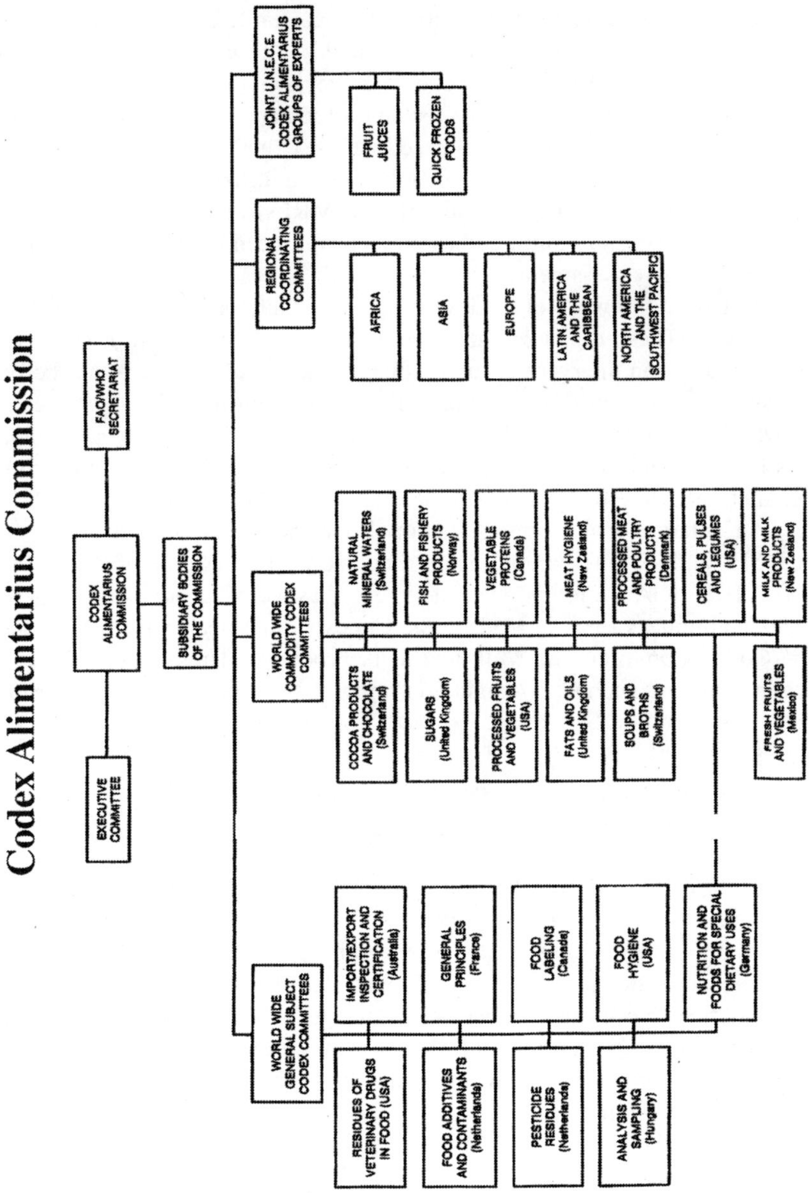

FIGURE 17.1 The structure of Codex Alimentarius.

Codex also utilizes the work and/or input of two other types of organizations. Joint WHO/FAO expert committees provide technical scientific risk assessment expertise for food additives, veterinary drugs, contaminants, and pesticide residues. The Joint Expert Committee on Food Additives (JECFA) provides this assistance for food additives, contaminants, and veterinary drugs, while the Joint Meeting on Pesticide Residues (JMPR) fills this role for pesticides. These committees are organizationally outside of Codex, but, in large measure, their work is determined by the agenda of the Codex committees responsible for food additives, veterinary drugs, and pesticides. Risk assessments are evaluated by these expert committees according to established protocols, and their recommendations are utilized by the appropriate committees to establish maximum residue limits (MRLs) or maximum permitted use levels. A similar expert committee has just recently been proposed for the field of microbiological risk assessment.

Also important to the work of Codex are international non-governmental organizations (INGOs). These organizations have no vote in the adoption of standards (see below), but they can intervene in, and influence the discussion on, issues before the Commission and its subsidiary bodies. There are many INGOs; examples of such organizations include the World Trade Organization, the European Commission, Consumers International, the International Organization for Standardization (ISO), AOAC International, International Council of Grocery Manufacturer Associations (ICGMA), the International Commission on Microbiological Specifications for Foods (ICMSF), and the International Dairy Federation (IDF). Some of the INGOs (e.g., ICMSF, IDF, AOAC International) have served as technical advisors to appropriate Codex committees (e.g., ICMSF to the Codex Committee on Food Hygiene, IDF to the Codex Committee on Milk and Milk Products). The role of INGOs is substantial in providing technical insight and important perspectives on the various issues under consideration by Codex.

V. THE STANDARDS DEVELOPMENT AND ADOPTION PROCEDURE OF CODEX

Codex utilizes an eight-step procedure to elaborate and adopt standards and other texts.[1] The process is a purposefully deliberative one that ensures all member countries are given opportunity to consider the issues coming before any individual Codex committee. The eight-step procedure is summarized in Table 17.1. New Codex standards, guidelines, codes of practice, and other texts may be proposed by any member government or by individual Codex committees. New work must be approved by the Commission or by the Executive Committee with subsequent confirmation by the Commission.

Steps 1 and 2 of the Codex procedure involve the initial development processes for the elaboration of Codex texts. Step 1 involves the initial approval by the Commission to proceed with new work. In some cases, especially in such areas as guidelines and codes of practice (including codes of hygienic practice), a discussion paper may be developed by the appropriate Codex committee on a proposed issue to determine whether the subject warrants further consideration by Codex. Step 1 also involves an assignment by the Executive Committee or the Commission to the

TABLE 17.1
The 8-Step Procedure of the Codex Alimentarius for Development and Adoption of Standards and Other Texts

Steps 1–4	Codex Commission approves proposal for new standard or other text and assigns it to Committee. Secretariat arranges for draft of proposed standard or text. Proposed draft standard or text sent to countries for comments. Draft proposed standard or text and country comments initially reviewed by assigned Codex committee.
Step 5	Initial review of proposed draft standard or text by the Codex Alimentarius Commission.
Steps 6–7	Draft proposed standard or text forwarded to member countries for second review. Second review, considering country comments, carried out by assigned Codex committee.
Step 8	Final review of proposed standard or text by Codex Alimentarius Commission. Standard or text adopted, modified and adopted, or rejected by Commission.
Fast Track	Proposed standard or text adopted at Step 5 when no objection exists.

appropriate Codex committee. Step 2 of the elaboration process involves the preparation of the draft proposed text, either by the member country or by the Secretariat. In the case of residues of pesticides or veterinary drugs, the Codex Secretariat distributes the recommendations for MRLs, when available, from JMPR or JECFA.

Step 3 of the Codex procedure involves the initial submission of the document to member countries for review and comment. Often, formal written comments on proposed Codex standards and texts are prepared by countries for consideration by the assigned Codex committee. Step 4 is the first formal review of the proposed standard or text by the assigned Codex committee. At this stage, standards and texts are normally reviewed in depth and revisions are made as appropriate.

Step 5 is the first review by the full Commission. This review is important, as normally only 25 to 30% of all member countries attend any specific meeting of a horizontal or vertical Codex committee. Total country participation at Commission meetings is substantially higher, hence this initial review by the Commission may be the first time that a country comments on a specific proposed Codex standard or text.

At Step 6, the now-revised standard, including any comments made by the Commission, is sent to all member countries for a second review. Again, as with Step 3, formal written comments are often prepared by a country for review by the assigned committee. Step 7 is the second opportunity for the assigned Codex committee to review the revised proposed Codex standard or text; the Committee has at hand the written country comments from Step 6. Step 8 is the second and final review by the Codex Commission. The Commission has the option of adopting the standard at this point, amending and adopting the standard, sending the standard back to Committee for further work, or rejecting the proposal. Normally, this review results in the adoption of the standard.

At any time during the process, a Codex committee or the Commission can retain a document at any specific step or can return a document to a previous step. Both of these instances frequently occur, particularly for proposed standards or texts that have difficult technical issues or in situations where consensus on a document cannot be reached. The desire in Codex is for decisions to be reached by consensus, although provisions exist within the formal Codex operating procedures for voting on proposed standards and texts. Voting on decisions relating to the adoption of Codex texts has occurred very infrequently and is usually associated with highly controversial issues (e.g., the elaboration of MRLs for growth-promoting animal hormones).

An accelerated or "fast track" procedure is provided for, in which Steps 6 and 7 are omitted. Thus, Steps 5 and 8 are combined, and the proposed Codex standard or text may be adopted at the first consideration by the Commission. Consensus or two thirds majority vote is required to "fast track" a Codex standard or text. Normally, this process is used for non-controversial items, such as pesticide residues MRLs.

It is important to note that the Commission meets only once every 2 years, while Codex committees meet, on the average, once every 12 to 18 months. Hence, it takes some period of time to elaborate and adopt a Codex standard. As previously noted, such deliberations are important to ensure the development of appropriate and correct standards or texts, particularly in light of the impact of Codex under the SPS and TBT Agreements and the ramifications that these interrelationships have on international trade.

VI. NATIONAL DELEGATIONS AND DECISION-MAKING IN CODEX

Most member countries of Codex maintain Codex contact points, which are listed on the Codex Internet Website. They are the focal point for Codex operations within a country. Functions of national Codex contact points vary from country to country, but activities include the management of national delegations to Codex committees, interaction with stakeholders, coordination and/or development of country positions on Codex documents, and receipt and distribution of Codex texts. Some Codex contact points maintain a library of Codex documents.

Representation at meetings of Codex committees and the Commission is made through national delegations from each country. Because Codex is an intergovernmental body, the country Delegates (and Alternate Delegates, if appointed) to any specific Codex committee and to meetings of the Commission must be governmental officials. Delegations may also contain both governmental and non-governmental advisors. Advisors are normally selected on the basis of the expertise and the assistance they can provide to the Delegate on the technical and/or policy issues before the Codex committee or Commission. Selection of non-governmental advisors may also be based on the constituencies they represent. Countries limit the size of delegations and normally provide balance with respect to issues or needed expertise and constituency representations (e.g., industry, consumers).

In preparation for specific Codex meetings, countries prepare positions on issues (proposed standards and other texts) coming before the Committees. These positions normally form the basis for a country's intervention (that is, verbally presenting a position or point of view) on the issue at the meeting of the Committee. Countries often present draft positions at public meetings to ensure transparency and the broadest representation possible in determining final country positions. Countries may also share their national positions with other countries to develop support for their position on an issue. Countries may involve interested parties through public meetings and other mechanisms in the development, review, and comment activities carried out for proposed Codex standards and texts.

Codex meetings normally proceed through a set agenda, discussing all issues assigned to the Committee. Countries intervene in the discussion in an orderly fashion, providing their position. Consensus among the Delegates is normally reached after several rounds of intervention and debate. As noted previously, Codex operates by intent and practice on a consensus basis, although procedures for voting are maintained. Once discussion or debate is concluded on an issue, consensus on the standard or text is reached within the context of the Step procedure. In the early stages of consideration, countries and the Committee may agree to a preliminary consensus position, recognizing that the issue will be revised at a future meeting where the Committee's final consensus decision may change. Ultimately, the Committee reaches a final consensus on the issue, a vote is taken, or (very rarely) the Committee agrees to cease work on the item. Countries wishing to record their opposition to a decision of a Committee or the Commission may do so. Should voting on an issue occur, each country is allotted one vote.

For the U.S., the Codex contact point is the U.S. Codex office, currently located within the U.S. Department of Agriculture, Food Safety Inspection Service, phone: 202-205-7760; e-mail: uscodex@usda.gov; Website address: http://www.fsis.usda.gov/OA/codex/index.htm.

VII. CODEX AND SCIENCE

From its beginning, the work of Codex has been science based. Experts and specialists in a wide range of scientific disciplines have contributed to every aspect of Codex activities to ensure that Codex standards withstand scientific scrutiny. It is fair to say that the work of the CAC, and that of FAO and WHO in their supportive roles, have provided a focal point for food-related scientific research and investigation, and the Commission itself has become a most important medium for the exchange of information about food.

The SPS Agreement, to which many countries are signatory, provides the basis and driving force for the current policy on international food standards. The SPS agreement, which went into effect in 1995, includes provisions that require the use of science, including risk assessment, in standards setting. The specific provisions are

- Article 2.2. Members shall ensure that any sanitary and phytosanitary measure is based on scientific principles.
- Article 5.1. Members shall ensure that their sanitary and phytosanitary measures are based on an assessment, as appropriate to the circumstances, of the risks to human, animal, or plant life or health.

TABLE 17.2
The Codex Alimentarius Statements of Principle Concerning the Role of Science in the Codex Decision-Making Process and the Extent to Which Other Factors Are Taken into Account

The food standards, guidelines, and other recommendations of Codex Alimentarius shall be based on the principle of sound scientific analysis and evidence, involving a thorough review of all relevant information, in order that the standards assure the quality and safety of the food supply.

When elaborating and deciding upon food standards, Codex Alimentarius will have regard, where appropriate, to other legitimate factors relevant to the health protection of consumers and for the promotion of fair practices in food trade.

In this regard it is noted that food labeling plays an important role in furthering both of these objectives.

When the situation arises that members of Codex agree on the necessary level of protection of public health but hold differing views about other considerations, members may abstain from acceptance of the relevant standard without necessarily preventing the decision by Codex.

Codex, recognizing its role under the SPS Agreement as the reference international organization for food safety standards, initiated a process to strengthen its policies and operational practices with respect to the use of science in standards setting. In 1995, the CAC established a set of four "Statements of Principle Relating to the Role of Food Safety Risk Assessment"[1] (Table 17.2), or "Sound Science Principles," as they are commonly referred to by those involved with Codex. These four principles established the fundamental policy by which Codex will undertake its standards-setting work. The first principle clearly articulates that food standards, guidelines, and other recommendations of the Codex Alimentarius shall be based on the principle of sound scientific analysis and evidence.

The second principle, relating to what are termed "other legitimate factors" (OLFs), is more complex and states that Codex, in its standards-setting activities, will have regard, where appropriate, to other legitimate factors relevant for the health protection of consumers and for the promotion of fair practices in food trade. Determination of what comprises OLFs is currently being undertaken by the CAC. There are differences of opinion among countries as to what constitutes an OLF. Countries generally accept that OLFs relating to the risk assessment process (e.g., good veterinary practices, good agricultural practices) are appropriate. Acceptance may be somewhat more difficult with factors such as economic feasibility, environmental impact, impact on health status, and consumer perception of risk. Acceptance will be most difficult with, and will likely be inappropriate for, such factors as animal welfare, consumer right to know, cost increases (or decreases) associated with technology, and relative risk (that is, the risks associated with a product compared to other risks to which a consumer is exposed) that some countries believe ought to be OLFs. Provisions of the SPS Agreement provide guidance as to factors countries can legitimately take into account when establishing sanitary and phytosanitary

TABLE 17.3
The Codex Alimentarius Draft Proposed Working Principles for Risk Analysis

Health and safety aspects of Codex decisions and recommendations should be based on a risk assessment, as appropriate to the circumstances.

Risk analysis should be soundly based on science, should incorporate the four steps of the risk assessment process, and should be documented in a transparent manner. There should be a general understanding to identify and systematically document the elements of risk assessment and risk management that are incorporated in the elaboration of Codex standards.

Risk analysis procedures used by Codex should be harmonized with those of other bodies and national governments to the greatest extent possible.

The risk analysis needs of developing countries should be specifically identified and addressed. Adequate flexibility should be provided where appropriate to meet any special needs of developing countries.

measures (e.g., economic factors including the relative cost effectiveness of alternative approaches to limiting risks, relevant processes and production methods, and relevant environmental conditions). These SPS provisions are likely to assist in clarifying the determination of acceptable OLFs for Codex standards setting. WTO jurisprudence in trade dispute settlement is ultimately the probable method by which the most controversial and difficult OLFs will be determined.

Codex, in conjunction with FAO and WHO, has also undertaken a series of Joint Expert Consultations[5–7] to elucidate the basic principles of risk analysis. Working from the findings of three Joint Expert Consultations dealing with the three components of risk analysis (risk assessment, risk management, risk communication) and from the discussions of several key Codex horizontal committees (Food Hygiene, Pesticide Residues, Residues of Veterinary Drugs in Foods, and Food Additives and Contaminants), Codex is in the process of delineating a comprehensive series of *Working Principles of Risk Analysis*[8] which will lay a further scientific foundation for the elaboration of Codex standards, guidelines, and recommendations. The current *Working Principles of Risk Analysis* and its component parts under consideration are presented in Tables 17.3 through 17.6.

This Codex effort to establish the "Sound Science Principles" and the *Working Principles of Risk Analysis* is formulating a comprehensive scientific framework for the work Codex itself undertakes in elaborating standards. This effort is also providing a framework for countries to use in their own individual food safety standards setting. As such, the work of Codex extends far beyond the organization itself. This comprehensive sound science/risk analysis approach to standards setting, which is compatible with the provisions of the SPS Agreement, is beginning to be seen in new draft Codex standards. A good example is the proposed draft *Code of Hygienic Practice for Milk and Milk Products* that is discussed below.

TABLE 17.4
The Codex Alimentarius Draft Proposed Working Principles for Risk Assessment

Risk assessments should use available quantitative information to the greatest extent possible and risk characterizations should be presented in a readily understandable form.

Risk assessments should be based on realistic exposure scenarios, with consideration of "worst-case" situations being defined by risk assessment policy.

Risk estimates should wherever possible include a numerical expression of uncertainty, and this should be conveyed to risk managers in a readily understandable form. The responsibility for resolving the impact of uncertainty on the risk management decision lies with the risk manager, not the risk assessor.

There should be increased recognition of differences between acute and chronic adverse health effects in carrying out risk assessment and establishing food standards.

There should be a functional separation of risk assessment and risk management, while recognizing that some interactions are essential for a pragmatic approach.

VIII. CODEX AND FOOD HYGIENE

The Codex Committee on Food Hygiene (CCFH) is primarily responsible for food hygiene matters within Codex. Food hygiene generally encompasses hygiene as it relates to good production, manufacturing, distribution, and marketing practices; the hazard analysis and critical control point (HACCP) system; microbiological criteria for foods; and microbiological risk assessment and risk management. The terms of reference for CCFH[1] are essentially threefold. First is to draft basic provisions on food hygiene applicable to all food. In practice, this takes the form of the development (and revision when needed) of Codex Codes of Hygienic Practice for various food commodities. A detailed discussion of these Codes of Hygienic Practice is presented below. The second responsibility is to consider or review food hygiene provisions of Codex commodity standards or other texts. A detailed discussion of this responsibility is also presented below. A third generic responsibility of the CCFH is to consider specific hygiene problems assigned to it by the Commission. Codex committees generally serve as a resource to the Commission in their area of expertise (e.g., Codex Committee on Food Labeling, for labeling issues; Codex Committee on Food Import and Export Inspection and Certification Systems, for import/export control issues). CCFH serves this role for the field of food hygiene.

A. Codes of Hygienic Practice

Codex Codes of Hygienic Practice provide guidance on the hygienic production and processing of foods. The base reference document in this area is the *Recommended International Code of Practice: General Principles of Food Hygiene*.[9] In addition to

TABLE 17.5
The Codex Alimentarius Draft Proposed Working Principles for Risk Management

Risk management should follow a structured approach.

Protection of human health should be the primary consideration in risk management decisions.

Risk management should be focused on agreed outcomes rather than processes.

Risk management policies should be documented and, where appropriate, acknowledged in individual Codex standards so as to foster a wider understanding of risk management concepts and the particular risk policy used in the elaboration of individual Codex standards.

Guidelines should be available for inclusion in risk management decisions of "other legitimate factors relevant for the health protection of consumers and for the promotion of fair practices in food trade."

If economic analysis is used in support of risk management decisions, the process should be subject to consistent and transparent decision-making criteria.

Where risk management involves selection of options other than (or in addition to) quantitative food standards for the prevention, elimination, or control of hazards, each available option should be evaluated according to a relevant risk management framework.

"Horizontal" issues in the elaboration of food standards and related texts according to risk analysis principles should be clearly identified and consistently addressed.

introductory sections on objectives and scope, this general food hygiene code of practice contains detailed recommended food hygiene practices for the following areas:

- Primary production
- Establishment: design and facilities
- Control of operation
- Establishment: maintenance and sanitation
- Establishment: personal hygiene
- Transportation
- Product information and consumer awareness
- Training

This "General Principles" document also contains a HACCP annex, which presents the recommendations of Codex for the application of HACCP to food production. The Codex approach to HACCP utilizes the seven internationally recognized principles of HACCP. The annex presents a detailed discussion on the application of each principle, including the establishment of critical control points (CCP) and

TABLE 17.6
The Codex Alimentarius Draft Proposed Working Principles for Risk Communication and for Documentation

Risk Communication	Risk analysis should include clear, interactive communication with consumers and other interested parties in all aspects of the process.
	A risk communication strategy should be proactive and include a plan specifying how information is to be communicated.
	Risk managers should include an assessment of uncertainty in risk estimates in their communication with the public.
Documentation	Risk assessment and risk management should be fully documented in a transparent manner. Risk management should be transparent, flexible, objective, and repeatable, and this requires full documentation.
	Risk management should be a continuing process that takes into account all newly generated data in the evaluation and review of risk management decisions. Food standards must be consistent with new scientific knowledge and other information relevant to risk analysis.

critical limits for each CCP. Also presented is a flow diagram for the application of HACCP ("Logic Sequence for Application of HACCP"), an example of a decision tree to identify CCPs, and an example of a HACCP worksheet.

The CCFH also undertakes the development of commodity-specific codes of hygienic practice. These commodity-specific codes reference the *Recommended International Code of Practice: General Principles of Food Hygiene* and are constructed so that only hygiene provisions supplemental to those present in the General Principles code and specific for the commodity type covered in the code are given. Examples of codes of hygienic practice that have been or are being developed by Codex include those for the following product areas:

- Canned fruit and vegetable products
- Processed meat products
- Milk and milk products
- Production, harvesting and packaging of fresh produce
- Fresh-cut raw fruits and vegetables
- Packaged/bottled water other than natural mineral waters
- Refrigerated packaged foods with extended shelf-life
- Transport of foodstuffs in bulk and semi-packaged foods

The current development of a proposed draft *Code of Hygienic Practice for Milk and Milk Products* deserves special mention because the construction of this Code is particularly reflective of the new responsibilities of Codex under the SPS Agreement. The development of this Code arose from prior work of the CCFH dealing with soft cheeses and the difficulty experienced by the Committee in dealing with complex microbiological risk issues associated with the production of soft, unripened, uncured cheeses made from raw milk. Countries differed as to the acceptable microbiological risk associated with this product type. The result to date has been the development of a draft proposed Code that permits countries to manufacture products in multiple ways (in which the microbiological risks may be different). Countries have the right to accept or reject product based on their scientifically established appropriate level of protection (ALOP). This Code, for the first time, employs SPS terminology and concepts, relating the construct of a specific Codex document to the provisions of the SPS Agreement, and reflects the new and enhanced role of Codex under the SPS Agreement.

B. Hygiene Provisions of Commodity Codes

Codex develops recommended standards for a multitude of specific food commodities. As noted earlier, provisions of these standards include sections on essential composition (a basic standard of identity), packaging and labeling, maximum permitted residue levels for pesticides, veterinary drugs, contaminants, and maximum permitted use levels for food additives; each of these sections refers to general broad Codex standards for pesticide residue MRLs, veterinary drug MRLs, and food additive permitted use levels. These commodity standards also contain a general section on food hygiene. The current recommended wording for the hygiene section of all commodity standards follows:[10]

- It is recommended that the products covered by the provisions of this standard be prepared and handled in accordance with the appropriate sections of the *Recommended International Code of Practice — General Principles of Food Hygiene* (CAC/RCP 1-1969, Rev. 3-1997) and other relevant Codex texts, such as Codes of Hygienic Practice and Codes of Practice.
- The products should comply with any microbiological criteria established in accordance with the *Principles for the Establishment and Application of Microbiological Criteria for Foods* (CAC/GL 21-1997).

As the above-mentioned Codex commodity committees develop individual commodity standards, the CCFH has the responsibility to review the food hygiene section to ensure that the Section's provisions are adequate for the product. This is normally a routine activity, and the recommended wording noted above is approved. CCFH has the capability to add specific additional provisions to the food hygiene section of commodity standards should the committee deem them necessary to ensure the safe production and/or processing of the products covered by the specific commodity standard. The review of eight generic dairy product standards, including that for cheese, is a case in point.[11] Because of concern with potential microbiological risk issues associated with some product types (e.g., soft, unripened, uncured cheeses made from raw milk), the CCFH approved the following additional provision for the hygiene section of the eight dairy product standards:

> From raw material production to the point of consumption, the products covered by this standard should be subject to a combination of control measures, which may include, for example, pasteurization, and these should be shown to achieve the appropriate level of public health protection.

C. Other Food Hygiene Activities

In the context of its broad food hygiene responsibilities, the work of CCFH extends beyond the confines of specific Codex codes of hygienic practice and the hygiene provisions of specific Codex commodity codes. The work of CCFH is wide ranging and significant, providing guidance to both Codex and countries, and affects the field of food microbiology generally. Work of CCFH in this broader context includes efforts in microbiological risk assessment and microbiological risk management (see below), the establishment of guidance for the use of microbiological criteria in foods (refer to *Principles for the Establishment and Application of Microbiological Criteria for Foods*[12]), the consideration of application of HACCP to small business, antibiotic resistance in bacteria in food, and guidelines for the control of viruses in foods. These last two items (antibiotic resistance and control of viruses) are new work items for CCFH, recommended for development by CCFH at its 1998 meeting. The antibiotic resistance topic is of interest to a great many countries, and the work of the CCFH in this area will be important to follow.

IX. Microbiological Risk Assessment and Risk Management in Codex

Codex, through the work of the CCFH, has developed a fundamental guidance document on microbiological risk assessment and is in the process of developing a companion document on microbiological risk management.

A. Microbiological Risk Assessment Guidance Document

Codex has established the first international intergovernmental guidance document for microbiological risk assessment, *Principles and Guidelines for the Conduct of Microbiological Risk Assessment.*[13] The scope of this document applies generally to risk assessment of microbiological hazards in food and is designed to ensure that sound science principles and practices are applied uniformly in evaluating microbiological hazards in foods. The document recognizes that microbiological risk assessment is a developing science and that, while the process should include quantitative information to the greatest extent possible, qualitative judgments will have to be made.

Eleven general principles applicable to the conduct of microbiological risk assessment (Table 17.7) are presented in the document. These principles state that microbiological risk assessment should be conducted according to the structured four-step approach that is common to risk assessment generally (hazard identification, hazard characterization, exposure assessment, and risk characterization). The principles also incorporate provisions relating to performing a microbiological risk assessment, including the need for:

TABLE 17.7
The Codex Alimentarius General Principles of Microbiological Risk Assessment

Microbiological risk assessment should be soundly based on science.
There should be a functional separation between risk assessment and risk management.
Microbiological risk assessment should be conducted according to a structured approach that includes hazard identification, hazard characterization, exposure assessment, and risk characterization.
A microbiological risk assessment should clearly state the purpose of the exercise, including the form of risk estimate that will be the output.
The conduct of a microbiological risk assessment should be transparent.
Any constraints that impact the risk assessment, such as costs, resources, or time, should be identified and their possible consequences described.
The risk estimate should contain a description of uncertainty and where the uncertainty arose during the risk assessment process.
Data should be such that uncertainty in the risk estimate can be determined; data collection systems should, as far as possible, be of sufficient quality and precision that uncertainty in the risk estimate is minimized.
A microbiological risk assessment should explicitly consider the dynamics of microbiological growth, survival, and death in foods and the complexity of the interaction (including sequelae) between humans and agents following consumption, as well as potential for further spread.
Wherever possible, risk estimates should be re-assessed over time by comparison with independent human illness data.
A microbiological risk assessment may require re-evaluation, as new relevant information becomes available.

- Functional separation between risk assessment and risk management
- A requirement for transparency in what is being done
- Identification of the limitation of data and resource constraints
- Consideration of the dynamics of microbial growth
- Specifying the uncertainties associated with each step of the risk assessment process
- Reassessment as new data (e.g., new quantitative information on the occurrence of microorganisms in food, new dietary intake information) become available

Guidance is provided for the steps required to carry out a microbiological risk assessment. Each of the four structured steps of risk assessment is discussed in detail. Helpful information specific to the application of microbiological risk assessment is given. For example, influences that may affect the hazard characterization are detailed (e.g., replication rates of microorganisms, virulence and infectivity, delay of onset of clinical symptoms, effect of food attributes such as fat content). Similar information is described for the exposure assessment step (e.g., characteristics of the pathogenic agent, the microbial ecology of the food, and the methods of packaging, processing, distribution, and storage). The need for proper documentation and reassessment as new data develop (an important consideration in the comparatively new field of microbiological risk assessment) is stressed in the guidance document.

B. Microbiological Risk Management

The CCFH is developing a new document on *Principles and Guidelines for the Conduct of Microbiological Risk Management.*[14] This document, when adopted by the CAC, will be the first international guidance to both Codex committees and to countries on how to undertake and implement microbiological risk management activities. The document builds on a series of risk management principles developed by the FAO/WHO Joint Expert Consultation on Risk Management[6] and provides a framework for undertaking microbiological risk management. In this document, microbiological risk evaluation is outlined as involving several steps, including the following:

- Identification of a microbiological food safety problem
- Establishment of risk profile
- Identification of the microbiological risk management goals and risk managers
- Ranking of hazards for microbiological risk management
- Establishment of microbiological risk assessment policy for the conduct of risk assessment
- Commissioning of the microbiological risk assessment
- Consideration of the results of the microbiological risk assessment

The document separately discusses each of these steps in microbiological risk evaluation. In addition, it reviews the area of microbiological option assessment, including the identification of available microbiological risk management options, selection of the preferred option, and the final management decision. Sections are also presented on implementation, monitoring, and review.

An important aspect of this paper is the incorporation of a new entity in risk analysis that can be used in risk management programs, the concept of food safety objectives (FSOs).[14,15] FSOs have been developed to describe the link between a sanitary measure and the appropriate level of protection that a country chooses for its consumers for a specific foodborne risk. FSOs are likely to become an important operational element in future food safety programs.

X. OTHER CODEX WORK AREAS

This presentation will conclude with a brief mention of some other key Codex work areas, some of which lie outside of the food hygiene area, that are important and may, in some cases, influence the work in the field of food hygiene.

A. Other Legitimate Factors

As noted earlier, work is being undertaken within the Codex Committee on General Principles in regard to other legitimate factors (OLFs). These are factors, other than science, directly related to food safety and quality that may be legitimately used as a basis to establish Codex standards. Such factors as technological and economic feasibility and impact on the environment are within this area. While the debate is still going on, its final outcome is important, as it will delineate the type and nature of future Codex standards. This will be an important area to watch.

B. Judgment of Equivalence of Food Regulatory Inspection Systems

The concept of equivalence is a fundamental provision of both the WTO SPS and TBT Agreements. Equivalence is the concept of two separate food regulatory control systems (of different countries), using, at least in part, different sets of control measures, to achieve the same level of protection or conformity. The concept is important to facilitate trade and conserve scarce resources. The Codex Committee on Food Import and Export Inspection and Certification Systems (CCFICS) is beginning new work on the judgment of equivalence for both the safety and non-safety areas.

C. Guidelines for the Utilization and Promotion of Quality Assurance Systems

The CCFICS is initiating work on this subject. Key elements that can have impact on food production and control systems include the relationship of HACCP to quality assurance systems, the relationship of quality assurance systems to the frequency of regulatory control inspections, and the use of third-party certification.

D. Guidelines for Food Import Control Systems

The CCFICS is also initiating work in this important area. The Guidelines are intended to give guidance to countries on the proper operation of food-import control systems. Such areas as critical elements of a food-import control system (aims and priorities, legal framework, regulations and standards, defined roles of authorities), administrative requirements (risk analysis, application of performance history to inspection, point of control, data exchange), and the operation and mechanism of the control system (documentation, training, system verification) are covered in this document.

E. Acceptance of Codex Texts

The Codex Committee on General Principles is in the process of considering whether current procedures as to how countries consider the acceptance of Codex standards and texts should be revised. At issue is the new relationship of Codex to the WTO and specifically to the SPS and TBT Agreements. Because the WTO now recognizes Codex as the reference international organization for food safety and quality standards, the question has arisen as to whether Codex needs to maintain acceptance procedures at all and, if so, in what form. The discussion is continuing at the moment, but it is important as it relates at least indirectly to the actual ultimate use of Codex standards by countries.

F. Review of Codex Commodity Standards

Most commodity-specific standards, such as those for processed fruits and vegetables, dairy products, and fishery products, were originally drafted many years ago. Many of these standards are now technologically outdated, do not incorporate newer product types, and do not reflect the new relationship of Codex to the WTO. Consequently, Codex has begun to undertake a comprehensive review of these commodity standards to update them with respect to all of these areas. The process is expected to take some time to complete.

G. New Codes of Hygienic Practice

It would be a fitting conclusion to this presentation to note that the CCFH is undertaking new work on two new important codes of hygienic practice, the *Proposed Draft Code of Hygienic Practice for the Primary Production, Harvesting, and Packaging of Fresh Produce* and the *Proposed Draft Code of Hygienic Code of Practice for Pre-Cut Raw Fruits and Vegetables.* With the apparent increase in the level of foodborne outbreaks associated with these internationally traded products, both codes of hygienic practice have significant international importance.

REFERENCES

1. Codex Alimentarius Commission. *Procedural Manual*, 10th ed. Food and Agriculture Organization of the United Nations, Rome, 1997, 150 pp.
2. FAO. *This Is Codex Alimentarius*, 2nd ed. Food and Agriculture Organization of the United Nations, Rome, [undated], 20 pp.
3. FAO/WHO Food Standards Programme. *Practices of the Codex Alimentarius Commission in the Context of the Code of Good Practice for the Preparation, Adoption, and Application of Standards by Standardizing Bodies.* Prepared for the World Trade Organization Agreement on Technical Barriers to Trade Committee, November, 1998, 16 pp.
4. WTO. *The Results of the Uruguay Round of Multilateral Trade Negotiations: The Legal Texts.* World Trade Organization, Geneva, Switzerland, 1995, 558 pp.
5. FAO. *Application of Risk Analysis to Food Standards Issues: Report of the Joint FAO/WHO Consultation*, Geneva, Switzerland, March 13–17, 1995.

6. FAO. *Risk Management and Food Safety: Report of the Joint FAO/WHO Consultation.* Rome, Italy, January 27–31, 1997.
7. FAO. *Risk Communication and Food Safety: Report of the Joint FAO/WHO Consultation*, report in process.
8. Codex Alimentarius Commission. *Risk Analysis.* 2. *Working Principles for Risk Analysis* (CX/GP 98/4). Rome, Italy, 1998.
9. FAO. Recommended International Code of Hygienic Practice: General Principles of Food Hygiene. In *Codex Alimentarius, General Requirements (Food Hygiene),* Vol. 1B (suppl.), 2nd ed. Food and Agriculture Organization of the United Nations, Rome, Italy, 1997, pp. 1–26.
10. Codex Alimentarius Commission. Draft Amendment to the Food Hygiene Provisions, Appendix VI in *Relations Between Commodity Committees and General Committees*, Section K, of the Procedural Manual. Report of the Thirtieth Session of the Codex Committee on Food Hygiene (ALINORM 99/13), Washington, D.C., October 20–24, 1997.
11. Codex Alimentarius Commission. Hygiene Provisions of the Draft Revised Standards for Butter, Milkfat Products, Evaporated Milks, Sweetened and Condensed Milks, Milk and Cream Powders, Cheese, Whey Cheese, and the Draft Standard for Cheeses in Brine, Appendix VII. Report of the Thirtieth Session of the Codex Committee on Food Hygiene (ALINORM 99/13), Washington, D.C., October 20–24, 1997.
12. FAO, Principles for the Establishment and Application of Microbiological Criteria for Foods. In *Codex Alimentarius, General Requirements (Food Hygiene),* Vol. 1B, (suppl.), 2nd ed. Food and Agriculture Organization of the United Nations, Rome, Italy, 1997, pp. 27–30.
13. Codex Alimentarius Commission. Draft Principles and Guidelines for the Conduct of Microbiological Risk Assessment, Appendix II in *Report of the Thirty-First Session of the Codex Committee on Food Hygiene* (ALINORM 99/13A), Orlando, FL, October 26–30, 1998.
14. Codex Alimentarius Commission, Joint FAO/WHO Food Standards Programme, Codex Committee on Food Hygiene. *Discussion Paper on Recommendations for the Management of Microbiological Hazards for Foods in International Trade*, CX/FH 97/10, 1998.
15. Van Schothorst, M., Principles for the establishment of microbiological food safety objectives and related control measures, *Food Control,* 9(6):379–384, 1998.

Index

Index

B

D

E

F

G

H

I

J

K

L

M

T

U

V

W

X

Y

Z